VOLUME EIGHTY SIX

ADVANCES IN BOTANICAL RESEARCH

Transgenic Plants *and Beyond*

ADVANCES IN BOTANICAL RESEARCH

Series Editors

VOLUME EIGHTY SIX

ADVANCES IN BOTANICAL RESEARCH

Transgenic Plants *and Beyond*

Volume Editor

MARCEL KUNTZ
Univ. Grenoble Alpes, CNRS, CEA, INRA,
Laboratoire de Physiologie Végétale & Cellulaire,
Grenoble, France

ELSEVIER

ACADEMIC PRESS
An imprint of Elsevier

Academic Press is an imprint of Elsevier
125 London Wall, London, EC2Y 5AS, United Kingdom
The Boulevard, Langford Lane, Kidlington, Oxford OX5 1GB, United Kingdom
50 Hampshire Street, 5th Floor, Cambridge, MA 02139, United States
525 B Street, Suite 1800, San Diego, CA 92101-4495, United States

First edition 2018

ISBN: 978-0-12-809447-1
ISSN: 0065-2296

For information on all Academic Press publications
visit our website at https://www.elsevier.com/books-and-journals

Publisher: Zoe Kruze
Acquisition Editor: Sam Mahfoudh
Editorial Project Manager: Joanna Collett
Production Project Manager: James Selvam
Cover Designer: Victoria Pearson

Typeset by SPi Global, India

CONTENTS

CONTRIBUTORS

Serife Akkoyunlu
AgroParisTech, Paris, France

Mahaletchumy Arujanan
Malaysian Biotechnology Information Centre, Petaling Jaya, Malaysia

Andrew Bartholomaeus
Faculty of Health, University of Canberra, Canberra, ACT; Diamantina Institute, Faculty of Medicine, University of Queensland, Brisbane, QLD, Australia

Alejandro Chaparro-Giraldo
Grupo de Ingeniería Genética de Plantas, Universidad Nacional de Colombia, Bogotá, Colombia

Qiang Chen
School of Life Sciences; Center for Immunotherapy, Vaccines and Cancer Therapy, Arizona State University, Tempe, AZ, United States

Nayla Farouki
Philosopher and Historian of Science, Grenoble, France

Ioannis Ganopoulos
Institute of Plant Breeding and Genetic Resources, Hellenic Agricultural Organization DEMETER (ex NAGREF), Thermi, Greece

Adriana Castaño Hernández
GMO Biosafety and AgroBiotech Consultant, Bogotá, Colombia

Aliki Kapazoglou
Institute of Olive Tree, Subtropical Crops and Viticulture, Hellenic Agricultural Organisation "DEMETER" (HAO-Demeter), Lykovrisi, Athens, Greece

Marcel Kuntz
Univ. Grenoble Alpes, CNRS, CEA, INRA, Laboratoire de Physiologie Végétale & Cellulaire, Grenoble, France

Silvio Alejandro López-Pazos
Facultad de Ciencias, Universidad Antonio Nariño, Bogotá, Colombia

Jacqueline Martin-Laffon
Univ. Grenoble Alpes, CNRS, CEA, INRA, Laboratoire de Physiologie Végétale & Cellulaire, Grenoble, France

Henry I. Miller
Hoover Institution, Stanford University, Stanford, CA, United States

Julián Mora-Oberlaender
Grupo de Ingeniería Genética de Plantas, Universidad Nacional de Colombia, Bogotá, Colombia

Piero A. Morandini
University of Milan, Milan, Italy

Léon Otten
Institut de Biologie Moléculaire des Plantes du CNRS, Strasbourg, France

Maxence Pfeiffer
AgroParisTech, Paris, France

Francis Quétier
University of Evry Val d'Essonne; Génopole-Evry, Evry, France

Agnès Ricroch
University of Paris-Sud, Collège d'Etudes Interdisciplinaires, Sceaux; AgroParisTech, Paris, France

Eleni Tani
Laboratory of Plant Breeding and Biometry, Agricultural University of Athens, Athens, Greece

Paul P.S. Teng
National Institute of Education, Nanyang Technological University, Singapore, Singapore

Athanasios Tsaftaris
Institute of Applied Biosciences (INAB), CERTH, Thermi-Thessaloniki; Perrotis College, American Farm School, Thessaloniki, Greece

PREFACE

Transgenic plants constitute an extraordinary chapter of the History of Science and Technology. As mentioned by Léon Otten in Chapter 2 of this volume, Mary-Dell Chilton (Washington University), Jeff Schell and Marc van Montagu (University of Ghent and Max Planck Institute for Plant Breeding in Cologne), and Rob Horsch (Monsanto) independently reported on the transfer and expression of a marker gene in tobacco plants at a Symposium in Miami in 1983. Transgenic plants were officially born! These achievements owe a lot to the long history of studies on bacteria from the genus Agrobacterium, which perform natural genetic engineering. This fascinating history is also included in Chapter 2.

Since then, transgenesis has become a major tool for plant biologists. And it still is for the more recent developments known as 'gene editing' which often requires transgenesis for transferring the editing tools into plants.

Transgenesis is viewed by many scientists as a crucial tool for modern plant breeding. Actually, this volume goes further back, to the invention of agriculture which would not have been without plant domestication, which in turn was dependent on genetic modification in a broad sense. Thus, in Chapter 1, Henry Miller and Piero Morandini argue that crop genetic engineering resulted in improved domestication traits.

Highly promising applications for human health, namely, the production of recombinant therapeutic molecules in plants, have also been developed using various plant-based expression systems, as detailed in Chapter 7 by Qiang Chen.

This volume is also about innovations beyond transgenesis: gene editing, as mentioned above (see Chapter 8 by Maxence Pfeiffer, Francis Quétier and Agnès Ricroch), and also epigenetics. In Chapter 9, Aliki Kapazoglou, Ioannis Ganopoulos, Eleni Tani and Athanasios Tsaftaris take us in the world of epigenetics, its regulatory role during plant development, its involvement in environmental stress tolerance and hybrid vigour, not forgetting graft-induced epigenetic changes. This chapter also discusses how epigenetics can be a source of phenotypic variation for plant breeding. It presents the main experimental techniques for DNA methylation profiling in plants and future trends such as third-generation sequencing.

Transgenic plants are not only about science and innovation. In 1990, the European Union legally defined the concept of 'genetic modification' in a Directive. This Directive first listed all existing genetic modification techniques, but then exempted them all from regulation apart from the most recent one, transgenesis. This has had enormous consequences worldwide regarding the use of biotechnology in agriculture. In this context, Arujanan Mahaletchumy and Paul P.S. Teng explain in Chapter 3 the legal, regulatory and labelling status of biotech crops.

This legal concept of 'genetically modified organisms' (GMO) has been adopted by many countries around the world and has triggered heavy regulation, and sometimes bans. Thus, the future of gene editing is strongly dependent on whether these 'edited' organisms will fall, or not, within this legal concept of 'GMO' as discussed in Chapter 8.

As another consequence of regulation, in most countries, transgenic plants are subjected to mandatory risk assessments before marketing is allowed. In Chapter 4, Andrew Bartholomaeus presents the basic principles of ethical regulation and the plausible risks borne by plant genetic modifications which need to be tested for food/feed safety. This chapter also warns against the damaging consequences of the current regulatory rules applied in a number of jurisdictions. Recombinant DNA technology applied to plants has actually been subjected to an impressive amount of risk assessment studies, both from industries seeking marketing authorization and from public laboratories, regarding the environmental impacts of these plants. In Chapter 5, as an emblematic case for nontarget organisms, Agnès Ricroch, Serife Akkoyunlu, Jacqueline Martin-Laffon and Marcel Kuntz summarize the risk assessment studies performed on honeybees.

As a case study for developing countries, the needs for genetic engineering of crop plants in Colombia are explained by Julián Mora-Oberlaender, Adriana Castaño Hernández, Silvio Alejandro López-Pazos and Alejandro Chaparro-Giraldo in Chapter 6. This country has taken the challenge of developing technical and institutional capacities to ensure the safe use of transgenic crops, and its experience could be highly valuable for many other countries. This chapter also gives advice on how to perform a 'Freedom to Operate Analysis' in order to increase the chance of bringing a product to the market despite a heavily regulated and patented landscape.

This volume would not have been complete without asking the question 'Why products, although highly diverse, just because they fall into the "GMO" classification, have struck a powerful emotion called fear?' Nayla Farouki analyses this phenomenon in Chapter 10 and the consequences of decades of politicization of the scientific domain.

I believe this volume will be highly useful to researchers but also to students and teachers (many of its illustrations will come in handy for teachers), as well as science journalists and regulators, especially in countries wishing to develop or improve existing biosafety regulations. Most chapters contain valuable Internet links to regulatory articles, international agreements or GMO databases.

Finally, I would like to thank all the authors and contributors who made it possible to produce this volume.

MARCEL KUNTZ

CHAPTER ONE

Plant Domestication, the Brave *Old* World of Genetic Modification

Henry I. Miller*, Piero A. Morandini[†,1]
*Hoover Institution, Stanford University, Stanford, CA, United States
[†]University of Milan, Milan, Italy
[1]Corresponding author: e-mail address: piero.morandini@unimi.it

Contents

Abstract

The genetic improvement of crop plants via the newer techniques of biotechnology to produce "genetically modified" crops is a significant driver of progress in agriculture. However, progress has not been unimpeded: various controversies swirl around the benefits, uniqueness, supposed risks and other aspects of "GMOs", or genetically modified organisms—which, as we explain, is a meaningless "category"—and the foods derived from them. In order to resolve the conundrums posed by those issues, it is important to understand the pedigree of genetic modification, which had its inception in the domestication of plants. In this chapter, we briefly introduce the crucial determinants of the "domestication syndrome" for cereals and legumes—that is, loss of seed shattering and reduced seed dormancy—and how it evolved through the ages into contemporary "genetic modification". We argue that the application of genetic engineering to crops within a few years brought a wave of improved domestication traits. Moreover, contrary to most of the early domestication traits, some of these novel traits are advantageous to the crop and not just to humans. The other chapters in this volume discuss current developments in technology, the promise of modern molecular genetic engineering, and the legal and regulatory landscape.

Advances in Botanical Research, Volume 86
ISSN 0065-2296
https://doi.org/10.1016/bs.abr.2017.11.001

1. INTRODUCTION

Worldwide, farmers are becoming an ever-smaller fraction of the population, and most city dwellers do not know where their food comes from, the technology involved, or the labour it requires to obtain it. And if they know little about the techniques, technologies and tenuousness of farming, they know even less about its history.

A critical part of that history, which was in effect the first step in genetic manipulation, old and new, was the domestication of plants, which humans have long depended upon for food, feed, fibre, fuel, ornamental flowers and beverages such as wine, beer, tea and coffee, and juices from fruits and vegetables. Since early domestication, there has been a seamless continuum of genetic improvement of crop plants, culminating in far more rapid, precise and predictable modern molecular "genetic engineering" (GE) over the past half century.

The latter has elicited a great deal of unscientific commentary, which frequently includes concerns about "tampering with Nature" or "playing God", as well as claims that the supposedly "natural" organisms are really better anyway—tastier heirloom tomatoes, healthier ancient grains and so on. Such claims show a lack of appreciation of the evolution of crop plants over millennia and their genetic modification by a seamless continuum of techniques.

2. THE "DOMESTICATION SYNDROME"

Domestication, the process that transforms a wild plant into a crop, is achieved by the selection (either conscious or as a by-product of farming practices) of traits advantageous to humans (for general references, see Harlan, 1992; Meyer, DuVal, & Jensen, 2012; Meyer & Purugganan, 2013; Zohary & Hopf, 2000). It is marked by a plant's stable acquisition (and therefore inheritance by its progeny) of a suite of traits, or characteristics, which are often shared across many species. (Reflecting the essential link between phenotype and genotype, the selection for and introduction of new heritable traits is, of course, accompanied by changes in DNA.)

The sum of all the traits is often referred to as "domestication syndrome", because it is, in a sense, a complex "illness" that usually makes the crop unable to survive without human care. The reason for this behaviour is evident from the fact that the most commonly selected traits of the syndrome, particularly for grains and pulses, are *loss of seed dispersal* (Arnaud, Lawrenson, Østergaard, & Sablowski, 2011; Dong et al., 2014; Funatsuki et al., 2014; Konishi et al., 2006; Li, Zhou, & Sang, 2006; Lin et al., 2012; Pourkheirandish et al., 2015; Zhou et al., 2012) exemplified in Fig. 1 for rice and *reduced seed dormancy* (Liu et al., 2015; Lush & Evans, 1980; Sugimoto et al., 2010) exemplified for maize in Fig. 2.

The impacts of those traits on the crop are critical. When seeds are not dispersed but are retained in tight groups (in an ear or a pod) on the mother plant, they have a difficult time reaching the ground. Even if they do, once germinated they produce seedlings that are very close to one other and experience strong competition for nutrients, light and water (an example of which is shown in Fig. 3), which causes most individuals to die before

Fig. 1 Seed dispersal. Shortly before harvest, seed dispersal is almost completed in some panicles of red rice (*left*), which is a form of weedy, semi-wild rice. Cultivated rice (*right*), on the contrary, retains all the seeds on the panicle and requires a harsh treatment (threshing) for the detachment of the seeds from the mother plant. The *inset* shows an enlargement showing no gaps in between seeds within the panicle. Pictures were taken in a cultivated field near Milan (Italy).

Fig. 2 Reduced seed dormancy. A maize cob that was partly buried in a field at harvest (late September) demonstrates that maize seeds have little dormancy when placed in favourable conditions. The part of the cob that was buried shows extensive and synchronous germination even if the seeds attained maturity just a few weeks before. Wild cereals in the same conditions, on the contrary, stay dormant in the soil for many months.

Fig. 3 The combined effect of the two main domestication trait. Voluntary maize seedlings (*left*) sprouted in October in a maize field, only a few days after harvest. The extreme high density and the patchiness are the obvious result of combining the lack of seed dispersal with the reduced seed dormancy. The high density results in strong competition for light, water and nutrients, further slowing growth at a time of the year that is already challenging for a cold-sensitive crop such as maize. After a few weeks (*right*), plants in similar situation have grown further, but are already showing some leaves burned by the low temperatures and will never make it to even produce flowers, let alone viable seeds, in temperate climates because they are soon going to be killed. The combination of the two domestication traits is thus a sort of "synthetic lethal" genotype for this crop in wild, temperate environments.

producing any offspring. Moreover, the reduced seed dormancy promotes germination soon after seed maturity is attained, which often means during an unfavourable season (e.g. just before an extended drought period) or at the wrong place (on the ear itself). The combination of the two key domestication traits is thus almost always lethal to the plant in wild environments (although their effects are compensated for during the farming of crop plants), which is why most crop plants are rarely found outside of cultivated fields.

Other traits commonly selected during domestications include changes in shape or size, timing of flowering, and nutrient and toxin content of the edible parts; these, in turn, affect appearance, reproductive strategies, edibility, resistance to pests and, most important, yield. Depending on the specific plant or its use, many other traits also could be selected for, such as colour, flavour, increased winter hardiness and seedlessness. More recently, resistance to herbicides or pests has been obtained by various means.

The domestication of a plant usually entails only a few crucial mutations, so domestication did (and still does to a significant extent) exploit spontaneous changes in the genetic material and their subsequent selection in new, useful combinations. Genetic modification (in a broad, generic sense) is therefore intrinsic to agriculture because crops would not exist without it. (Readers willing to delve into the scientific literature of old and new domestications are referred to Pigna & Morandini, 2017).

3. THE MODERN ERA

During the past century, to shorten the time required to identify and introduce useful traits, scientists and plant breeders have developed a panoply of techniques, including the treatment of seeds with radiation or mutagenic chemicals in order to increase significantly the frequency of mutations, and the screening of large numbers of plants with automated and sophisticated analyses. More recently, several powerful molecular techniques for precise site-directed mutagenesis, gene deletion/inactivation, and gene replacement or insertion in genomes have become available. These tools have markedly accelerated crop improvement and development.

Consider, for example, herbicide tolerance (HT), which simplifies weed management and, in certain cases, allows the control of parasitic weeds that can be serious pests. This successful trait has been achieved not only by moving genes from one organism to another via molecular techniques (a process sometimes called "transgenesis") but also through spontaneous mutation and random and site-directed mutagenesis. In 2016 alone, transgenic HT crops were cultivated on 161.9 million hectares or 87% of the 185.1 million hectares of transgenic crops planted globally. *Non*transgenic HT crops—the most important of which are rice, maize, soybean, sunflower and lentil—are also grown on many millions of hectares worldwide (Tan, Evans, Dahmer, Singh, & Shaner, 2005).

For about a century, plant breeders have performed "wide cross" hybridizations, in which hundreds or thousands of genes are moved across what were once considered "natural breeding boundaries", giving rise to plant varieties that cannot and do not occur spontaneously in nature (Fuentes, Stegemann, Golczyk, et al., 2014; Goodman, Hauptli, Crossway, & Knauf, 1987; Mason & Batley, 2015; Randhawa, Bona, & Graf, 2015; Sharma, 1995). In these hybridizations between different species or genera, the parental plants may be sufficiently compatible to produce a viable zygote that will not, however, develop into a fertile plant. To overcome this obstacle, scientists have devised mechanical and biochemical ways to "rescue" the embryos and produce viable, fertile plants from them. Commercial crops derived from wide crosses include common varieties of tomato, potato, oat, rice, wheat, corn and pumpkin, among others. Illustrating the continuum of genetic modification that exists, the products of wide crosses might well be called "nonmolecular transgenics".

Thus, attempts to create meaningful categories of "GMOs", "transgenics", and the like, particularly for purposes of regulatory review or labelling, are fraught, and shed more heat than light (Miller, 2014).

The products of genetic modification, whatever the technique(s) used, are not wholly without problems, but they are not referable to the technology—except perhaps in the sense that the premolecular methods are imprecise and often introduce unwanted genetic changes. A good example is the backdrop to the last major crop epidemic in the United States, which occurred in 1970: A fungus that causes a disease called Southern corn leaf blight claimed approximately 15% of the corn crop, costing farmers to lose 20 million metric tons worth about a billion dollars (Levings, 1990). The culprit was the relative imprecision of older techniques of genetic modification. For several years, most of the corn in the country had been grown from hybrid lines that contained a trait called "Texas cytoplasm male sterility". The extensive use of male sterility obviates the need to remove plants' tassels mechanically or by hand in order to produce hybrid seed, but unknown to plant breeders, those hybrid varieties were more sensitive to Southern corn leaf blight, and 1970 offered climatic conditions (warm and wet) especially favourable to the fungus (Levings, 1990). When genetic changes are introduced with more precise molecular genetic modification techniques, by contrast, the number of genes moved or altered is small, the predictability of the final plant is greater and the likelihood of unwanted changes is less.

A common generic problem with certain crops is gene flow between the domesticated crop and wild or weedy relatives. This happens, for example, with herbicide-tolerant crops, whether they are transgenic or crafted by mutagenesis. Indeed, resistance to the herbicide imidazolinone has moved from cultivated rice to weedy red rice in many places, including Costa Rica, Brazil, the United States and Italy (Busconi, Rossi, Lorenzoni, Baldi, & Fogher, 2012; Gressel & Valverde, 2009). Similarly, the continued use of any pesticide (herbicide, insecticide, fungicide, etc.) or other mean of selection will eventually favour resistant individuals, if given enough time and a sufficient number of individuals.

Thus, the emergence of herbicide-resistant weeds is a predictable result of selection pressure, especially in the absence of good agronomic practices such as crop rotation or the switching of active herbicidal agents. However, although gene flow that spreads herbicide resistance does increase the fitness of weeds, it does so, for that trait, only in the fields. It is therefore not an ecological or safety problem, but an agronomic one.

In fact, although extensive gene flow between HT crops and weedy relatives may reduce the usefulness of the specific herbicide, it does not affect wild environments because herbicides are employed mainly on cultivated fields and, to a lesser extent, on other areas managed by humans (e.g. roadsides and golf courses). In this case, therefore, gene flow does not make the plant more fit in the wild, and, since the trait comes from a crop, gene flow is very likely to confer a *dis*advantage to any recipient plant in the wild because of the genetic burden of all the domestication alleles that are transferred along with the resistance.

Transgenic approaches have been developed to mitigate gene flow from crops (whether for HT or other traits). This is achieved by linking the specific trait to a "mitigator" gene, a gene that is integral to domestication (e.g. conferring loss of shattering or reduced seed dormancy) and which is therefore already present in crops but which is missing in wild or weedy relatives. By tightly linking the two genes, it is possible to reduce the fitness of individuals and prevent their thriving in the wild if gene flow has occurred. Such mitigation is only achievable using molecular genetic engineering techniques (Al-Ahmad, Galili, & Gressel, 2005; Al-Ahmad & Gressel, 2006; Rose et al., 2009).

GE herbicide-resistant plants make possible the use of no-till farming techniques, in which the soil is not ploughed, which reduces soil erosion, runoff of agricultural chemicals, and fuel consumption and carbon emissions by mechanized farm equipment. In 2014 alone, this was equivalent to removing 22.4 billion kg of carbon dioxide from the atmosphere or to removing 10 million cars from the road for 1 year (Barfoot & Brookes, 2016).

The use of modern genetic techniques to improve plants also both enhances sustainability and food security. The United Nations' Food and Agriculture Organization estimates that some 20%–40% of global crop totals are lost annually due to disease and pests. Much of these losses are preventable, and resistance to insect and viral pests has been among plant biotechnology's most striking success stories.

The common bacterium *Bacillus thuringiensis* (*Bt*) has played an important role in agriculture and in the development of GE crop plants. The insecticidal activity of certain proteins in these bacteria has made them an important pest control tool over the past century. Paradoxically, while the use of the intact bacteria as a natural biopesticide sprayed on crops is widely accepted and approved for conventional and organic applications, the engineering of *Bt* genes into major crops has been more controversial—due largely to organized, self-interested opposition to genetic engineering in general.

Since 1996 various recombinant DNA-modified plants have been crafted to contain short sequences of genes from *Bt* to express the crystal proteins that are toxic to certain insect pests (Koch et al., 2015). The plants synthesize the proteins and thereby protect themselves from insects without any external spraying with *B. thuringiensis*. *Bt* crops are protected specifically against the European corn borer, southwestern corn borer, tobacco budworm, cotton bollworm, pink bollworm or Colorado potato beetle. More than 90% of corn and cotton cultivated in the United States are now recombinant *Bt* varieties.

Because GE crops require less pesticide sprays, farmers and their families are at lower risk of poisoning as a result of handling them in large quantities. From 1996 to 2014, the cultivation of GE crops has reduced pesticide spraying by 581 million kg (−8.2%), equal to the total amount of pesticide active ingredient applied to crops in China for more than a year (Barfoot & Brookes, 2016). This has decreased the environmental impact associated with herbicide and insecticide use on the area planted to recombinant DNA-modified crops by 18.5%. A second-order benefit of reduced insect damage to crops is lower levels of specific mycotoxins in *Bt* corn, which means fewer birth defects such as spina bifida, and less toxicity to livestock (Wu, 2006).

By the mid-1990s, papaya ringspot virus, which devastates the trees and for which there is no cure, was laying waste to Hawaii's papaya industry. American scientists produced two varieties of papayas resistant to the virus via an ingenious strategy (Tripathi, Suzuki, & Gonsalves, 2007). They modified the papaya trees with the addition to the genome of a gene that expresses a coat protein of the virus, the presence of which interferes with viral propagation after infection.

Those papaya varieties saved Hawaii's papaya industry, its second most important crop (after pineapple). Similar commercialized products include virus-resistant squash and potatoes.

It must be stressed that it is difficult to obtain insect- or virus-resistant varieties, as well as many other traits, by conventional means and, when achieved, the new variety is normally the outcome of a random process. Moreover, the presence of concomitant undesirable genetic changes could cause the kinds of problems described earlier for US corn farmers afflicted with Southern corn leaf blight in 1970. By contrast, molecular genetic engineering techniques are versatile tools that precisely, predictably and relatively easily make possible the creation of plants resistant to pests and disease.

Genetic engineering provides consumer-friendly as well as agronomically important traits. Commercialized products include GE potato and apple varieties that are bruise resistant because the enzymes that mediate bruising have been downregulated, reducing the high percentage of waste for those crops (Healy, 2015). The potatoes also contain much less asparagine, a chemical that is converted to acrylamide, a probable carcinogen, when heated to high temperatures.

The domestication of many crops has required thousands of generations for the selection of desirable traits in the farmers' fields and for the genetic material expressing those traits to become fixed within crop genomes. With the benefit of long experience, the time required to domesticate a species has shortened, with several recognizable phases. The first one is a fairly rapid fixation of a few key traits because of strong selection by farmers, then diffusion of the crop to several places together with a long period of minor alterations and variety diversification.

In the second phase, often the crop experiences yield stasis due to inefficient selection. Traits such as fruit colour are easy to select, segregate and multiply in a new breed by farmers, but higher-yielding individuals are more difficult to identify in the absence of easily visible markers, and their identity is more difficult to preserve. Such superior genotypes are present in the population but remain at low frequency because seeds are propagated in bulk. This means that it takes a long time for such traits to undergo positive selection and to become sufficiently frequent in the population to show their effect.

By contrast, the use of modern genetic engineering techniques has allowed wider benefits to more farmers: the access to new, superior genotypes that usually provide large yield increases. (In the case of hybrid maize a fivefold increase compared to open-pollinated varieties is a common outcome.) The development of such genotypes would be extremely difficult or impossible to simple farmers, because the actual "breeding", or genetic modification, requires complex know-how, advanced tools and large investments in time, materials and, therefore, money. This is now most often performed by specialists working in large companies—which may be the source of much of the ideological opposition to modern genetic engineering—rather than by the farmers themselves. However, the benefits of the use of the new techniques have been widely distributed; in fact, over the past 20 years, the cumulative financial benefits from recombinant DNA-modified plants have been equally divided between farmers in industrialized and developed countries (Barfoot & Brookes, 2016).

The present phase, which began with the application of genetics, breeding, sterile tissue culture and more recently, molecular biology, has improved the yields and the pest and disease resistance of several crops far beyond expectations.

4. REALIZING GENETIC ENGINEERING'S POTENTIAL

We now have a far deeper understanding of plant growth, physiology, biochemistry and development than ever before, and that knowledge and the availability of new techniques are being exploited to improve our crops more precisely and predictably, via the transfer and manipulation of genes. This last domestication phase represents an open-ended process and holds great promise for the creation of novel crops and the incremental improvement of existing ones. Various aspects of this knowledge and new technologies are explored in the following chapters.

Agrobacterium is well known for its capacity to transfer specific fragments of DNA (transferred DNA or T-DNA) into plant cells, leading to the formation of tumours (crown galls) by *Agrobacterium tumefaciens* and to abundant root growth (hairy roots) by *Agrobacterium rhizogenes*. The discovery of this natural plant transformation system about a century ago, which can stably introduce foreign DNA into plants, has led to a revolution in agriculture. The chapter by Otten explains "How Agrobacterium, a Natural Genetic Engineer, Became a Tool for Modern Agriculture".

In spite of their enormous contribution to mankind, biotech crops continue to face excessive, discriminatory regulatory scrutiny. The chapter by Arujanan and Teng discusses the "Legal, Regulatory and Labelling Status of Biotech Crops".

The principle of proportionality, which embodies concepts of fairness, equity and consistency, is fundamental not only to government regulation but also to human rights and national and international law. In chapter "Regulating Safety of Novel Food and Genetically Modified Crops", Bartholomaeus explains why current regulatory burdens on recombinant DNA technology and the newer biotechnologies are applied in a discriminatory way, are disproportionate to the known (lack of) plausible food safety risks and are ignorant of the broader knowledge of natural plant genome plasticity.

Bees, which play an important role in the pollination of a wide range of plants, are likely to encounter GE crops during their foraging period—especially insect-resistant crops, which have been cultivated worldwide. It is important, therefore, to assess potential impacts of these crops on the nontarget organism honey bee (*Apis mellifera* L.), the most important pollinator species worldwide. By examining the available scientific literature, Ricroch et al. conclude in the chapter "Assessing the Environmental Safety of Transgenic Plants: Honey Bees as a Case Study" that the studied protease inhibitors, Cry proteins, RNAi and herbicide-tolerance proteins do not negatively affect the survival of honey bees and have no potential sublethal effect in controlled laboratory conditions or in actual or simulated field trials.

In chapter "Genetic Engineering of Crop Plants: Colombia as a Case Study", Mora-Oberlaender et al. describe the path followed by Colombia for the deployment of GE crops in agriculture. Colombia participated in the formulation of international regulatory policies such as the Cartagena protocol and established a national framework to deal with environmental release of GE crops and their use as food/feed. Within this framework, events for several crops (maize, cotton, flowers and soybeans) have been authorized for cultivation, while there are 104 events from 7 crops approved for food use since 2003 and 59 for feed. While the national framework is still far from a scientifically sound, risk-based approach, the experience gained in Columbia can be of value for other developing and developed countries.

Plant-based expression systems that produce protein-based therapeutic molecules offer several production advantages: low-cost, rapidity, scalability and a significantly lower chance of contamination with prion or mammalian viruses. In chapter "Recombinant Therapeutic Molecules Produced in Plants", Chen explains the current status and recent advancement of such systems, important clinical products and challenges and future directions.

Genome editing with engineered nucleases (e.g. CRISPR-Cas9, TALENs) represents a highly precise and efficient tool to generate useful novel phenotypes in crops by base additions, deletions, gene replacement or transgene insertion. These techniques generate phenotypic variation in plants that can be indistinguishable from those obtained through natural means or conventional mutagenesis. In chapter "Genome Editing in Agricultural Biotechnology", Pfeiffer et al. discuss issues surrounding the regulatory status of plants edited by engineered nucleases.

In chapter "Epigenetics, Epigenomics and Crop Improvement", Kapazoglou et al. discuss how intense breeding over the past centuries has eroded genetic diversity. Heritable epigenetic variation based on chromatin variation that does not involve alterations in DNA sequence can expand the sources of phenotypic diversity and affect important agronomical traits. Hence, epigenetic variation and the emerging field of epi-breeding can be exploited towards crop improvement in the context of enhancing yield, quality and adaptation to the ongoing climatic changes.

In concluding chapter "Biotechnologies: The Ideal Victim ?" Farouki, a philosopher and historian of science, presents a brief history of the way fear and mistrust about science and technology have been built over the past few decades. She analyses the attacks on new biotechnologies in the context of two complementary phenomena: the suspicion that surrounds everything that is related to food, and the questioning of scientific knowledge, method and truth in the intellectual postmodern sphere.

5. CONCLUSION

There is a long, venerable and seamless continuum of agricultural advances that began with the domestication of plants and has continued with incremental refinements as our experience and knowledge have accumulated and new technologies have become available. Transgenesis and genome-editing techniques have great potential and less risk of unintended effects when compared to spontaneous mutation or random mutagenesis. We need further advances to ensure that our food supply has sufficient resilience—the ability to recover from or adapt to adversity—in the face of the stresses of weather, pestilence and disease.

However, many of those advances—the use and diffusion of the newest technologies for genetic improvement, in particular, many of which are discussed elsewhere in this volume—are being hindered by extrascientific factors. These include government overregulation and antagonism from activists opposed to modern agriculture (for the various reasons discussed earlier) and industries with competing financial interests (e.g. organic agriculture). An important element of these problems is the myth that "GMOs" are a genuine "category" with unique characteristics (Tagliabue, 2016).

If we are to unleash the ingenuity of plant scientists in order to make farming more resilient and to improve food security for a growing world population, we will need enlightened public policy that takes into consideration the seamless continuum from domestication to 21st century genetic engineering. This must include sweeping regulatory reform to make regulation scientifically defensible and risk based. Many scientists have made credible proposals to achieve this (Conko, Kershen, Miller, & Parrott, 2016; McHughen, 2012; Potrykus, 2013), but as yet, there is nary a hint of that on the political horizon.

ACKNOWLEDGEMENT

Many thanks to Luca Mizzi for invaluable help with the figures.

REFERENCES

Al-Ahmad, H., Galili, S., & Gressel, J. (2005). Poor competitive fitness of transgenically mitigated tobacco in competition with the wild type in a replacement series. *Planta, 222*, 372–385.

Al-Ahmad, H., & Gressel, J. (2006). Mitigation using a tandem construct containing a selectively unfit gene precludes establishment of *Brassica napus* transgenes in hybrids and backcrosses with weedy *Brassica rapa*. *Plant Biotechnology Journal, 4*, 23–33.

Arnaud, N., Lawrenson, T., Østergaard, L., & Sablowski, R. (2011). The same regulatory point mutation changed seed-dispersal structures in evolution and domestication. *Current Biology, 21*, 1215–1219.

Barfoot, P., & Brookes, G. (2016). *GM crops: Global socio-economic and environmental impacts 1996–2014*. PG Economics Ltd., *http://biotechbenefits.croplife.org/paper/gm-crops-global-socio-economic-and-environmental-impacts-1996-2014/*. Accessed 30 May 2017.

Busconi, M., Rossi, D., Lorenzoni, C., Baldi, G., & Fogher, C. (2012). Spread of herbicide-resistant weedy rice (red rice, *Oryza sativa* L.) after 5 years of Clearfield rice cultivation in Italy. *Plant Biology (Stuttgart, Germany), 14*, 751–759.

Conko, G., Kershen, D. L., Miller, H., & Parrott, W. A. (2016). A risk-based approach to the regulation of genetically engineered organisms. *Nature Biotechnology, 34*(5), 493–503.

Dong, Y., Yang, X., Liu, J., Wang, B. H., Liu, B. L., & Wang, Y. Z. (2014). Pod shattering resistance associated with domestication is mediated by a NAC gene in soybean. *Nature Communications, 5*, 3352.

Fuentes, I., Stegemann, S., Golczyk, H., et al. (2014). Horizontal genome transfer as an asexual path to the formation of new species. *Nature*, *511*, 232–235.

Funatsuki, H., Suzuki, M., Hirose, A., Inaba, H., Yamada, T., Hajika, M., et al. (2014). Molecular basis of a shattering resistance boosting global dissemination of soybean. *Proceedings of the National Academy of Sciences of the United States of America*, *111*, 17797–17802.

Goodman, R. M., Hauptli, H., Crossway, A., & Knauf, V. C. (1987). Gene transfer in crop improvement. *Science*, *236*, 48–54.

Gressel, J., & Valverde, B. E. (2009). A strategy to provide long-term control of weedy rice while mitigating herbicide resistance transgene flow, and its potential use for other crops with related weeds. *Pest Management Science*, *65*, 723–731.

Harlan, J. R. (1992). *Crops and man*. Madison: American Soc. Agronomy, Madison, WI.

Healy, M. (2015). FDA says GMO apples, potatoes are 'safe for consumption'. In *Los Angeles Times*. March 20, 2015.

Koch, M. S., Ward, J. M., Levine, S. L., Baum, J. A., Vicini, J. L., & Hammond, B. G. (2015). The food and environmental safety of Bt crops. *Frontiers in Plant Science*, *6*, 283.

Konishi, S., Izawa, T., Lin, S. Y., Ebana, K., Fukuta, Y., Sasaki, T., et al. (2006). An SNP caused loss of seed shattering during rice domestication. *Science*, *312*, 1392–1396.

Levings, C. S., 3r. (1990). The Texas cytoplasm of maize: Cytoplasmic male sterility and disease susceptibility. *Science*, *250*, 942–947.

Li, C., Zhou, A., & Sang, T. (2006). Rice domestication by reducing shattering. *Science*, *311*, 1936–1939.

Lin, Z., Li, X., Shannon, L. M., Yeh, C.-T., Wang, M., Bai, G., et al. (2012). Parallel domestication of the Shattering1 genes in cereals. *Nature Genetics*, *44*, 720–724.

Liu, S., Sehgal, S. K., Lin, M., Trick, H. N., Gill, B. S., & Bai, G. (2015). Independent mis-splicing mutations in TaPHS1 causing loss of preharvest sprouting (PHS) resistance during wheat domestication. *The New Phytologist*, *208*, 928–935.

Lush, W. M., & Evans, L. T. (1980). The seed coats of cowpeas and other grain legumes: Structure in relation to function. *Field Crops Research*, *3*, 267–286.

Mason, A. S., & Batley, J. (2015). Creating new interspecific hybrid and polyploid crops. *Trends in Biotechnology*, *33*, 436–441.

McHughen, A. (2012). Introduction to the GM crops special issue on biosafety, food and GM regulation. *GM Crops & Food*, *3*, 6–8.

Meyer, R. S., DuVal, A. E., & Jensen, H. R. (2012). Patterns and processes in crop domestication: An historical review and quantitative analysis of 203 global food crops. *The New Phytologist*, *196*, 29–48.

Meyer, R. S., & Purugganan, M. D. (2013). Evolution of crop species: Genetics of domestication and diversification. *Nature Reviews. Genetics*, *14*, 840–852.

Miller, H. I. (2014). *The GMO stigma*. Project Syndicate. https://www.project-syndicate.org/commentary/henry-i--miller-on-the-meaningless-distinction-between-genetically-modified-organisms-and-their-conventional-counterparts. Accessed 6 April 2017.

Pigna, G., & Morandini, P. (2017). Domestication of new species. In R. Pilu & G. Gavazzi (Eds.), *More food: Road to survival* (pp. 255–320). Sharja, United Arab Emirates: Bentham Science Publisher.

Potrykus, I. (2013). Unjustified regulation prevents use of GMO technology for public good. *Trends in Biotechnology*, *31*(3), 131.

Pourkheirandish, M., Hensel, G., Kilian, B., Senthil, N., Chen, G., Sameri, M., et al. (2015). Evolution of the grain dispersal system in barley. *Cell*, *162*, 527–539.

Randhawa, H. S., Bona, L., & Graf, R. J. (2015). Triticale breeding—Progress and prospect. In F. Eudes (Ed.), *Triticale* (pp. 15–32). New York: Springer International Publishing.

Rose, C. W., Millwood, R. J., Moon, H. S., Rao, M. R., Halfhill, M. D., Raymer, P. L., et al. (2009). Genetic load and transgenic mitigating genes in transgenic *Brassica rapa* (field mustard) x *Brassica napus* (oilseed rape) hybrid populations. *BMC Biotechnology*, *9*, 93.
Sharma, H. C. (1995). How wide can a wide cross be? *Euphytica*, *82*, 43–64.
Sugimoto, K., Takeuchi, Y., Ebana, K., Miyao, A., Hirochika, H., Hara, N., et al. (2010). Molecular cloning of Sdr4, a regulator involved in seed dormancy and domestication of rice. *Proceedings of the National Academy of Sciences of the United States of America*, *107*, 5792–5797.
Tagliabue, G. (2016). The meaningless pseudo-category of "GMOs" *EMBO Reports*, *17*, 10–13.
Tan, S., Evans, R. R., Dahmer, M. L., Singh, B. K., & Shaner, D. L. (2005). Imidazolinone-tolerant crops: History, current status and future. *Pest Management Science*, *61*, 246–257.
Tripathi, S., Suzuki, J., & Gonsalves, D. (2007). Development of genetically engineered resistant papaya for papaya ringspot virus in a timely manner: A comprehensive and successful approach. *Methods in Molecular Biology*, *354*, 197–240.
Wu, F. (2006). Mycotoxin reduction in Bt corn: Potential economic, health, and regulatory impacts. *Transgenic Research*, *15*, 277–289.
Zhou, Y., Lu, D., Li, C., Luo, J., Zhu, B.-F., Zhu, J., et al. (2012). Genetic control of seed shattering in rice by the APETALA2 transcription factor shattering abortion1. *Plant Cell*, *24*, 1034–1048.
Zohary, D., & Hopf, M. (2000). *Domestication of plants in the old world* (3rd ed.). Oxford: Oxford University Press.

CHAPTER TWO

How *Agrobacterium*, a Natural Genetic Engineer, Became a Tool for Modern Agriculture

Léon Otten[1]
Institut de Biologie Moléculaire des Plantes du CNRS, Strasbourg, France
[1]Corresponding author: e-mail address: leon.otten@ibmp-cnrs.unistra.fr

Contents

Abstract

Agrobacterium is well known for its capacity to transfer specific fragments of DNA (transferred DNA or T-DNA) into plant cells, leading to the formation of tumours (crown galls) by *A. tumefaciens* and to abundant root growth (hairy roots) by *A. rhizogenes*. The T-DNA contains genes which change the growth of plant cells in various ways and lead to the production of special metabolites (called opines) used by the bacterium for its growth. The discovery of this natural plant transformation system started about one hundred years ago, and the adaptation of *A. tumefaciens* as a vector to stably introduce foreign DNA into plants has led to a revolution in plant research and in agriculture. The potential of *A. rhizogenes* is not yet fully exploited and much remains to be learnt about its root-inducing properties. Recent research has shown that apart from tumours and hairy roots, *Agrobacterium* can also produce transgenic plants in at least three different plant genera (*Nicotiana*, *Linaria* and *Ipomoea*), with stable transmission of T-DNA genes across species. In the case of *Nicotiana tabacum*, some cultivars express the TB-*mas2′* T-DNA gene to high levels in roots and produce the corresponding opine. The possible growth-modifying role of T-DNA genes in natural transformants remains to be studied.

Advances in Botanical Research, Volume 86
ISSN 0065-2296
https://doi.org/10.1016/bs.abr.2017.11.002

1. INTRODUCTION

Plant molecular biology would be very different without the use of *Agrobacterium* transformation vectors. *Agrobacterium* is a ubiquitous soil bacterium that induces tumours (called crown galls) or profusely growing roots (hairy roots) on susceptible plants. *Agrobacterium* is a member of the Rhizobiaceae family within the α-proteobacteria group. In nature, crown galls are mostly found on raspberry, fruit trees and grapevine, but the experimental host range of the bacterium is very wide. When it was discovered that *Agrobacterium* acts by transferring part of its DNA to plant cells, it was rapidly realized that this natural phenomenon could be used to introduce new DNA into plants. Soon after the creation of the first genetically transformed plants, *Agrobacterium* was adopted as a universal tool for plant genetic engineering. In basic science, *Agrobacterium*-based gene transfer mainly serves to study the expression and biological function of plant genes, for example, by overexpression or silencing or by the use of so-called reporter genes in which easily detectable proteins like the green fluorescent protein (GFP) are expressed under control of the gene promoter under study. Recent advances in plant gene silencing with the CRISPR-Cas9 technique often also rely on *Agrobacterium*-mediated transformation. Among the many practical applications of the *Agrobacterium* technology, crop improvements, horticultural innovations, synthesis of medically important proteins and synthesis of plant-derived chemicals have been achieved. Corn, cotton and soybean are among the most important genetically transformed plant species worldwide, and other major crops like wheat and rice are likely to follow. In view of the considerable importance of *Agrobacterium* in present-day basic and applied science it is worthwhile to take a step back in time and to present a short history of the field.

2. BASIC *AGROBACTERIUM* DISCOVERIES AND EARLY BIOTECHNOLOGY DEVELOPMENTS

Agrobacterium was discovered as a gall-inducing bacterium at the end of the 19th century. In 1907, Smith and Townsend isolated the bacterium and induced crown gall under experimentally controlled conditions (Smith & Townsend, 1907), providing the start of a scientific approach to the elucidation of the gall-inducing mechanism. Early on, parallels were established between *Agrobacterium*-induced tumours and animal tumours. In 1941,

White and Brown discovered that crown galls could be grafted onto normal plants and maintained their tumourous growth in the absence of the bacteria, showing a permanent change (White & Brown, 1941). Braun and Mandle therefore proposed the transfer of a tumour-inducing principle (TIP) from the bacterium to the plant, but could not isolate such an agent. They also showed that the tumour-inducing process, but not tumour growth, became blocked at temperatures of 32°C, thereby obtaining a way to establish the time it takes to complete the transformation process (Braun & Mandle, 1948). In 1956, Lioret detected a large amount of an unusual amino acid derivative in sterile crown gall tissues (Lioret, 1956). This compound was later found to be lysopine, a condensation product of pyruvic acid and lysine (Bielmann, Lioret, Asselineau, Lederer, & Polonski, 1960). The name lysopine is derived from octopine (derived from pyruvate and arginine, and originally found in *Octopus vulgaris*). Lysopine was subsequently found to be degraded by *Agrobacterium* (Lejeune & Jubier, 1967). Compounds similar to lysopine (among those octopine and nopaline) were soon discovered in other crown gall tissues and collectively called opines (Ménagé & Morel, 1964). The opines are conjugation products of widely occurring metabolites like amino acids, α-keto acids and sugars. At the end of the 1960s, the group of Tempé and Morel discovered that both the synthesis of a given opine type in a tumour and its use by *Agrobacterium* were strain dependent (Petit, Delhaye, Tempé, & Morel, 1970). This, and the fact that tumours continued to produce opines in the absence of the bacterium, was strong evidence for Braun's TIP hypothesis which postulated genetic transfer from the bacterium to the plant. In 1969, a major discovery was made by Kerr. He showed (Kerr, 1969) that agrobacteria can transfer their virulent properties (and therefore the TIP) to nonvirulent strains by conjugation in plants, but not in vitro. This unusual type of bacterial conjugation in planta was called a Kerr cross. Subsequently it became clear that conjugation between agrobacteria required the presence of opines, explaining why the cross worked well in tumours. Two years later, Hamilton described a method (Hamilton, 1971) to eliminate virulence by heat treatment (heat curing). Kerr cross and heat curing now made it possible to transfer or remove the still hypothetical TIP at will. At that time the most likely TIP candidate was a bacteriophage, and a report had appeared (based on the use of liquid DNA/DNA hybridization) that apparently showed bacteriophage PS8 DNA in crown gall DNA (Schilperoort, Van Sittert, & Schell, 1973), but later this result was shown to be due to an impurity in the crown gall DNA samples. In the early 1970s, the *Agrobacterium* group of van Montagu and Jeff Schell in Ghent started to look for phages with an electron microscope. To their surprise, they detected extrachromosomal DNA molecules (plasmids) in virulent *Agrobacterium* strains (Zaenen, van Larebeke, Teuchy, van Montagu, & Schell, 1974). The unusually large size of these plasmids (250 kb) made it difficult to isolate them, but techniques were developed to do so. Thus, instead of being a phage, the TIP might be the large plasmid, or part of it. Kerr crosses and heat curing soon showed that the presence of the large plasmids was correlated with virulence, opine synthesis and opine utilization (Van Larebeke et al., 1975; Watson, Currier, Gordon, Chilton, & Nester, 1975), and the plasmid was called tumour-inducing or pTi plasmid. However, it still had to be demonstrated that crown gall cells contained bacterial DNA. This demonstration was provided by Chilton in a series of brilliant experiments, involving pTi isolation, pTi restriction, restriction fragment purification and liquid DNA/DNA hybridization with crown gall DNA. The crown gall DNA was shown to contain a specific, relatively small pTi fragment, called the T-DNA (Chilton et al., 1977). This successful, but cumbersome, technique was rapidly replaced by the much easier Southern DNA blotting technique. Different groups (mainly those of Rob Schilperoort in Leiden, Gene Nester in Seattle, Marc Van Montagu and Jeff Schell in Ghent and Cologne) then established genetic methods like transposon insertion and homologous recombination to mutate pTi plasmids and to identify their functions (Garfinkel et al., 1981; Holsters et al., 1980; Ooms, Hooykaas, Moolenaar, & Schilperoort, 1981). In one of these studies the bacterial Tn7 transposon was found to be inserted in the T-DNA and to be transferred to plants, clearly showing the potential of the T-DNA to transfer foreign sequences (Hernalsteens et al., 1980). Among the Tn7 insertion mutants were Ti plasmids that had lost their tumour-inducing capacity but retained opine synthesis genes, and were still able to transfer T-DNA to plant cells. Plants obtained from such mutants were genetically transformed, but morphologically normal and transmitted their new sequences to the F1 progeny, showing stable incorporation into the germline (Otten et al., 1981). Interestingly, some natural pTi/pRi plasmids also carry bacterial transposons on their T-DNAs (like *IS870*, Fig. 1A) that can be transferred to plant cells. Transposon mutagenesis studies led to the discovery of the tumour-induction genes, genes for opine synthesis and degradation, genes required for virulence (*vir* genes), genes for plasmid transfer (as occurred in a Kerr cross) and genes for pTi replication. Some Ti plasmids (like the octopine Ti plasmids) were found to contain two T-DNA fragments that could be transferred independently (Thomashow, Nutter, Montoya, Gordon, & Nester, 1980), or even

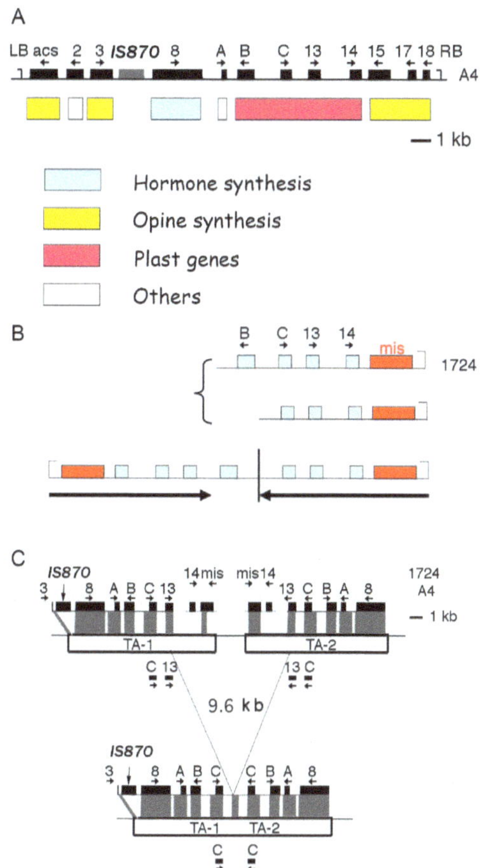

Fig. 1 *A. rhizogenes* T-DNA and cellular T-DNA (cT-DNA) in natural transgenic plants. (A) Map of the *A. rhizogenes* A4 TL-DNA region. IS*870*: bacterial insertion sequence IS*870*. (B) Map of the *N. glauca* T-DNA. *Top* and *middle*: two fragments of the pRi1724 DNA similar to the cT-DNA of *N. glauca*. *Bottom*: the two pRi1724-like fragments combined in the partial inverted repeat (*arrows*) found in *N. glauca*. The mikimopine synthase gene (*mis*) has been found in the pRi1724 T-DNA and replaces the *orf15-17-18* region from pRiA4. (C) *Top*: map of the tobacco TA region. *Bottom*: 9.6 kb internal deletion found in some tobacco cultivars. 1724 and A4 indicate regions with DNA homology to pRi1724 and pRiA4. *Panel (C) adapted from Chen, K., Dorlhac de Borne, F., Szegedi, E., & Otten, L. (2014). Deep sequencing of the ancestral tobacco species* Nicotiana tomentosiformis *reveals multiple T-DNA inserts and a complex evolutionary history of natural transformation in the genus* Nicotiana. The Plant Journal, 80, *669–682.*

three (Canaday, Gérard, Crouzet, & Otten, 1992). The *Agrobacterium tumefaciens* T-DNAs contain three genes responsible for tumour induction: the isopentenyl transferase (*ipt*) gene, leading to cytokinin synthesis (Buchmann, Marner, Schröder, Waffenschmidt, & Schröder, 1985) and two genes (*iaaM*, coding for indoleacetamide synthesis from tryptophan, and *iaaH*, coding for the conversion of indolacetamide into indoleacetic acid) required for indoleacetic acid synthesis (Schröder, Waffenschmidt, Weiler, & Schröder, 1984). Similar genes were found in other plant-associated bacteria (Morris, 1986) and are therefore most likely of bacterial origin. In the case of the *Agrobacterium* T-DNA genes, their promoters probably evolved to function in plants. DNA sequencing studies, still in their infancy, showed that the T-DNA was surrounded by two similar, short sequences in direct repeat (called left and right border sequences), and it was shown that they were sufficient for DNA transfer (Wang, Herrera-Estrella, van Montagu, & Zambryski, 1984; Yadav, Vanderleyden, Bennett, Barnes, & Chilton, 1982). When it was demonstrated that the natural T-DNA could be replaced by an artificial sequence (Hernalsteens et al., 1980), this opened up the new era of plant genetic engineering. The pTi was modified to allow efficient and easy DNA transfer, and two vector types were developed, cointegration vectors (Zambryski et al., 1983) and so-called binary vectors (De Framond, Barton, & Chilton, 1983; Hoekema, Hirsch, Hooykaas, & Schilperoort, 1983), the latter has been universally adopted. It separates the pTi in two parts: a large helper plasmid carries the *vir* genes, whereas a smaller plasmid (the binary vector) carries the T-DNA and can be easily isolated, modified and reintroduced in *Agrobacterium* strains carrying the helper plasmid. However, genetic engineering not only required DNA transfer but also the development of other tools, like selectable marker genes and plant gene promoter cassettes (Bevan, Flavell, & Chilton, 1983; Fraley et al., 1983; Herrera-Estrella, Depicker, Van Montagu, & Schell, 1983) and of course genes that could confer agronomically important traits, like insect resistance (Perlak, Fuchs, Dean, McPherson, & Fischhoff, 1991; Vaeck et al., 1987) or herbicide tolerance (Thompson et al., 1987). The first selectable marker genes for plants were derived from a bacterial kanamycin resistance gene placed under the control of a nopaline synthase (*nos*) promoter. Plant regeneration techniques for so-called recalcitrant species became very important and allowed the regeneration of the first transgenic major crops, like maize (Rhodes, Pierce, Mettler, Mascarenhas, & Dettmer, 1988) and rice (Hiei, Ohta, Komari, & Kumashiro, 1994; Zhang & Wu, 1988). An early development of these applications happened in Ghent at a newly founded plant biotechnology company, Plant Genetic Systems (PGS). PGS was set up in 1982 through a joint venture between Advanced Genetic Sciences (AGS) and european investment capital. Large companies also became active at an early stage. Dr. Ernie Jaworski, working at Monsanto, contacted the Ghent group as early as 1977, looking for ways to protect crops against the Monsanto herbicide glyphosate (Roundup, first commercialized in 1975) by genetic transformation. In 1980, when it became clear that plant genetic engineering with *Agrobacterium* vectors was indeed possible (Hernalsteens et al., 1980), Monsanto set up a special lab with Rob Horsch, Steve Rogers and Robb Fraley, while maintaining close contacts with the various *Agrobacterium* groups. In 1983, at a Symposium in Miami (Advances in Gene Technology: Molecular Genetics of Plants and Animals), three groups independently announced the production of the first genetically engineered plants. Mary-Dell Chilton (Washington University), Jeff Schell and Marc van Montagu (University of Ghent and Max Planck Institute for Plant Breeding in Cologne) and Rob Horsch (Monsanto) all reported on the transfer and expression of antibiotic resistance genes in plants. These results showed not only that these new genes could be introduced in plants but also provided an easy way to select transgenic tissues. With the further development of herbicide and insect resistant crops, plant biotechnology became a worldwide business. The basic discoveries in the *Agrobacterium* field have revealed a remarkable and still unique case of horizontal gene transfer, where a bacterium transfers part of its own DNA to plant cells, in order to manipulate cell division (leading to tumours or hairy roots) and plant metabolism (leading to opines). The combined result is the forced production of large amounts of conjugated compounds that only the crown gall-inciting bacterium can use for its growth, thanks to specific opine uptake and catabolism genes on the Ti plasmid. This unique system has aptly been called "genetic colonization" (Schell et al., 1979). It was later shown that other types of bacteria also acquired the possibility to use opines (Baek et al., 2005), and thus act as pirates attacking the colonists.

3. *AGROBACTERIUM RHIZOGENES* AND ITS USE IN BIOTECHNOLOGY

Although the major discoveries in the *Agrobacterium* field were made with *A. tumefaciens*, other types of *Agrobacterium* species are known and also merit attention. *A. rubi* has been isolated from blueberries, but very little is known about this species. *A. vitis* has only been found on grapevine where it can cause considerable damage. One of the major factors for its adaptation is its unique capacity to use tartaric acid, most often encoded by the Ti plasmid (Burr & Otten, 1999). *A. rhizogenes* causes aberrant root growth (hairy roots, abbreviated HR) on a large number of plant species (Riker, 1930). These root-inducing bacteria basically use the same system as *A. tumefaciens*, i.e., transfer of a T-DNA fragment (Chilton et al., 1982) surrounded by border sequences and located on a large plasmid (called the pRi or root-inducing plasmid; Moore, Warren, & Strobel, 1979). Several strains have been isolated with different T-DNA structures, and they are presently classified as agropine (Gyon, Chilton, Petit, & Tempé, 1980), mannopine (Hansen et al., 1991; Koplow, Byrne, Jen, Tempé, & Chilton, 1984), cucumopine (Serino, Clérot, Brevet, Costantino, & Cardarelli, 1994) and mikimopine (Kiyokawa, Kobayashi, Kikuchi, Kamada, & Harada, 1994) strains. *A. rhizogenes* T-DNAs carry different opine synthesis genes, some of which are shared with *A. tumefaciens* and *A. vitis*. Among these, the *mas1′*, *mas2′* and *ags* genes encode the synthesis of deoxyfructosylglutamine (DFG), mannopine, mannopinic acid, agropine and agropinic acid (Baek et al., 2005). As in the case of the Ti plasmids, the corresponding catabolic genes are carried by the pRi plasmid, outside the T-DNA region(s).

The root-inducing *A. rhizogenes* T-DNA genes were called root locus (*rol*) genes. Full hairy root development requires a minimum of three *rol* genes: *rolA*, *rolB* and *rolC* (Aoki & Syono, 1999b; Capone et al., 1989; Cardarelli et al., 1987; Costantino et al., 1994; Gidoni, Bar, & Gilboa, 2001; Schmülling, Schell, & Spena, 1988; Spena, Schmülling, Koncz, & Schell, 1987; White, Taylor, Huffmann, Gordon, & Nester, 1985), additional genes can modify the efficiency of the *rol* genes: *orf13* and *orf14*. A map of the TL-DNA region from the model strain *A. rhizogenes* A4 is shown in Fig. 1A.

Some *A. rhizogenes* strains carry *iaaM* and *iaaH* genes, which also contribute to root induction. More than 30 years after their discovery the precise function of the *rol* genes is still unclear. Even more astonishing is the fact that practically no research has been carried out to establish what exactly makes hairy roots different from normal roots. Contrary to crown gall tumours, hairy roots can easily regenerate into plants, either spontaneously, or by culture on artificial media (Ackermann, 1977; Costantino, Spanò, Pomponi, Benvenuto, & Ancora, 1984; David, Chilton, & Tempé, 1984; Spanò & Costantino, 1982; Tepfer, 1984). Such plants show an "hairy root or HR phenotype", generally characterized by wrinkled leaves, a high rate of root growth, reduced apical dominance in roots and stems, and broader leaves (Tepfer, 1984). HR roots are affected in gravitropism (Legué et al., 1996). Hairy roots of many plant species grow indefinitely in vitro in simple media and have been used in several biotech applications (Giri & Narasu, 2000; Hu & Du, 2006). Their main use is in the production of high-value secondary metabolites (Georgiev, Pavolov, & Bley, 2007). Another interesting application of hairy roots is the production of transformed roots on plant species that are otherwise difficult to regenerate, leading to composite plants (Hansen, Jørgensen, Stougaard, & Marcker, 1989). This allows rapid testing of different root-related phenomena like induction of nitrogen-fixing nodules by *Rhizobium* and *Frankia*, study of mycorrhizal interactions, or uptake and secretion of various compounds. Hairy roots have also been used in horticulture to improve rooting (Strobel & Nachmias, 1985; Zhu, Holefors, Ahlman, Xue, & Welander, 2001) and in floriculture (Casanova, Trillas, Moysset, & Vainstein, 2005). It is clear that a better knowledge of the genes that allow hairy roots to grow so efficiently will be of interest for fundamental plant science and for breeders. It should notably be investigated at which level the *rol* genes modify root growth and anatomy, whether they act on the cell cycle, and at which particular phase, what are the possible molecular targets, and whether they change cell expansion and root cell differentiation. Since *rol* genes also affect leaf morphogenesis, their action is not limited to roots. Studies aimed at finding the *rol* gene targets should be carried out with single *rol* genes but also with different *rol* gene combinations, preferably with inducible promoters like the dexamethasone-inducible gene promoter (Aoyama & Chua, 1997; Grémillon, Helfer, Clément, & Otten, 2004), and using techniques like confocal microscopy, GFP reporter genes and wild-type and mutant *Arabidopsis* lines. The combination of these approaches can provide levels of molecular and anatomical detail that were not available at the time of the initial studies, over 30 years ago.

4. THE *PLAST* GENE FAMILY

Interestingly, many T-DNA genes belong to a gene family that was identified in 1987 on the basis of weak protein similarities (Levesque, Delepelaire, Rouzé, Slightom, & Tepfer, 1988). Surprisingly, this family is extremely diverged in spite of the fact that it seems mostly limited to the *Agrobacterium* T-DNAs. Because several of its members cause strong phenotypic changes, it was called the *plast* family (for morphogenetic plasticity). The *rolB*, *rolC*, *orf13* and *orf14* genes of *A. rhizogenes* and several genes from *A. tumefaciens* and *A. vitis* (genes *5*, *6a*, *6b*, *d* and others), belong to this group. A puzzling case is the 5′ fragment of the *iaaM* (and homologous *orf8*) gene, which strongly resembles the *rolB* gene. This part is not present in the bacterial *iaaM* genes (like the *Pseudomonas savastanoi iaaM* gene; Morris, 1986) and not required for IAM synthesis. When expressed separately in plants, the

5′ *rolB* part of *orf8* leads to starch accumulation (Otten & Helfer, 2001) and retention of sucrose in source leaves (Umber, Voll, Weber, Michler, & Otten, 2002). Still other genes on the pRi T-DNAs like *rolA* (Sinkar, Pythoud, White, Nester, & Gordon, 1988) or *orf3* (Lemcke & Schmülling, 1998), not related to the *plast* genes, await more detailed studies (Meyer, Tempé, & Costantino, 2000; Nilsson & Olsson, 1997). T-DNAs consistently contain opine genes and *plast* genes, suggesting the *plast* genes are of great importance to the bacterium.

The *6b plast* gene causes very strong and highly unusual growth changes (Chen & Otten, 2015; Helfer, Clément, Michler, & Otten, 2003; Hooykaas, den Dulk-Ras, & Schilperoort, 1988) and stimulates sucrose uptake, while favouring retention of sucrose at places where it is expressed, showing a similar effect as for the 5′ part of *orf8* (Clément et al., 2007). This may be an important factor in the observed growth modifications and could convert crown gall tumours into strong sinks. However, other models have been put forward (Ishibashi, Kitakura, Terakura, Machida, & Machida, 2014; Ito & Machida, 2015; Kakiuchi, Gális, Tamogami, & Wabiko, 2006; Kakiuchi, Takahashi, & Wabiko, 2007; Kitakura et al., 2002; Kitakura, Terakura, Yoshioka, Machida, & Machida, 2008; Takahashi et al., 2013; Terakura et al., 2007; Wang et al., 2011). In the case of the *rolC* gene, it was originally assumed that it coded for a β-glucosidase hydrolysing cytokinin glucosides (Estruch, Chriqui, Grossmann, Schell, & Spena, 1991), but this was later shown to be unlikely (Faiss et al., 1996). Similarly, the *rolB* gene has been reported to be an β-indole-glucosidase acting on auxin derivatives (Estruch, Schell, & Spena, 1991), but later results have put this in doubt (Nilsson, Moritz, Imbault, Sandberg, & Olsson, 1993). Others reported that RolB is a tyrosine phosphatase (Filippini et al., 1996), but mutations of the putative phosphatase motif CX5R did not abolish the biological activity of the RolB protein (Mohajjel-Shoja, 2010).

5. *AGROBACTERIUM* AS A NATURAL GENETIC ENGINEER

In the early days of *Agrobacterium* research, when the first Southern blots showed that sterile crown gall and hairy root tissues contained T-DNA sequences, it was discovered that untransformed plants of several species carried T-DNA-like sequences in their genomes (called cellular T-DNA or cT-DNA). These sequences were homologous to *A. rhizogenes* sequences (White, Garfinkel, Huffman, Gordon, & Nester, 1983). This immediately suggested that such plants were natural transformants that had acquired their cT-DNA by an ancient transformation event. It was postulated that *A. rhizogenes*-induced hairy roots had spontaneously regenerated into transgenic plants. Initially, a cT-DNA was found in *N. glauca* (a member of the *Noctiflorae* section in the Solanaceae family), and called Ng cT-DNA. A systematic search by Southern analysis in the *Nicotiana* genus (Furner et al., 1986) revealed cT-DNA sequences in other species, like *N. benavidesii*, *N. tabacum*, *N. tomentosa*, *N. tomentosiformis* and *N. otophora*. Only several years later, a complete sequence of the *N. glauca* cT-DNA was obtained. For a list of cT-DNA genes, see Table 1. Small partial sequences were published for other *Nicotiana* species and an attempt was made to study their distribution (Intrieri & Buiatti, 2001; Suzuki, Yamashita, & Tanaka, 2002). It was originally assumed that the *Nicotiana* cT-DNA sequences had been introduced during a rare transformation event in an ancestral *Nicotiana* species and were then transmitted during the speciation events that led to the various modern *Nicotiana* species with cT-DNAs. A large-scale search for cT-DNA sequences in other plant genera using PCR by the group of Tatiana Matveeva and Ludmila Lutova in Saint-Petersburg led to the discovery of another natural transformant: *Linaria vulgaris*, a member of the Plantaginaceae family (Matveeva, Bogomaz, Pavlova, Nester, & Lutova, 2012). Similar cT-DNA sequences were also found in other *Linaria* species (Matveeva & Lutova, 2014), suggestive for an ancient cT-DNA introduction in a *Linaria* ancestor with vertical transmission to different derived species. Recently, cT-DNAs were reported in a well-known member of the Convolvulaceae, *Ipomoea batatas* (sweet potato), widely used as an agricultural crop (Kyndt et al., 2015; Quispe, 2012). Two cT-DNAs, *Ib*T-DNA1 and *Ib*T-DNA2, were found, and they contain several genes that are expressed in different organs of the "Huachano" cultivar, indicating possible functions (Table 1). *Ib*T-DNA1 consists of a partial inverted repeat with the following genes: *acs* (agrocinopine synthase), gene *c*, *iaaH* and *iaaM*. *Ib*T-DNA2 was not fully sequenced, but contains sequences related to *orf14*, *orf17n*, *rolB/rolC*, *orf13* and *orf18/orf17n* in an order that is usual for *A. rhizogenes* T-DNAs. Most of the open reading frames are intact. Gene *c* has no known function and was previously only known for *A. tumefaciens*. Recently, it has been found in several organisms other than *Agrobacterium* (Chen, Dorlhac de Borne, Szegedi, & Otten, 2014), mainly in Fungi (Fig. 2).

In the cultivar "Huachano" each cT-DNA occurs in four copies. All tested cultigens (291) of cultivated sweet potato, but no related wild species, contained *Ib*T-DNA1, whereas *Ib*T-DNA2 was found in many, but not all cultivars and in the related *Ipomoea* species *I. trifida* (considered as an ancestor

Table 1 cT-DNA Genes in Natural Transgenic Plants

Species	cT-DNA	cT-DNA Genes	Intact	Expression	Biological Activity
N. glauca	T-DNA	*rolB*	–	+	+ (after restoration)
		rolC	+	+	+
		orf13	+	+	+
		orf14	+	+	+
		mis	+	+	+ (in *E. coli*)
Linaria vulgaris	T-DNA	*acs*	–	nt	nt
		orf2	–	nt	nt
		orf3	–	nt	nt
		orf8	–	nt	nt
		rolA	–	nt	nt
		rolB	–	–	nt
		rolC	+	–	nt
		orf13	–	–	nt
		orf14	–	–	nt
		mis	–	nt	nt
Ipomoea batatas	*Ib*T-DNA1	*acs*	+	+	nt
		Gene *c*	+	+	nt
		iaaH	+	+	nt
		iaaM	+	+	nt
	*Ib*T-DNA2	*orf14*	–	nt	nt
		orf17n	–	nt	nt
		rolB-like	+	+	nt
		orf13	+	+	nt
		orf18/ orf17n	+	nt	nt

Table 1 cT-DNA Genes in Natural Transgenic Plants—cont'd

Species	cT-DNA	cT-DNA Genes	Intact	Expression	Biological Activity
N. tomentosiformis and *N. tabacum* (except TC)	TA	*orf8*	–	nt	nt
		rolA	–	nt	nt
		rolB	–	nt	nt
		rolC	+	+	+ (*N. tabacum*)
		orf13	+	+	+ (*N. tabacum*)
		orf14-like	–	+	nt
		mis	–	nt	nt
	TB	*orf14*	+	nt	nt
		mis	–	nt	nt
		ags	–	nt	nt
		mas1′	–	nt	nt
		mas2′	+	+	+ (*N. tabacum*)
	TC	*ocs-like*	+	nt	—
		orf2	–	nt	nt
		orf3	–	nt	nt
		orf8	–	nt	nt
		rolA	–	nt	nt
		rolB	–	nt	nt
		Gene *c*	–	nt	nt
	TD	*orf18*	–	nt	nt
		orf14-like	+	nt	nt
		orf15	+	nt	nt
		orf511	?	nt	nt
N. otophora	TC	*ocs-like*	?	nt	nt
		orf2	?	nt	nt
		orf3	?	nt	nt
		orf8	?	nt	nt
		rolA	?	nt	nt
		rolB	?	nt	nt
		Gene c	?	nt	nt

Continued

Table 1 cT-DNA Genes in Natural Transgenic Plants—cont'd

Species	cT-DNA	cT-DNA Genes	Intact	Expression	Biological Activity
N. otophora	TE	*vis*	?	nt	nt
		6b	+	nt	+
		mas1′	?	nt	nt
		mas2′	?	nt	nt
		rolB	?	nt	nt
		rolC	?	nt	nt
		orf13	?	nt	nt
		orf14	?	nt	nt
		iaaH	?	nt	nt
		iaaM	?	nt	nt
		acs	?	nt	nt
N. noctiflora	T-DNA	*rolC*	?	+	nt
		orf13	?	+	nt
		orf14	?	+	nt
		mis	?	+	nt
		iaaH	?	+	nt
		iaaM	?	+	nt

The copies of the internal repeats have not been separated. Because TD-*orf511* is unique, it is unknown whether it is intact or not. *N. otophora* and *N. noctiflora* cT-DNA sequences were obtained from collections of paired sequence reads (each read 100 nucleotides) and partial contigs. As these reads and contigs have not yet been assembled, it is still unknown whether they carry intact cT-DNA gene copies. *nt*, not tested.

Adapted from Matveeva, T., & Lutova, L. A. (2014). Horizontal gene transfer from *Agrobacterium* to plants. *Frontiers in Plant Science, 5*, 1–11, and extended for *N. otophora* and *N. noctiflora.*

of *I. batatas*). In the case of the naturally transgenic *Nicotiana* species, the idea of a single event spreading into different species was found to be incorrect when cT-DNA patterns were elucidated in *N. tomentosiformis* (the paternal ancestor of tobacco) and in related species of the Tomentosae section (Chen et al., 2014). Analysis of complete genomes showed that during the evolution of this *Nicotiana* subgroup, at least five different transformation events had occurred, carried out by different *Agrobacterium* strains, with different types of *A. rhizogenes*-like T-DNAs. Even more surprisingly,

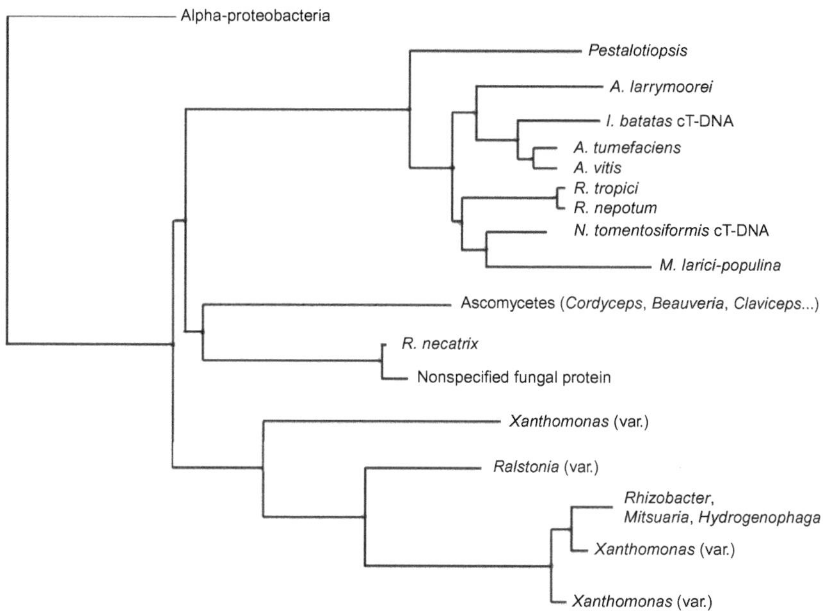

Fig. 2 Occurrence of T-DNA gene *c* in various organisms. The presence of gene *c* in different species suggests horizontal gene transfer. Tree was constructed for proteins from BLAST pairwise alignments with the following settings: Fast Minimum Evolution, Maximum Sequence Difference: 0.85, Distance: Grishin. Part of the tree was simplified to avoid redundancy. *A, Agrobacterium, I, Ipomoea; M, Melampsora; N, Nicotiana; R, Rhizobium.*

N. tomentosiformis was found to carry no less than four different cT-DNAs (TA, TB, TC and TD). Each of the cT-DNAs contained inverted, partially repeated sequences and was different from the *N. glauca* cT-DNA.

As it is highly unlikely that the four cT-DNAs of *N. tomentosiformis* were introduced in a single transformation event, several successive transformations must have occurred, leading to the accumulation of multiple cT-DNAs. The relative order of insertion could be established by using the sequence divergence between the cT-DNA repeats, more divergence indicating older events. This led to the hypothesis that the first natural transformant which had received the TC DNA was retransformed with the TB and TD region, and finally by the TA region, yielding a transgenic plant with about 65 kb of foreign DNA. Interestingly, cT-DNA loss was also observed. In tobacco, the complete TC region, and about 1 kb on both sides of it, are deleted. The tobacco genome also showed a partial loss of the TA region (Fig. 1C).

An ancestor of *N. otophora* (already carrying the TC region) was transformed by yet another *Agrobacterium*, introducing the TE region, as evidenced by sequences found in a collection of 44 nonassembled contigs of the *N. otophora* genome. The sequence of this TE region still remains to be pieced together. Interestingly, it contains *rol* genes, but also vitopine synthase (TE-*vis*) and *6b* gene sequences (TE-*6b*), normally found in *A. tumefaciens* or in *A. vitis* (Chen et al., 2014), indicating transformation by a new type of *Agrobacterium* strain that combines features of classical *A. rhizogenes* and *A. tumefaciens*/*A. vitis* strains. Recently, cT-DNA sequences of yet another nature have been found in the transcriptome of *N. noctiflora* (Long, Ren, Xiang, Wan, & Dong, 2016). The latter species is related to *N. glauca*, but the *noctiflora* sequences (Fig. 3) differ from the *glauca* sequences, suggesting that multiple transformations took place in the *Noctiflorae* section, as in the case of the Tomentosae section.

6. EXPRESSION OF CT-DNA GENES

An important question about cT-DNAs concerns the possible function of their genes in natural transgenics. Several such genes have been studied in detail. Among these, the mikimopine synthase gene (*mis*) from *N. glauca* was shown to encode an active enzyme (by expression in *Escherichia coli*), but no mikimopine was found in plants (Suzuki, Tanaka, Kamada, & Yamashita, 2001; Suzuki et al., 2002). *rolB* from *N. glauca* is mutated but could be restored to an active form (Aoki, 2004). The expression patterns of several cT-DNA genes in *N. glauca* were similar to those of *A. rhizogenes* T-DNA genes (Nagata, Kosono, Sekine, Shinmyo, & Syono, 1995). *NgrolC* (Aoki & Syono, 1999a) and *Ngorf13* (Aoki & Syono, 1999c) were shown to be biologically active. In the case of *N. tabacum*, *torf13* (Fründt, Meyer, Ichikawa, & Meins, 1998) and *trolC* (Meyer, Ichikawa, & Meins, 1995; Mohajjel-Shoja, Clément, Perot, Alioua, & Otten, 2011) were found to induce growth modifications like those found for *A. rhizogenes* equivalents. However, it could not yet be shown whether *NgrolC*, *Ngorf13*, *trolC* and *torf13* influence the growth of natural transformants. It has been proposed that *N. glauca* cT-DNA genes play a role in the growth of the spontaneous tumours observed in *N. glauca* × *N. langsdorffi* hybrids, since the genes show high expression levels in such tumours (Aoki et al., 1994). Recently, it was shown that the TB-*mas2′* cT-DNA gene is little or not expressed in *N. tomentosiformis* and most tobacco cultivars, but a few tobacco cultivars express it to high levels (Chen et al., 2016). Reporter gene experiments

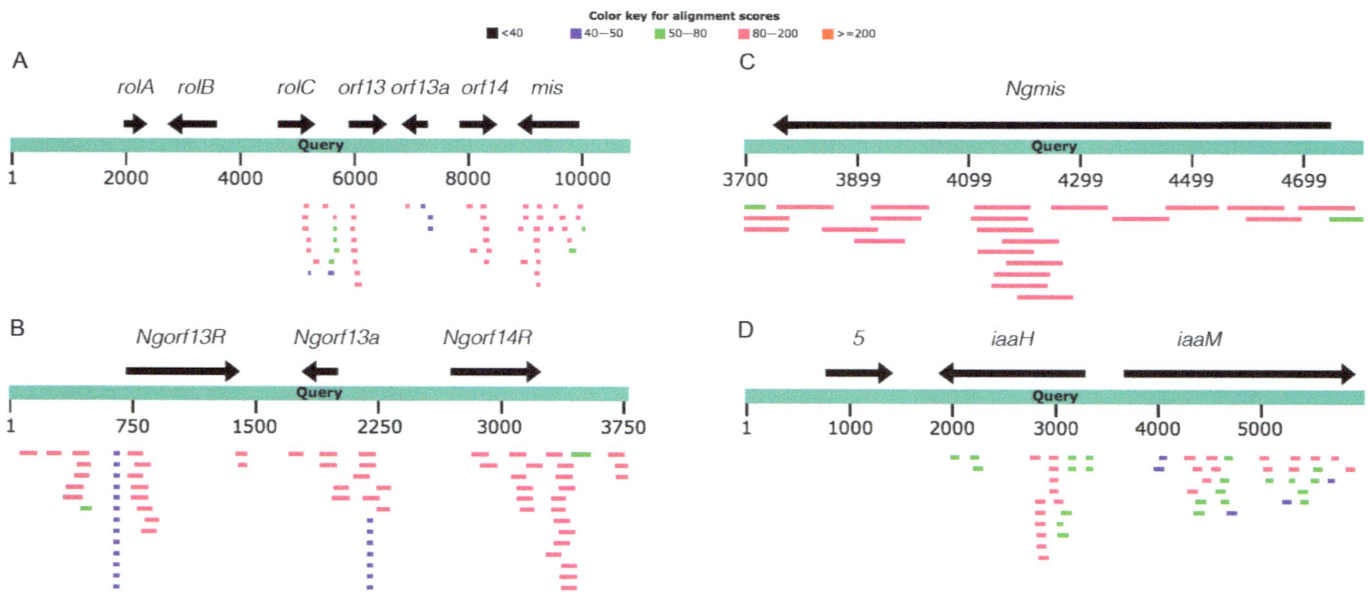

Fig. 3 Transcriptome-derived reads from *N. noctiflora* (Long et al., 2016), mapped on four different (c)T-DNA query sequences. (A) Query: 1724 T-DNA, partial. (B) Query: *N. glauca* cT-DNA, right arm, partial. (C) Query: *Ngmis* gene from *N. glauca*. (D) Query: C58 T-DNA, partial.

showed that the TB-*mas2′* gene is strongly expressed in root tips and leads to easily detectable levels of DFG in the roots of high-expression cultivars. DFG synthesis in tobacco was the first proof for a change caused by a cT-DNA gene in a natural transgenic plant. It remains to be investigated what could be the effects of TB-*mas2′*-encoded DFG synthesis. In high expression cultivars it may influence root growth by sequestering amino acids and sugars. Opine production might also affect the growth of bacteria that live in the rhizosphere of these plants. In experiments on artificial symbiosis, opine synthesis genes were introduced in various plant species and shown to lead to opine production and thereby change the composition of the rhizosphere by an accumulation of opine-degrading soil bacteria (Guyon, Petit, Tempé, & Dessaux, 1993; Savka, Dessaux, Oger, & Rossbach, 2002). Natural large-scale DFG synthesis in tobacco fields could be of benefit to agrobacteria or to other DFG-degrading microorganisms. In view of these findings, it even seems possible that the major function of the *A. rhizogenes* T-DNA is the creation of transgenic plants rather than hairy roots, as such plants would provide much higher quantities of opines than hairy roots. This interesting possibility will require quantitative analysis on opine synthesis, secretion, stability and uptake by agrobacteria in soil, using isogenic, opine-negative plants as controls. Only removal of the cT-DNA genes by homologous recombination techniques like CRISPR-Cas9, or by silencing of the cT-DNA gene expression with interfering RNA techniques, can address the question on the importance of the cT-DNA genes for the plants and the agrobacteria that created them.

7. ON THE ORIGIN OF NATURAL TRANSFORMANTS

With the knowledge we have so far, it is worthwhile to discuss a few more general points about natural transformants. One concerns the evolutionary origins of the transformants. Several steps have to be envisaged to explain these origins. First, a transformation event had to occur by an *A. rhizogenes* strain, presumably leading to hairy roots. Second, the hairy roots had to give rise to fertile plants, possibly through selection of the right combination of cT-DNA genes and their expression levels. Plants have been regenerated from hairy roots in a number of early hairy root experiments, but very little is known about the cT-DNA combinations that allow plant regeneration. More detailed studies are needed to address this question. The regenerant plants were probably different from their immediate nontransformed ancestors, because they carried T-DNA genes capable of inducing growth changes (as shown by the analysis of *NgrolB*, *NgrolC*, *Ngorf13*, *trolC* and *torf13* genes). It seems difficult to imagine that a very rare transformed plant could have survived among the nontransformed ancestor plants, without being significantly different. It is possible that the transformation event provided a selective advantage to the transformed plant. However, there exists another possibility. The degree of phenotypic change induced by *A. rhizogenes* T-DNA genes can be quite considerable, as shown by various experiments with individual T-DNA genes or with regenerated hairy roots. It is therefore possible that the morphological changes were so drastic that hybridization with the nontransformed ancestors became impossible, thus effectively creating a new species. In that case, no special selective advantages have to be postulated to assure survival of the rare transformants. Speciation can occur by different mechanisms like geographical isolation (leading to allopatric speciation) or genetic incompatibility (sympatric speciation), but horizontal gene transfer by *Agrobacterium* should now also be considered as a potential mechanism for speciation. What happened to the nontransformed ancestors? In the Tomentosae section only one species, *Nicotiana setchellii*, lacks cT-DNA, as shown by analysis of its transcriptome (Long et al., 2016). Thus, *N. setchellii* seems to be a descendant of the nontransformed ancestor. Can the different evolutionary possibilities for the origin of natural transgenic plants be experimentally investigated? In order to do so, we first must identify the original *A. rhizogenes* strains. All cT-DNAs seem to be incomplete, making full reconstruction of the original T-DNAs impossible. Although some cT-DNA genes are very similar to *A. rhizogenes* T-DNA genes, others on the same cT-DNA are quite different. It has been known for some time that T-DNAs are chimeric and evolve rapidly, probably by conjugation and Ti plasmid recombination between different strains, making it difficult to reconstruct the evolutionary history of these structures (Nuenen, de Ruffray, & Otten, 1993). The relatively low nucleotide divergence between the cT-DNA repeats (1%–6%, Chen et al., 2014) suggests that if the original *A. rhizogenes* strain survived, it should be easy to recognize. However, unless an *A. rhizogenes* strain is found with the same T-DNA structure as one of the cT-DNAs, it will be difficult to experimentally reproduce the postulated transformation event. Such a reconstruction will also require knowledge about the putative target plants. An ancestor of *N. setchellii* could be a candidate, but it certainly continued to evolve since the first transformation event occurred. In the meantime, one can hope to deduce some conclusions from the infection of different plant species with different *A. rhizogenes* strains, by investigating which cT-DNA combinations allow regeneration of fertile plants. Natural transformants lack intact *rolB* genes, suggesting that *rolB* is detrimental to their survival. Knockout of the different cT-DNA genes in natural transformants will not only provide important data on their growth effects but could also lead to clues about the phenotype of the original nontransformed plants.

Possibly, natural transformation can be observed as it occurs in nature. However, it might take place at such a modest scale (both in time and in space, and possibly underground) that it would be difficult to detect.

The remarkable accumulation of cT-DNAs within a single species like in *N. tomentosiformis*, also merits further investigation. Indeed, if natural transformation is rare, then secondary, tertiary and quaternary transformations would be increasingly unlikely. However, it could be that the nontransformed ancestor was prone to regeneration from roots. Some plants are known to easily regenerate shoots from roots under natural conditions, among those *Linaria* and *Ipomoea*. Thus, it would be interesting to see whether *Linaria* hairy roots can also regenerate easily. Alternatively, hairy roots obtained by the introduction of a first cT-DNA could have led to HR plants with an increased regeneration ability from roots, thus enabling further transformation events. The roots of tobacco HR regenerants spontaneously form shoots in pots, whereas normal tobacco roots do not (Tepfer, 1984).

Finally, the evolution of cT-DNAs after their initial establishment in natural transformants should be further investigated. If cT-DNA introduction led to speciation, this could have assured the survival of such new plants, but they possibly did not have optimal growth properties, so that negative selection would tend to remove some of the cT-DNA genes. Also, useless cT-DNA genes would certainly degenerate over time, either by point mutations or by partial or full deletion. Different environmental conditions might select for different cT-DNA expression patterns. It would therefore be interesting to compare cT-DNA structures and their expression, for example, by studying the relatively "old" TC region, both in different *Nicotiana* species (*N. otophora*, *N. tomentosiformis*, *N. kawakami*, *N. tomentosa*), and within the same species, between different accessions. In the case of cultivated natural transformants like tobacco and sweet potato, which are grown in different parts of the world, cT-DNA genes could have adapted to local conditions. The apparent activation of the TB-*mas2′* gene in certain tobacco cultivars (see above) could be such a case. Adaptation of cT-DNA genes can also be studied experimentally. Potato HR with artificially introduced T-DNAs easily produce spontaneous deletions (Hänisch ten Cate et al.,

1990). Another report describes a case of spontaneous transcriptional inactivation leading to a reversal from the HR to the normal phenotype (Sinkar, White, et al., 1988).

Finally, natural transformation may also occur in other plant species. Additional natural transformants might be most easily detected by automatically updated searches for T-DNA-related sequences within plant genomes and transcriptomes.

8. CONCLUSION

Agrobacterium has developed an extraordinary and unique capacity to exploit plants in order to assure its own survival. It uses genetic manipulation to efficiently introduce T-DNA genes in plant cells. The stable integration of T-DNA into the nuclear DNA of the transformed cells coupled with T-DNA-induced cell division ensures the creation of tumours and hairy roots, whereas T-DNA-encoded opine synthesis produces food for the bacterium. Hairy roots can lead to the regeneration of fertile naturally transgenic plants with new properties. Although *Agrobacterium* is now firmly established as a gene vector for both fundamental and applied research, much remains to be learnt about the way this natural genetic engineer changes the growth of plants, either in tumours, hairy roots or naturally transgenic plants.

ACKNOWLEDGEMENT

Work in my laboratory has been supported by basic funding from the CNRS.

REFERENCES

Ackermann, C. (1977). Pflanzen aus *Agrobacterium rhizogenes*-Tumoren an *Nicotiana tabacum*. *Plant Science Letters*, *8*, 23–30.

Aoki, S. (2004). Resurrection of an ancestral gene: Functional and evolutionary analyses of the Ngrol genes transferred from *Agrobacterium* to *Nicotiana*. *Journal of Plant Research*, *117*, 329–337.

Aoki, S., Kawaoka, A., Sekine, M., Ichikawa, T., Fujita, T., Shinmyo, A., et al. (1994). Sequence of the cellular T-DNA in the untransformed genome of *Nicotiana glauca* that is homologous to ORFs 13 and 14 of the Ri plasmid and analysis of its expression in genetic tumours of *N. glauca* x *N. langsdorffii*. *Molecular and General Genetics*, *243*, 706–710.

Aoki, S., & Syono, K. (1999a). Horizontal gene transfer and mutation: Ngrol genes in the genome of *Nicotiana glauca*. *Proceedings of the National Academy of Sciences of the United States of America*, 96, 13229–13234.

Aoki, S., & Syono, K. (1999b). Synergistic function of *rolB*, *rolC*, *ORF13* and *ORF14* of TL-DNA of *Agrobacterium rhizogenes* in hairy root induction in *Nicotiana tabacum*. *Plant and Cell Physiology*, *40*, 252–256.

Aoki, S., & Syono, K. (1999c). Function of *Ngrol* genes in the evolution of *Nicotiana glauca*: Conservation of the function of *NgORF13* and *NgORF14* after ancient infection by an *Agrobacterium rhizogenes*-like ancestor. *Plant and Cell Physiology*, *40*, 222–230.

Aoyama, T., & Chua, N.-H. (1997). A glucocorticoid-mediated transcriptional induction system in transgenic plants. *The Plant Journal*, *11*, 605–612.

Baek, C. H., Farrand, S. K., Park, D. K., Lee, K. E., Hwang, W., & Kim, K. S. (2005). Genes for utilization of deoxyfructosyl glutamine (DFG), an amadori compound, are widely dispersed in the family Rhizobiaceae. *FEMS Microbiology Ecology*, *53*, 221–233.

Bevan, M. W., Flavell, R. B., & Chilton, M.-D. (1983). A chimaeric antibiotic resistance gene as a selectable marker gene for plant cell transformation. *Nature*, *304*, 184–187.

Bielmann, K., Lioret, C., Asselineau, K., Lederer, E., & Polonski, J. (1960). Sur la structure chimique de la lysopine, nouvel acide aminé isolé du tissu de crown-gall. Bulletin de la Société de Chimie Biologique, *42*, 979–991.

Braun, A. C., & Mandle, R. J. (1948). Studies on the inactivation of the tumor-inducing principle in crown gall. *Growth*, *12*, 255–269.

Buchmann, I., Marner, F.-J., Schröder, J., Waffenschmidt, S., & Schröder, J. (1985). Tumour genes in plants: T-DNA encoded cytokinin biosynthesis. *The EMBO Journal*, *4*, 853–859.

Burr, T. J., & Otten, L. (1999). Crown Gall of grape: Biology and Disease Management. *Annual Review of Phytopathology*, *37*, 53–80.

Canaday, J., Gérard, J. C., Crouzet, P., & Otten, L. (1992). Organization and functional analysis of three T-DNAs from the vitopine Ti plasmid pTiS4. *Molecular and General Genetics*, *235*, 292–303.

Capone, I., Spanò, L., Cardarelli, M., Bellincampi, D., Petit, A., & Costantino, P. (1989). Induction and growth properties of carrot roots with different complements of *Agrobacterium rhizogenes* T-DNA. *Plant Molecular Biology*, *13*, 43–52.

Cardarelli, M., Mariotti, D., Pomponi, M., Spanò, L., Capone, I., & Costantino, P. (1987). *Agrobacterium rhizogenes* T-DNA genes capable of inducing hairy root phenotype. *Molecular and General Genetics*, *209*, 475–480.

Casanova, E., Trillas, M. I., Moysset, L., & Vainstein, A. (2005). Influence of *rol* genes in floriculture. *Biotechnology Advances*, *23*, 3–39.

Chen, K., de Borne, F. D., Julio, E., Obszynski, J., Pale, P., & Otten, L. (2016). Root-specific expression of opine genes and opine accumulation in some cultivars of the naturally occurring genetically modified organism *Nicotiana tabacum*. *The Plant Journal*, *87*, 258–269.

Chen, K., Dorlhac de Borne, F., Szegedi, E., & Otten, L. (2014). Deep sequencing of the ancestral tobacco species *Nicotiana tomentosiformis* reveals multiple T-DNA inserts and a complex evolutionary history of natural transformation in the genus *Nicotiana*. *The Plant Journal*, *80*, 669–682.

Chen, K., & Otten, L. (2015). Morphological analysis of the *6b* oncogene-induced enation syndrome. *Planta*, *243*, 131–148.

Chilton, M.-D., Drummond, M. H., Merlo, D. J., Sciaky, D., Montoya, A. L., Gordon, M. P., et al. (1977). Stable incorporation of plasmid DNA into higher plant cells: The molecular basis of crown gall tumorigenesis. *Cell*, *11*, 263–271.

Chilton, M.-D., Tepfer, D., Petit, A., David, C., Casse-Delbart, F., & Tempé, J. (1982). *Agrobacterium rhizogenes* inserts T-DNA into the genomes of host plant root cells. *Nature*, *295*, 432–434.

Clément, B., Perot, J., Geoffroy, P., Legrand, M., Zon, J., & Otten, L. (2007). Abnormal accumulation of sugars and phenolics in tobacco roots expressing the *Agrobacterium* T-*6b* oncogene and the role of these compounds in *6b*-induced growth. *Molecular Plant-Microbe Interactions*, *20*, 53–62.

Costantino, P., Capone, I., Cardarelli, M., De Paolis, A., Mauro, M. L., & Trovato, M. (1994). Bacterial plant oncogenes: The *rol* genes' saga. *Genetica*, *94*, 203–211.

Costantino, P., Spanò, L., Pomponi, M., Benvenuto, E., & Ancora, G. (1984). The T-DNA of *Agrobacterium rhizogenes* is transmitted through meiosis to the progeny of hairy root plants. *Journal of Molecular and Applied Genetics*, *2*, 465–770.
David, C., Chilton, M.-D., & Tempé, J. (1984). Conservation of T-DNA in plants regenerated from hairy root cultures. *Nature Biotechnology*, *2*, 73–76.
De Framond, A. J., Barton, K. A., & Chilton, M.-D. (1983). Mini-Ti: A new vector strategy for plant genetic engineering. *Bio/Technology*, *1*, 262–269.
Estruch, J. J., Chriqui, D., Grossmann, K., Schell, J., & Spena, A. (1991). The plant oncogene *rolC* is responsible for the release of cytokinins from glucoside conjugates. *The EMBO Journal*, *10*, 2889–2895.
Estruch, J. J., Schell, J., & Spena, A. (1991). The protein encoded by the *rolB* plant oncogene hydrolyses indole glucosides. *The EMBO Journal*, *10*, 3125–3128.
Faiss, M., Strnad, M., Redig, P., Dolezal, K., Hanus, J., van Onckelen, H., et al. (1996). Chemically induced expression of the *rolC* encoded β-glucosidase in transgenic tobacco plants and analysis of cytokinin metabolism: RolC does not hydrolyze endogenous cytokinin glucosides in planta. *The Plant Journal*, *10*, 33–46.
Filippini, F., Rossi, V., Marin, O., Trovato, M., Costantino, P., Downey, P. M., et al. (1996). A plant oncogene as a phosphatase. *Nature*, *379*, 499–500.
Fraley, R. T., Rogers, S. G., Horsch, R. B., Sanders, P. R., Flick, J. S., Adams, S. P., et al. (1983). Expression of bacterial genes in plants. *Proceedings of the National Academy of Sciences of the United States of America*, *80*, 4803–4807.
Fründt, C., Meyer, A. D., Ichikawa, T., & Meins, F. (1998). A tobacco homologue of the Ri-plasmid *orf13* gene causes cell proliferation in carrot root disks. *Molecular and General Genetics*, *259*, 559–568.
Furner, I. J., Huffman, G. A., Amasino, R. M., Garfinkel, D. J., Gordon, M. P., & Nester, E. W. (1986). An *Agrobacterium* transformation in the evolution of the genus Nicotiana. *Nature*, *319*, 422–427.
Garfinkel, D., Simpson, R. B., Ream, L. W., White, F., Gordon, M. P., & Nester, E. W. (1981). Genetic analysis of crown gall: Fine structure map of the T-DNA by site-directed mutagenesis. *Cell*, *27*, 143–153.
Georgiev, M. I., Pavolov, A. I., & Bley, T. (2007). Hairy root type plant *in vitro* systems as sources of bioactive substances. *Applied Microbiology and Biotechnology*, *74*, 1175–1185.
Gidoni, D., Bar, M., & Gilboa, N. (2001). FLP/FRT-mediated restoration of normal phenotypes and clonal sectors formation in *rolC* transgenic tobacco. *Transgenic Research*, *10*, 317–328.
Giri, A., & Narasu, M. L. (2000). Transgenic hairy roots: Recent trends and applications. *Biotechnology Advances*, *18*, 1–22.
Grémillon, L., Helfer, A., Clément, B., & Otten, L. (2004). New plant growth-modifying properties of the *Agrobacterium* T-*6b* oncogene revealed by the use of a dexamethasone-inducible promoter. *The Plant Journal*, *37*, 218–228.
Guyon, P., Petit, A., Tempé, J., & Dessaux, Y. (1993). Transformed plants producing opines specifically promote growth of opine-degrading agrobacteria. *Molecular Plant-Microbe Interactions*, *6*, 92–98.
Gyon, P., Chilton, M.-D., Petit, A., & Tempé, J. (1980). Agropine in "null-type" crown gall tumors: evidence for generality of the opine concept. *Proceedings of the National Academy of Sciences of the United States of America*, 77, 2693–2696.
Hamilton, R. H. (1971). The loss of tumor-inducing ability in *Agrobacterium tumefaciens* by incubation at high temperature. *Experientia*, *27*, 229–230.
Hänisch ten Cate, C. H., Loonen, A. E., Ottaviani, M. P., Ennik, L., van Eldik, G., & Stiekema, W. J. (1990). Frequent spontaneous deletions of Ri T-DNA in *Agrobacterium rhizogenes* transformed potato roots and regenerated plants. *Plant Molecular Biology*, *14*, 735–741.

Hansen, J., Jørgensen, J. E., Stougaard, J., & Marcker, K. A. (1989). Hairy roots—A short cut to transgenic root nodules. *Plant Cell Reports*, *8*, 12–15.

Hansen, G., Larribe, M., Vaubert, D., Tempé, J., Biermann, B. J., Montoya, A. L., et al. (1991). *Agrobacterium rhizogenes* pRi8196 T-DNA: Mapping and DNA sequencing of functions involved in mannopine synthesis and hairy root differentiation. *Proceedings of the National Academy of Sciences of the United States of America*, *88*, 7763–7767.

Helfer, A., Clément, B., Michler, P., & Otten, L. (2003). The *Agrobacterium* oncogene AB-*6b* causes a graft-transmissible enation syndrome in tobacco. *Plant Molecular Biology*, *52*, 483–493.

Hernalsteens, J.-P., Van Vliet, F., De Beuckeleer, M., Depicker, A., Engler, G., Lemmers, M., et al. (1980). The *Agrobacterium tumefaciens* Ti plasmid as a host vector system for introducing foreign DNA in plant cells. *Nature*, *287*, 654–656.

Herrera-Estrella, L., Depicker, A., Van Montagu, M., & Schell, J. (1983). Expression of chimaeric genes transferred into plant cells using a Ti-plasmid-derived vector. *Nature*, *303*, 209–213.

Hiei, Y., Ohta, S., Komari, T., & Kumashiro, T. (1994). Efficient transformation of rice (*Oryza sativa* L.) mediated by *Agrobacterium* and sequence analysis of the boundaries of the T-DNA. *The Plant Journal*, *6*, 271–282.

Hoekema, A., Hirsch, P. R., Hooykaas, P. J. J., & Schilperoort, R. A. (1983). A binary plant vector strategy based on separation of vir- and T-region of the *Agrobacterium tumefaciens* Ti-plasmid. *Nature*, *303*, 179–180.

Holsters, M., Silva, B., van Vliet, F., Genetello, C., De Block, M., Dhaese, P., et al. (1980). The functional organization of the nopaline *Agrobacterium tumefaciens* plasmid pTiC58. *Plasmid*, *3*, 212–230.

Hooykaas, P. J. J., den Dulk-Ras, H., & Schilperoort, R. A. (1988). The *Agrobacterium tumefaciens* T-DNA gene *6b* is an oncogene. *Plant Molecular Biology*, *11*, 791–794.

Hu, Z.-B., & Du, M. (2006). Hairy root and its application in plant genetic engineering. *Journal of Integrative Plant Biology*, *48*, 121–127.

Intrieri, M. C., & Buiatti, M. (2001). The horizontal transfer of *Agrobacterium rhizogenes* genes and the evolution of the genus Nicotiana. *Molecular Phylogenetics and Evolution*, *20*, 100–110.

Ishibashi, N., Kitakura, S., Terakura, S., Machida, C., & Machida, Y. (2014). Protein encoded by oncogene *6b* from *Agrobacterium tumefaciens* has a reprogramming potential and histone chaperone-like activity. *Frontiers in Plant Sciences*, *5*, 1–7.

Ito, M., & Machida, Y. (2015). Reprogramming of plant cells induced by 6b oncoproteins from the plant pathogen *Agrobacterium*. *Journal of Plant Research*, *128*, 423–435.

Kakiuchi, Y., Gális, I., Tamogami, S., & Wabiko, H. (2006). Reduction of polar auxin transport in tobacco by the tumorigenic *Agrobacterium tumefaciens* AK-*6b* gene. *Planta*, *223*, 237–247.

Kakiuchi, Y., Takahashi, S., & Wabiko, H. (2007). Modulation of the venation pattern of cotyledons of transgenic tobacco for the tumorigenic 6b gene of Agrobacterium tumefaciens AKE10. *Journal of Plant Research*, *120*, 259–268.

Kerr, A. (1969). Transfer of virulence between isolates of *Agrobacterium*. *Nature*, *223*, 1175–1176.

Kitakura, S., Fujita, T., Ueno, Y., Terakura, S., Wabiko, H., & Machida, Y. (2002). The protein encoded by oncogene *6b* from *Agrobacterium tumefaciens* interacts with a nuclear protein of tobacco. *The Plant Cell*, *14*, 451–463.

Kitakura, S., Terakura, S., Yoshioka, Y., Machida, C., & Machida, Y. (2008). Interaction between *Agrobacterium tumefaciens* oncoprotein 6b and a tobacco nucleolar protein that is homologous to TNP1 encoded by a transposable element of *Antirrhinum majus*. *Journal of Plant Research*, *121*, 425–433.

Kiyokawa, S., Kobayashi, K., Kikuchi, Y., Kamada, H., & Harada, H. (1994). Root-inducing region of mikimopine type Ri plasmid pRi1724. *Plant Physiology*, *104*, 801–802.

Koplow, J., Byrne, M. C., Jen, G., Tempé, J., & Chilton, M. D. (1984). Physical map of the *Agrobacterium rhizogenes* strain 8196 virulence plasmid. *Plasmid, 11*, 17–27.
Kyndt, T., Quispe, D., Zhai, H., Jarret, R., Ghislain, M., Liu, Q., et al. (2015). The genome of cultivated sweet potato contains *Agrobacterium* T-DNAs with expressed genes: An example of a naturally transgenic food crop. *Proceedings of the National Academy of Sciences of the United States of America, 112*, 5844–5849.
Legué, V., Driss-Ecole, D., Maldiney, R., Tepfer, M., & Perbal, G. (1996). The response to auxin of rapeseed (*Brassica napus* L.) roots displaying reduced gravitropism due to transformation by *Agrobacterium rhizogenes*. *Planta, 200*, 119–124.
Lejeune, B., & Jubier, M. F. (1967). Etude de la dégradation de la lysopine par *Agrobacterium tumefaciens*. *Comptes Rendus de l'Académie des Sciences de Paris, 264*, 1803–1805.
Lemcke, K., & Schmülling, T. (1998). Gain of function assays identify non-*rol* genes from *Agrobacterium rhizogenes* TL-DNA that alter plant morphogenesis or hormone sensitivity. *The Plant Journal, 15*, 423–433.
Levesque, H., Delepelaire, P., Rouzé, P., Slightom, J., & Tepfer, D. (1988). Common evolutionary origin of the central portion of the Ri TL-DNA of *Agrobacterium rhizogenes* and the Ti T-DNAs of *Agrobacterium tumefaciens*. *Plant Molecular Biology, 11*, 731–744.
Lioret, C. (1956). Sur la mise en évidence d'un acide aminé non-identifié particulier aux tissus de crown-gall. *Bulletin de la Société Française de Physiologie Végétale, 2*, 76.
Long, N., Ren, X., Xiang, Z., Wan, W., & Dong, Y. (2016). Sequencing and charcterization of leaf transcriptomes of six diploid *Nicotiana* species. *Journal of Biological Research (Thessaloniki), 23*, 1–12.
Matveeva, T. V., Bogomaz, D. I., Pavlova, O. A., Nester, E. W., & Lutova, L. A. (2012). Horizontal gene transfer from genus *Agrobacterium* to the plant *Linaria* in nature. *Molecular Plant-Microbe Interactions, 25*, 1542–1551.
Matveeva, T., & Lutova, L. A. (2014). Horizontal gene transfer from *Agrobacterium* to plants. *Frontiers in Plant Science, 5*, 1–11.
Ménagé, A., & Morel, G. (1964). Sur la présence d'octopine dans les tissus de crown-gall cultivés in vitro. *Comptes Rendus de l'Académie des Sciences de Paris, 259*, 4795–4796.
Meyer, A. D., Ichikawa, T., & Meins, F. (1995). Horizontal gene transfer: Regulated expression of a tobacco homologue of the *Agrobacterium rhizogenes rolC* gene. *Molecular and General Genetics, 249*, 265–273.
Meyer, A., Tempé, J., & Costantino, P. (2000). Hairy root: A molecular overview. Functional analysis of *Agrobacterium rhizogenes* T-DNA genes. In G. Stacey & N. T. Keen (Eds.), *Plant-microbe interactions: Vol. 5* (pp. 93–139). St. Paul, MN: American Phytopathological Society Press.
Mohajjel-Shoja, H. (2010). Contribution to the study of the *Agrobacterium rhizogenes plast* genes *rolB and rolC* and their homologs in Nicotiana tabacum. University of Strasbourg (Ph. D. Thesis).
Mohajjel-Shoja, H., Clément, B., Perot, J., Alioua, M., & Otten, L. (2011). Biological activity of the *Agrobacterium rhizogenes*-derived *trolC* gene of *Nicotiana tabacum* and its functional relationship to other plast genes. *Molecular Plant-Microbe Interactions, 24*, 44–53.
Moore, L., Warren, G., & Strobel, G. (1979). Involvement of a plasmid in the hairy root disease of plants caused by *Agrobacterium rhizogenes*. *Plasmid, 2*, 617–626.
Morris, R. O. (1986). Genes specifying auxin and cytokinin biosynthesis in phytopathogens. *Annual Review of Plant Physiology, 37*, 509–538.
Nagata, N., Kosono, S., Sekine, M., Shinmyo, A., & Syono, K. (1995). The regulatory functions of the *rolB* and *rolC* genes of *Agrobacterium rhizogenes* are conserved in the homologous genes (*Ngrol*) of *Nicotiana glauca* in tobacco genetic tumors. *Plant Cell Physiology, 36*, 1003–1012.
Nilsson, O., Moritz, T., Imbault, N., Sandberg, G., & Olsson, O. (1993). Hormonal characterization of transgenic tobacco plants expressing the *rolC* gene of *Agrobacterium rhizogenes* TL-DNA. *Plant Physiology, 102*, 363–371.

Nilsson, O., & Olsson, O. (1997). Getting to the root: The role of the *Agrobacterium rhizogenes rol* genes in the formation of hairy roots. *Physiologia Plantarum, 100*, 463–473.
Nuenen, M., de Ruffray, P., & Otten, L. (1993). Rapid divergence of *Agrobacterium vitis* octopine-cucumopine Ti plasmids from a recent common ancestor. *Molecular and General Genetics, 240*, 49–57.
Ooms, G., Hooykaas, P. J. J., Moolenaar, G., & Schilperoort, R. A. (1981). Crown gall plant tumours of abnormal morphology, induced by *Agrobacterium tumefaciens* carrying mutated octopine Ti plasmids; analysis of T-DNA functions. *Gene, 14*, 33–50.
Otten, L., De Greve, H., Hernalsteens, J.-P., Van Montagu, M., Schieder, O., Straub, J., et al. (1981). Mendelian transmission of genes introduced into plants by the Ti plasmids of *Agrobacterium tumefaciens*. *Molecular and General Genetics, 183*, 209–213.
Otten, L., & Helfer, A. (2001). Biological activity of the *rolB*-like 5′ end of the A4-*orf8* gene from the *Agrobacterium rhizogenes* TL-DNA. *Molecular Plant-Microbe Interactions, 14*, 405–411.
Perlak, F. J., Fuchs, R. L., Dean, D. A., McPherson, S. L., & Fischhoff, D. A. (1991). Modification of the coding sequences enhances plant expression of insect control protein genes. *Proceedings of the National Academy of Sciences of the United States of America*, 88, 3324–3328.
Petit, A., Delhaye, S., Tempé, J., & Morel, G. (1970). Recherches sur les guanidines des tissus de Crown Gall. Mise en évidence d'une relation biochimique spécifique entre les souches d'*Agrobacterium tumefaciens* et les tumeurs qu'elles induisent. *Physiologie Végétale, 2*, 205–213.
Quispe, D. (2012). *Identificacion y caracterizacion molecular de secuencias homologas al T-DNA del plasmido Ti de* Agrobacterium *spp. insertados en el genoma de* Ipomoea batatas *(L.) Lam « camote » y especies silvestres relacionadas*. Master Thesis, Lima: National University of San Marcos.
Rhodes, C. A., Pierce, D. A., Mettler, I. J., Mascarenhas, D., & Dettmer, J. J. (1988). Genetically transformed maize plants from protoplasts. *Science, 240*, 204–207.
Riker, A. J. (1930). Studies on infectious hairy root of nursery apple trees. *Journal of Agricultural Research, 41*, 507–540.
Savka, M. A., Dessaux, Y., Oger, P., & Rossbach, S. (2002). Engineering bacterial competitiveness and persistence in the phytosphere. *Molecular Plant-Microbe Interactions, 15*, 866–874.
Schell, J., Van Montagu, M., De Beuckeleer, M., De Block, M., Depicker, A., De Wilde, M., et al. (1979). Interactions and DNA transfer between *Agrobacterium tumefaciens*, the Ti-plasmid and the plant host. *Proceedings of the Royal Society of London, Series B, Biological Sciences, 204*, 251–266.
Schilperoort, R. A., Van Sittert, N. J., & Schell, J. (1973). The presence of both phage PS8 and *Agrobacterium tumefaciens* A6 DNA base sequences in A6-induced sterile crown-gall tissue cultured in vitro. *European Journal of Biochemistry, 15*, 1–7.
Schmülling, T., Schell, J., & Spena, A. (1988). Single genes from *Agrobacterium rhizogenes* influence plant development. *The EMBO Journal*, 7, 2621–2629.
Schröder, G., Waffenschmidt, S., Weiler, E. W., & Schröder, J. (1984). The T-region of Ti plasmids codes for an enzyme synthesizing indole-3-acetic acid. *European Journal of Biochemistry, 138*, 387–391.
Serino, G., Clérot, D., Brevet, J., Costantino, P., & Cardarelli, M. (1994). *rol* genes of *Agrobacterium rhizogenes* cucumopine strain: Sequence, effects and pattern of expression. *Plant Molecular Biology, 26*, 415–422.
Sinkar, V. P., Pythoud, F., White, F. F., Nester, E. W., & Gordon, M. P. (1988). *rolA* locus of the Ri plasmid directs developmental abnormalities in transgenic tobacco plants. *Genes and Development, 2*, 688–697.

Sinkar, V. P., White, F. F., Furner, I. J., Abrahamsen, M., Pythoud, F., & Gordon, M. P. (1988). Reversion of aberrant plants transformed with *Agrobacterium rhizogenes* is associated with the transcriptional inactivation of the T(L)-DNA genes. *Plant Physiology, 86*, 584–590.

Smith, E. F., & Townsend, C. O. (1907). A plant-tumor of bacterial origin. *Science, 25*, 671–673.

Spanò, L., & Costantino, P. (1982). Regeneration of plants from callus cultures of roots induced by *Agrobacterium rhizogenes* on tobacco. *Zeitschrift für Pflanzenphysiologie, 106*, 87–92.

Spena, A., Schmülling, T., Koncz, C., & Schell, J. S. (1987). Independent and synergistic activity of *rol A*, *B* and *C* loci in stimulating abnormal growth in plants. *The EMBO Journal, 6*, 3891–3899.

Strobel, G. A., & Nachmias, A. (1985). *Agrobacterium rhizogenes* promotes the initial growth of bare root stock almond. *Journal of General Microbiology, 131*, 1245–1249.

Suzuki, K., Tanaka, N., Kamada, H., & Yamashita, I. (2001). Mikimopine synthase (*mis*) gene on pRi1724. *Gene, 263*, 49–58.

Suzuki, K., Yamashita, I., & Tanaka, N. (2002). Tobacco plants were transformed by *Agrobacterium rhizogenes* infection during their evolution. *The Plant Journal, 32*, 775–787.

Takahashi, S., Sato, R., Takahashi, M., Hashiba, N., Ogawa, A., Toyofuku, K., et al. (2013). Ectopic localization of auxin and cytokinin in tobacco seedlings by the plant-oncogenic AK-*6b* gene of *Agrobacterium tumefaciens* AKE10. *Planta, 238*, 753–770.

Tepfer, D. (1984). Transformation of several species of higher plants by *Agrobacterium rhizogenes*: sexual transmission of the transformed genotype and phenotype. *Cell, 37*, 959–967.

Terakura, S., Ueno, Y., Tagami, H., Kitakura, S., Machida, C., Wabiko, H., et al. (2007). An oncoprotein from the plant pathogen *Agrobacterium* has histone chaperone-like activity. *The Plant Cell, 19*, 2855–2865.

Thomashow, M. F., Nutter, R., Montoya, A. L., Gordon, M. P., & Nester, E. W. (1980). Integration and organization of Ti plasmid sequences in crown gall tumors. *Cell, 19*, 729–739.

Thompson, C. J., Movva, N. R., Tizard, R., Crameri, R., Davies, J. E., Lauwereys, M., et al. (1987). Characterization of the herbicide-resistance gene bar from *Streptomyces hygroscopicus*. *The EMBO Journal, 6*, 2519–2523.

Umber, M., Voll, L., Weber, A., Michler, P., & Otten, L. (2002). The *rolB*-like part of the *Agrobacterium rhizogenes orf8* gene inhibits sucrose export in tobacco. *Molecular Plant-Microbe Interactions, 15*, 956–962.

Vaeck, M., Reynaerts, A., Höfte, H., Jansens, S., De Beuckeleer, M., Dean, C., et al. (1987). Transgenic plants protected from insect attack. *Nature, 328*, 33–37.

Van Larebeke, N., Genetello, C., Schell, J., Schilperoort, R. A., Hermans, A. K., Hernalsteens, J. P., et al. (1975). Acquisition of tumour-inducing ability by non-oncogenic agrobacteria as a result of plasmid transfer. *Nature, 255*, 742–743.

Wang, K., Herrera-Estrella, L., van Montagu, M., & Zambryski, P. (1984). Right 25 bp terminus sequence of the nopaline T-DNA is essential for and determines direction of DNA transfer from *Agrobacterium* to the plant genome. *Cell, 38*, 455–462.

Wang, M., Soyano, T., Machida, S., Yang, J. Y., Jung, C., Chua, N. H., et al. (2011). Molecular insights into plant cell proliferation disturbance by *Agrobacterium* protein 6b. *Genes and Development, 25*, 64–76.

Watson, B., Currier, T. C., Gordon, M. P., Chilton, M.-D., & Nester, E. W. (1975). Plasmid required for virulence of *Agrobacterium tumefaciens*. *Journal of Bacteriology, 123*, 255–264.

White, P. R., & Brown, A. C. (1941). Crown gall production by bacteria-free tumor tissues. *Science, 94*, 239–241.

White, F., Garfinkel, D., Huffman, G. A., Gordon, M., & Nester, E. W. (1983). Sequences homologous to *Agrobacterium rhizogenes* T-DNA in the genomes of uninfected plants. *Nature, 301*, 348–350.

White, F., Taylor, B. H., Huffmann, G. A., Gordon, M. P., & Nester, E. W. (1985). Molecular and genetic analysis of the transferred DNA regions of the root-inducing plasmid of *Agrobacterium rhizogenes*. *Journal of Bacteriology, 164*, 33–44.

Yadav, N. S., Vanderleyden, J., Bennett, D. R., Barnes, W. M., & Chilton, M.-D. (1982). Short direct repeats flank the T-DNA on a nopaline Ti plasmid. *Proceedings of the National Academy of Sciences of the United States of America, 79*, 6322–6326.

Zaenen, I., van Larebeke, N., Teuchy, H., van Montagu, M., & Schell, J. (1974). Supercoiled circular DNA in crown gall inducing *Agrobacterium* strains. *Journal of Molecular Biology, 86*, 109–127.

Zambryski, P., Joos, H., Genetello, C., Leemans, J., Montagu, M. V., & Schell, J. (1983). Ti plasmid vector for the introduction of DNA into plant cells without alteration of their normal regeneration capacity. *The EMBO Journal, 2*, 2143–2150.

Zhang, W., & Wu, R. (1988). Efficient regeneration of transgenic plants from rice protoplasts and correctly regulated expression of the foreign gene in plants. *Theoretical and Applied Genetics, 76*, 835–840.

Zhu, L., Holefors, A., Ahlman, A., Xue, Z., & Welander, M. (2001). Transformation of the apple rootstock M.9/29 with the *rolB* gene and its influence on rooting and growth. *Plant Science, 160*, 433–439.

CHAPTER THREE

Legal, Regulatory and Labelling Status of Biotech Crops

Mahaletchumy Arujanan*,[1], Paul P.S. Teng†
*Malaysian Biotechnology Information Centre, Petaling Jaya, Malaysia
†National Institute of Education, Nanyang Technological University, Singapore, Singapore
[1]Corresponding author: e-mail address: maha@bic.org.my

Contents

Advances in Botanical Research, Volume 86
ISSN 0065-2296
https://doi.org/10.1016/bs.abr.2017.11.019

Abstract

Biotech crops provide food, feed, fuel and fibre and increasingly contribute towards global food security, alleviation of poverty, addressing malnutrition and mitigation of environmental impact caused by agricultural practices and production of pharmaceutical proteins. They are also climate resilient to tolerate harsher weather conditions. As of 2016, 18 million farmers grow biotech crops on 185 million ha of land in 26 countries. In spite of their enormous contribution to mankind, biotech crops continue to face regulatory scrutiny. These crops are heavily regulated around the world and in recent years, there have been calls from various parties to label food ingredients derived from biotech crops. A number of countries have national laws on labelling, while discussion is ongoing on low-level presence and adventitious presence of GMOs. There are also talks on harmonizing regulations to ease trade and enforcement activities. This chapter will discuss the global status of biotech crops, labelling regimes and the need for, challenges to and consequences of biotech crops. It is noteworthy that biotech crops have now accumulated a flawless record of human and environmental safety since their first release in 1996.

1. INTRODUCTION: A RECAP OF CROP BREEDING

In the last 21 years, GM (genetically modified), transgenic or biotech crops have emerged to be the fastest adopted crop technology in the history of modern agriculture, although they are also the most scrutinized. The public reluctance in accepting GM crops and the call for stringent policies and regulation mainly stem from the misconception that all the foods that are commercially available are natural unless they are genetically modified through molecular technologies (Tagliabue, 2016). Another major impediment to GM crop adoption is that the regulations are process-based and not product-based.

Understanding the development of new crop varieties and the evolution of plant breeding techniques is important to be able to evaluate the need for GM crops, and the way they should be regulated. The science of plant breeding is a dynamic discipline and it is continuously evolving. It started with selection and cross breeding when man started domesticating plants. Hybrid seeds were introduced in the 1920s (Crow, 1998) and have had a large impact on food production. This was followed by mutant crop varieties in the 1960s, and since then, over 3200 mutant varieties from more than 200 crop species have been released into the market in more than 70 countries (Plant Breeding and Genetics Newsletter, 2015). All these methods are known as conventional plant breeding techniques but have a number of limitations. Before any improved seed can become a commercially viable variety, the undesirable traits that result from cross pollination must be bred out of the new plant by multiple backcrosses that involve the crossing of thousands of genes, which also requires a long period of time. Another major limitation is that this method is only viable when the desired trait can be found in plants that can sexually mate. In the case of mutation breeding, chemicals and irradiation are used to create the desired traits by inducing mutagenesis.

Advances in molecular biology and the advent of molecular technologies have resulted in plant breeders gaining access to the identity and function of specific genes. This has meant that plant breeders can now use plants for which they know possess the desired genes (and therefore traits conferred by the genes) to breed new varieties, instead of the slower, conventional approach of observing the phenotypes without knowing the genetic basis of the phenotypes and their properties. In the 1970s, genetic modification technology allowed scientists to break the species boundaries and enabled them to insert genes from any organism into plants without the need to cross breed or scrambling the genes via mutation techniques. Cross breeding and hybridization of plants are largely constrained by the availability of desired traits in closely related plant species. GM technology is an extension of conventional plant breeding but it gives additional precision, flexibility and speed, and allows desired traits to be introduced by transferring new genes into current plant types.

While all crops have their genes modified through various breeding methods, why are only certain crops termed 'genetically modified'? To address this, it is important to understand the definition used in the Cartagena Protocol on Biosafety (CPB) as explained by The World Conservation Union (IUCN)—'Genetic modification', which is also referred to as 'genetic engineering', uses a variety of methods to isolate single genes from one or more microorganisms, plants or animals, and insert them into the genetic material of the cells of another. These methods are collectively termed 'in vitro nucleic acid techniques' and have been developed since the 1970s (see in this volume, chapter 'How *Agrobacterium*, a Natural Genetic Engineer, Became a Tool for Modern Agriculture' by Otten). Through genetic modification, genes are transferred and modified in ways that are not possible in nature, i.e., between different species and between animals and plants and microorganisms. Once inserted, these genes may be transferred to offspring of the modified individual through normal reproductive processes (Mackenzie et al., 2003). It is this possibility of cross species transfer of genes, and additionally, the introduction of synthetic gene constructs by artificial techniques, which have prompted the demand for special regulations on genetically engineered organisms, and calls for their food products to be labelled.

In 2004, The International Service for the Acquisition of Agribiotech Applications (ISAAA) introduced the term 'Biotech Crops' as an alternative to 'GM crops' to put things in perspective as all crops have their genome modified by one way or another, and the term GM crops also gives a negative connotation that creates public fear. ISAAA was also farsighted to envisage that gene technology will give rise to other breakthroughs in plant breeding in the future and today we see the emergence of technologies such as gene editing (new breeding technologies) and gene drive. Collectively, all crops developed through these methods could be termed 'biotech crops'.

The term 'biotech crops' used in this chapter refers only to crops developed through GM technology and crops developed through gene-editing or new breeding technologies are not discussed here (see in this volume, chapter 'Genome Editing in Agricultural Biotechnology' by Pfeiffer et al.). These biotech crops provide agronomic, environmental, economic and social benefits to farmers and consumers. Table 1 provides information on the global value of the seed market for biotech crops between 1996 and 2014 and shows the demand for such seeds by farmers.

All biotech crops are subject to country-specific regulatory systems to assess risk and ensure their safety before they are approved for planting by farmers or for use as food or feed items in that country. The regulatory system is science-based and includes in some countries, socioeconomic evaluation criteria. Indeed it has been noted that the regulatory approval process for biotech crops is even more stringent and demanding than for nonbiotech crops (Teng, 2008). Regulatory control is exercised at different stages of the process to develop and commercialize a biotech crop variety; this will be discussed in detail in a later section of this chapter. Background on the rationale for and the mechanics of regulating biotech crops is described in the following sections. It is important to understand that while regulations are based on facts and science, political agenda takes over in decision-making process in many countries.

2. THE REGULATORY BACKDROP

All plant breeding techniques alter the genetic makeup of plants. However, only GM technology has attracted much controversy and public debate. It has raised complex policy and regulatory questions and

Table 1 The Global Value of the Biotech Seed Market, 1996–2014

Year	Value (Millions of US$)
1996	93
1997	591
1998	1560
1999	2354
2000	2429
2001	2928
2002	3470
2003	4046
2004	5090
2005	5714
2006	6670
2007	7773
2008	9045
2009	10,607
2010	11,780
2011	13,251
2012	14,840
2013	15,610
2014	15,690
2015	15,267
Total	148,808

Source: James, C. (2015). Global status of commercialized biotech/GM crops: 2015. ISAAA Brief No. 51. ISAAA, Ithaca, NY.

resulted in serious international negotiations and the development of national and international regulatory frameworks around the world. A brief history of regulations and the entities involved, and why and how these crops are singled out to be regulated will form a useful backdrop to this chapter.

2.1 Asilomar Conference

The need to regulate GM technology was mooted by scientists themselves. A number of scientists expressed concerns over the potential risks of developing new strains of bacteria as a result of transferring genetic material from viruses and other bacteria. The GM bacteria were feared to compete and outdo the natural microbial population or transmit viral genes that might cause cancer. These concerns resulted in two historic conferences among scientists in the mid-1970s. The first conference took place in January 1973, spearheaded by the National Science Foundation's Human Cell Biology Steering Committee and the National Cancer Institute (NCI) of the National Institutes of Health (NIH). The consensus of that meeting was that researchers should proceed cautiously, that they should attempt to quantify the potential risks of working with such genes, and that additional efforts should be expended to determine what safety precautions should be taken to avoid spreading a potential carcinogenic risk through the environment (Pew Initiative on Food and Biotechnology, 2001).

This conference was followed by the renowned 1975 Asilomar Conference which discussed the overall safety issues related to GM technology. Although most of the participants believed that the technology neither posed significant health risks nor created new hazards, they agreed to abide by a set of research guidelines for the safe use of the technology. Chief among these was the agreement to limit work to disabled bacteria that were not able to grow outside a laboratory environment. Thus, one of the first recognized risk management decisions applied to the technology was the adoption of voluntary controls by an otherwise unregulated community of scientists, primarily in academic laboratories (Pew Initiative on Food and Biotechnology, 2001).

2.2 The Cartagena Protocol on Biosafety

The CPB is a protocol to the Convention on Biological Diversity (CBD). Two articles under CBD explicitly describe modern biotechnology. Article 16 (*Access to and Transfer of Technology*) says access to and transfer of modern biotechnology are essential elements for the attainment of the objectives of the CBD. First two paragraphs of Article 19 (*Handling of Biotechnology and Distribution of its Benefits*) express that Parties shall promote and advance priority access to the results and benefits arising from biotechnologies. The third paragraph of article 19 instructs consideration of a protocol in the field of the safe transfer, handling and use of living modified organisms (LMOs) (Convention on Biological Diversity, 1992).

Article 19 gave rise to the adoption of the CPB in 2000. The preamble of CPB clearly states the benefits of modern biotechnology: *Recognizing that modern biotechnology has great potential for human well-being if developed and used with adequate safety measures for the environment and human health* (Cartagena Protocol on Biosafety to the Conventional on Biological Diversity, 2000).

The main elements of the CPB are:

- Procedures that allow a Party that does not yet have a regulatory framework for biosafety, to make informed decisions about the import of LMOs for environmental introduction in its territory.
- General principles and methodology for risk assessment.
- A mechanism for information sharing: the Biosafety Clearing House (BCH).

While the CPB acknowledges the benefits of modern biotechnology, this instrument is often misinterpreted by Parties due to lack of awareness and understanding; critics of GM technology exploit CPB to their convenience to enforce strict national regulations. The CPB serves as a reference in developing national regulatory framework to its Parties.

2.3 Codex Alimentarius

The Codex Alimentarius or 'Food Code' was established by the Food and Agricultural Organization of the United Nations (FAO) and the World Health Organization in 1963 to develop harmonized international food standards, which protect consumer health and promote fair practices in food trade (Codex Alimentarius, http://www.fao.org/fao-who-codexalimentarius/en/). The Codex Committee on Food Labelling (CCFL) has been considering the labelling issue of food derived from GM organisms in an effort to develop guidelines for international harmonization. However, the effort which commenced in 1994 has yielded no results and came to an end in May 2011 where CCFL was divided between countries proposing process-based GMO labelling and those proposing that GMOs should be declared on the label only when they are present in the final product. CCFL decided to discontinue its work on definitions related to GMOs, and instead will develop a compilation of Codex texts relevant to the labelling of foods derived from modern biotechnology (Asia Food Journal, 2011). This then leaves each interested country to develop its own position to the labelling of food derived from GMOs.

2.4 World Trade Organization

The World Trade Organization (WTO) was founded on 1 January 1995 and has several agreements that cover goods, services and intellectual property, spell out the principles of liberalization and the permitted exceptions. These agreements include individual countries' commitments to lower customs tariffs and other trade barriers, and to open and keep open services markets. WTO also sets procedures for settling disputes. It is driven by member countries and has 162 member countries as of 30 November 2015. Several of these agreements apply to GM products (Josling, 2015):

(i) The Sanitary and Phytosanitary Agreement (SPS) https://www.wto.org/english/tratop_e/sps_e/spsagr_e.htm: SPS measures are aimed to protect human or animal life from risks arising from additives, contaminants, toxins or disease-causing organisms in their food, beverages or feedstuffs, or to protect a country from damage caused by the entry, establishment or spread of pests. This is applied to GMOs and many countries trigger this agreement in regulating GMOs.

(ii) Technical Barriers to Trade (TBT) https://www.wto.org/english/tratop_e/tbt_e/tbt_e.htm: Agreement deals with technical regulations, standards, including labelling requirements and conformity assessment. TBT allows governments to take measures to protect health or environment, provided they are not more trade-restrictive than necessary. Often this agreement is also invoked when making decisions about GMOs.

(iii) The Agreement on Trade-Related Aspects of Intellectual Property Rights (TRIPS) https://www.wto.org/english/docs_e/legal_e/27-trips_03_e.htm: This agreement could also be relevant in cases where the issue of the patentability of GM products comes into question. TRIPS allows governments to exclude from patentability plants and animals and essentially biological processes for the production of plants and animals. It also allows temporary exclusion from patentability when necessary to protect human, animal or plant life or health or to avoid prejudice to the environment (Halle, 2008).

(iv) General Agreement on Tariffs and Trade (GATT) https://www.wto.org/english/docs_e/legal_e/gatt47_e.pdf: This was incorporated into the WTO as the GATT-1994, that apply to all trade in goods can have fundamental implications for national GM policies.

2.5 The International Plant Protection Convention (IPPC)

The IPPC is an international plant health agreement, established in 1952, that aims to protect cultivated and wild plants by preventing the introduction and spread of pests with a secretariat at the FAO. It has 182 contracting parties. It develops International Standards on Phytosanitary Measures (ISPMs) against plant pests and plant products including GMOs (Jindal & Sharma, 2010).

2.6 Organization for Economic Cooperation and Development (OECD)

A request from the G8 leaders at their summit in Cologne in June 1999 that the OECD undertake a study of the implications of biotechnology and other aspects of food safety, and an earlier endorsement of the OECD's biotechnology programme by OECD ministers in May 1999 gave rise to two conferences: OECD Edinburgh Conference on the Scientific and Health Aspects of Genetically Modified Foods in 2000, and OECD Consultation with Nongovernmental Organizations on Biotechnology and Other Aspects of Food Safety in 1999 (OECD, 2000).

Both conferences yielded mixed reactions but there was general agreement among the majority of the participants on the need for a more open, transparent and inclusive debate and for openness and transparency in the policy process, as well as an acknowledgement that there is potential benefit to be gained from GM technology. A key point of contention concerned the issue of whether individual countries should be allowed to develop GM technology for food production according to their own needs, or whether there should be a global moratorium on GM crops. In addition, there was a lack of complete agreement on issues such as the mandatory labelling of GM foods, the usefulness of feeding trials in animals of GM foods and on the process of assessing consumer concerns. Finally, a need for further work was identified in relation to the potential long-term effects of GM food on human health, worker safety and the environment.

3. FACTORS THAT DRIVE REGULATIONS

The regulations, legal status of biotech crops and their labelling are not only determined by prevailing international and national legal instruments but also driven by other factors such as public concerns and fear and NGO activism. Food retailers and producers are largely pressured to label GM foods due to concerns voiced by consumers and not for safety or scientific reasons. It is ironic that the same countries which ban the growing of GM crops are often the same countries which import large quantities of GM crop products for food, feed and processing (FFP) (James, 2015). The ban on growing biotech crops and the moratorium on field trials in several countries are due to the additional precautionary measures propagated by activist groups. This has led many governments to increase the burden of evidence for demonstrating safety and broadened the practical definition of the 'adverse effects' which must be prevented (Levidow, Carr, & Wield, 2000).

3.1 Public Concerns

Public concerns and debate play an important role in determining a country's legal and labelling status of biotech crops. The act of transferring genes from unrelated species to crops that have been historically used as food, feed and fibre have caused overhyped public concerns and debate. Food safety, environmental risks, trade monopoly and ethical and religious issues associated to this technology initiated the call to regulate biotech crops or even ban it in some countries.

3.2 Public vs Private Funded Research and Products

In the 1940s and 50s, the 'Green Revolution' radically changed the agriculture landscape and introduced new crop varieties, hybrid seeds and other modern practices. This transformed the economy of many developing countries, alleviated famine and also improved farmers' agronomic practices. The major difference between Green Revolution and Gene Revolution (the advent of GM technology) is that the former was driven by government-funded institutes, whereas GM crops are overwhelmingly in the hands of the private sector. The development of GM seeds requires huge investment that limits commercialization by the public sector because of the same onerous regulations which the private sector has to meet. However, there is much basic research that takes place in public institutions on developing biotech crops (James, 2015).

3.3 Trade Monopoly

A common controversy created by critics of biotech crops is the issue related to intellectual property rights. The huge investment required to develop biotech crops makes patenting these seeds by private sector companies essential to safeguard the investment, research and innovation. However, this is not the case just for biotech seeds as hybrid seeds have been patented long before GMOs. In fact, the EU regulations allow farmers to reuse seeds for their own needs even if they contain patented traits. See links to legal documents here: http://www.marcel-kuntz-ogm.fr/article-brevet-41058140.html. In addition, in many other countries farmers are required to buy hybrid seeds from these companies every planting season. However, this is perceived as trade monopoly by the critics, despite the higher yield, better farm management and quality of crop enjoyed by the farmers. Biotech crops are singled out and are accused for bringing an end to the seed saving tradition that has been practiced for generations and also the transfer of seed production into the hands of private companies, which was previously held by the public sector institutes. The evidence though is that farmers in developing countries actually prefer good quality seeds which assure a better harvest, even if these seeds require to be purchased, as long as the seeds result in higher or assured yields.

The cost of discovery, development and approval of a new biotech crop introduced between 2008 and 2012 was US$136 million, of which 26% (US$35.1 million) is spent on regulatory testing and registration process. It takes an average 13 years for a biotech crop to reach the farm for commercial cultivation from the initiation of research and the longest phase of product development is regulatory science and registration activities, at about 5.5 years (McDougall, 2011).

However, GM seeds are not the only ones that are patented. According to US Patent and Trademark Office (http://www.uspto.gov/patents-getting-started/patent-basics/types-patent-applications/general-information-about-35-usc-161), many seeds that are bred through conventional methods are also patented. Hybrid seeds started in corn in the 1920s and are highly priced because they produce uniform plants and higher yield when compared to inbred lines. Hybrid seeds are developed by the hybridization or crossing of parent lines that are 'pure lines' produced through inbreeding. Pure lines are plants that 'breed true' or produce sexual offspring that closely resemble their parents (Pocket K. No. 13, 2006). It also takes a very long period to produce hybrids seeds, so there is a need to recoup the profit from the investment made in research. Thus, hybrid seeds are also patented, although this is not allowed in all countries.

4. NATIONAL BIOSAFETY FRAMEWORKS (NBFs)

The CPB is the most important international instrument that has impacted on how countries cooperate towards the safe transfer, handling and use of LMOs that come from modern biotechnology. The CPB was adopted in 2000 and entered into force on 11 September 2003. As of June 2016, 170 countries have acceded to the CPB. It is important for Parties to CPB to develop national legislative measures that will enable them to comply with the minimum requirements of CPB. The legislative instrument at the national level is the NBFs and each is developed in a format that is deemed appropriate, workable and practical for the individual countries, and meets the national priorities and international obligations. Parties may adopt legislation, laws, acts, regulations, decrees or guidelines.

The NBFs represent "combination of policy, legal, administrative and technical instruments that are developed to ensure an adequate level of protection in the field of safe transfer, handling and use of LMOs resulting from modern biotechnology that may have adverse effect on the conservation and sustainable use of biological diversity taking into account risks to human health" (UNEP-GEF, 2012). NBFs broadly have five components: (i) a national biosafety policy; (ii) a regulatory regime comprising laws, acts, regulation, decrees or guidelines, or equivalent; (iii) an administrative system that includes the component authority(ies) responsible for receiving and handling requests for permits (import, export, domestic use, including placing on the market, intentional introduction into the environment, field trials, contained use, transit, etc.); (iv) mechanisms for public awareness, education and participation include public access to information on GMOs; and (v) systems for follow-up, including monitoring for environmental effects and effects on human, animal or plant life or health; enforcement of biosafety law (Gupta, Karihaloo, & Khetarpal, 2014).

Developing the NBFs poses a huge challenge, especially to developing countries as it requires institutional capacities, technical competence and funding to exercise regulatory oversight. Lack of NBFs not only stifles the development and commercialization of biotech crops in the country but also becomes a major barrier for the country to import biotech crop-derived products. International developmental agencies such as the FAO, the Global Environment Facility (GEF), United Nations Development Programme and the United Nations Environment Programme (UNEP) play an important role in capacity building in enabling developing countries to develop and implement NBFs (FAO, 2009; UNEP, 2006). The range of activities include: (i) the development of national policies and formulation of regulations; (ii) GMO detection and monitoring including equipping of laboratories and harmonizing protocols among countries; (iii) facilitating effective communication and public awareness; and (iv) human resource development in biosafety (Pertry et al., 2013).

Table 2 lists the NBFs in different countries and the year each came into force. This information was obtained from various sources such as the BCH, USDA-FAS GAIN reports, regulatory authorities' websites, ISAAA's Crop Biotech Update, papers presented at conference, etc.

Table 2 The National Biosafety Frameworks in Different Countries

Country	**Type of NBFs**
Asia Pacific	
Australia	Gene Technology Act 2000
Bangladesh	Biosafety Guidelines of Bangladesh (2005)
Cambodia	In development
China	Regulation on the Safety Administration of Agricultural Transgenic Organisms (2001)
India	Recombinant-DNA Safety Guidelines (1998)
Indonesia	Regulation No. 21 (2005)
Iran	National Biosafety Act (2009)
Japan	Law Concerning the Conservation and Sustainable Use of Biological Diversity through Regulations on the Use of Living Modified Organisms (Law No. 97 of 2011)
Laos Republic	In development
Malaysia	Biosafety Act 2007
Myanmar	In development
Pakistan	The Pakistan Biosafety Rules (2005)
Philippines	Biosafety Guidelines on Genetic Engineering (1998)
Thailand	Draft Biosafety Law (pending approval)
Vietnam	Regulation on management of biological safety of GMOs products and goods originating from GMOs (2005)
Latin America	
Argentina	Law on Seeds and Phytogenetic Creations and the Law on the Promotion of the Development and Production of Modern Biotechnology
Brazil	Presidential Decree No. 5591, Law No. 11.105 (2005)
Mexico	Law on Biosecurity of Genetically Modified Organisms
Paraguay	Draft Biosafety Bill (pending approval since 2003)
Europe	
Norway	Gene Technology Act No. 38 (1993)
European Union	Directive 2001/18/EC of the European Council and the Parliament

Continued

Table 2 The National Biosafety Frameworks in Different Countries—cont'd

Country	Type of NBFs
Americas	
Canada	Federal regulatory framework for biotechnology (1993)
United States	Coordinated Framework for Regulation of Biotechnology (1986)
Africa	
Cote d'Ivoire	Drafting the law
Burkina Faso	Biosafety Law 2006, Amended 2013
Egypt	Permissive approach to GMOs. No law in place
Ghana	Biosafety Act 2011
Kenya	Biosafety Act 2009
Malawi	Biosafety Act 2002, revised 2007
Mali	Biosafety Law 2008
Nigeria	Biosafety Law, 2015
Senegal	Biosafety Law 2009
South Africa	Genetically Modified Organisms Act, 1997 Act No. 15
Togo	Biosafety law 2009
Uganda	Biosafety Bill (pending approval since 2012)
Zambia	Biosafety Act (Amendment) Act 2007

5. THE REGULATORY PROCESS FOR BIOTECH CROPS FROM 'LAB TO TABLE'

There is much ongoing research in Asia using recombinant-DNA (R-DNA) technology to produce genetically engineered plants with improved traits, i.e., biotech crops (Asian Productivity Organization, 2007; Teng, 2006). However, only a small number of final products will make it to the marketplace and be available to farmers or consumers. This is partly because the cost of regulatory compliance to get a biotech crop from the lab to the market is high, and so much research unfortunately remains in the lab, especially in public institutions (Teng, 2006).

5.1 Stages in Developing a Biotech Crop Product

There are many stages in the process to commercialize a crop biotech product, i.e., moving it from 'lab to table'. This commonly starts with a 'product concept' and ends with a 'market product', and the required time ranges from 10 to 13 years. The individual stages vary by country, depending mainly on the regulatory framework (NBF) in place for biosafety and food/feed safety assessments. These stages may be illustrated by means of the 'stage gate' procedure (Fig. 1) that directly or indirectly is used by private companies to plan and 'drive' the commercialization of a biotech crop (Teng, 2008). Stage gating also provides a framework to assess how far a product is to being released into the marketplace. As indicated in Fig. 1, to get from one stage to the succeeding one requires that all requirements are met before the gate is cleared. Although the schematic shows a linear process, in practice there are concurrent stages going on and multiple responsibilities to ensure concurrent fulfilment of requirements.

In companies, through research and development, a Discovery Group commonly identifies a valuable protein which helps to confer a useful trait,

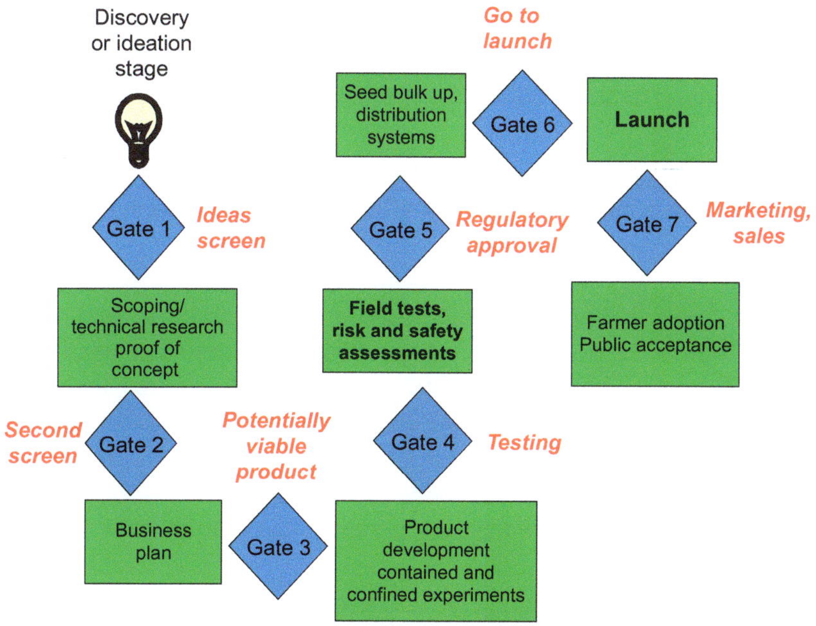

Fig. 1 'Stage Gating' of a biotech crop commercialization process. *After Teng, P.S. (2008). BioScience entrepreneurship—Creating value from biology.* World Scientific. *pp. 335.*

such as insect resistance. The B.t. protein is an example (in this volume, see chapter 'Assessing the Environmental Safety of Transgenic Plants: Honey Bees as a Case Study' by Ricroch et al.). At this stage too, many questions are asked about the history of the protein, especially its safe use in its conventional form. When the history of safety is ascertained, a next step commonly is then for molecular biologists to develop artificial constructs of the gene which makes the protein. These steps may take up to a year and commonly done without biosafety regulation as there is still no genetic consideration. The insertion of the artificial gene construct into target plant cells is next done through the process commonly called 'transformation' when the target gene is inserted into plant cells.

This stage is subject to regulatory oversight as most countries require approval to do such transformations. The target cells are from plant varieties with desirable commercial traits as well as being useful parents in a breeding program. The successful fraction of transformations is very low, hence in the same laboratories, facilities are developed for automation. Successfully transformed cells containing the desired gene are then grown to whole plants using tissue culture techniques. The regulatory entity at this time is commonly the Institutional Biosafety Committee (IBC), made up of representatives from research, government and society. Next comes testing the plantlets or whole plants to determine if the desired trait is expressed in strong enough levels to justify a useful product. For example, a plant purported to contain the B.t. gene may be tested by exposing it to insect larvae from the pest which it was developed to resist; technically this is called a bioassay. This screening process again is often done under biosafety supervision in greenhouses, and may take 1–3 years (Teng, 2008).

The 'Proof of Concept' stage is completed when the transgene in a new variety successfully expresses itself by producing proteins in high enough quantity to have the desired effect. Most likely at the time the 'proof of concept' stage is proceeding, another team would be developing the business plan, which includes both market analysis, and also 'table-top exercises' on any potential regulatory, intellectual property protection, safety or public acceptance issues (Teng, 2008). Next is the 'product development' stage, when enough seed is multiplied for the many tests required by regulatory authorities, commonly starting with greenhouse tests in a 'contained' environment with no air exchange with the external environment, and progressing to screenhouse in a 'confined' environment where there is air exchange but care is taken to prevent insect infestations or escape of biotech plant material to the external environment. If the new biotech plant at this time meets all the criteria it was developed for, it then proceeds to field testing, starting at a single location, and progressively, to multiple locations and crop seasons. All these are done under conditions which comply with the existing biosafety regulations (NBF) of that county, and only after regulators have approved the field trial design and location. Commonly, countries also require public notification and hearings for field trials, and for the technology developer to dialogue with communities. At the same time that this is going on, companies or public institutions proposing commercialization or release of the biotech crops in question are also conducting or accumulating evidence on the food/feed safety of the biotech crop. After several seasons over 2–3 years of field evaluations and food/feed safety testing, the government body empowered to make a decision to approve or disapprove the product must weigh all the evidence and render its decision. Assuming the biotech crop product passes all the gates (especially 'Gate 5') in Fig. 1 and is approved for wide area planting by farmers, the company or government institution (if it is a biotech crop developed by the public sector) then undertakes seed multiplication in preparation for sale or release to farmers.

In Asia, only Australia, the Philippines, Indonesia, China, Vietnam, Bangladesh and India have experienced the full cycle of commercialization. An important question that must be asked is whether governments will expect public institutions to be subject to the same long and expensive process of testing a biotech crop before it is approved. In North America, the experience shows that regulatory agencies are willing, on a scientific basis, to allow data sharing between products so that the process can become less cumbersome.

On top of regulatory requirements to obtain approval for release into the environment for use by farmers, producers of GM crops further have to assure the safe and sustained use of their products through product stewardship (Teng, 2008). Product stewardship is the provision of the subsequent after-sales support to ensure that the product is properly used. This includes resistance management schemes especially for the insect-protected products (B.t. corn, B.t. cotton), detection techniques and among others. In North America, insect resistance management strategies for B.t. crops rely on deployment of non-B.t. crops within specified geographic areas, a strategy commonly called refugia management (Gould, 1998). In some countries, product stewardship is stipulated by regulatory agencies as guidelines for using biotech crops.

5.2 Regulating the Safety of Biotech Crops in Relation to Commercialization

As noted earlier, before any biotech crop can be released for commercial planting or public use, it has to be approved through a regulatory process developed and operated by the government of a country using internationally recognized principles, techniques and technology. This is commonly done for biosafety (environmental safety), and food/feed safety, and should be science-based. The process to obtain regulatory approvals for food and feed safety include the submission of complete dossiers of information required by regulatory agencies, and the conduct of public hearings (Thomas & Fuchs, 2002). The status of biosafety approvals in Asia is shown in Table 3.

5.3 Safety Assessment Systems and Processes

The commercialization of any biotechnology product in agriculture produced using genetic engineering (R-DNA technology) requires that there be policies and procedures to ensure that these products are environmentally safe. Such policies and procedures have come to be known collectively as 'biosafety'. Biosafety is the subject of an international protocol, called the CPB under the Convention for Biological Diversity, both of which were described in detail in the preceding sections of this chapter.

5.4 Risk Assessment and Risk Management

Central to all biotechnology regulatory systems is a framework that leads to decision making. Science-based safety or risk assessment is central to this, but may not be the only factor that informs the decision to approve or not approve a GM food.

Risk assessment is a scientific process used to identify hazards, the likelihood of their occurrence, and their impacts on human and animal health and on the environment. It also leads to the identification of possible risk mitigation measures. The risk assessment of a GM food should be evidence-based, scientifically defensible and typically requires a diverse collection of expertise and a significant knowledge base.

At the scientific and technical levels, very similar questions are asked when evaluating the potential human health risks of GM products. In evaluating novel foods a comparative approach is taken that focuses on the defined differences between the novel food and its traditional counterpart,

Table 3 Status of Regulatory Approval and Regulatory Systems for Biosafety in Selected Asian Countries

Country	Status of Contained Experiments	Status of Confined Field Trials	Status of Open Field Trials, Precommercalization	Status of Biosafety Regulatory Systems	Status of Importation of GMOs for Food, Feed and Processing
China	+	+	+	+	+
India	+	+	+	+	–
Australia	+	+	+	+	+
Philippines	+	+	+	+	+
Thailand	+	+	(+)	+	+
Vietnam	+	+	+	+	+
Korea	+	+	–	+	+
Japan	+	+	+	+	+
Malaysia	+	(+)	–	+	+
Singapore	+	+	+	+	+
Cambodia	–	–	–	–	?
Indonesia	+	+	+	+	+
Bangladesh	+	+	+	+	+
Myanmar	(+)	(+)	(+)	–	+

Keys: + indicates current, existing strength and/or activity; – indicates no known activity; and ? indicates no data.

and the effect these differences have on composition, nutritional quality, toxicology and potential allergenicity.

Those charged with undertaking a risk assessment must ensure that information requirements are scientifically justifiable. In the context of a GM food safety assessment, this means that the information and data required by the risk assessor must be safety related; risk assessors should be able to identify the biological significance of statistically significant differences where these exist and the context must be safety related.

In the context of novel food safety assessment, risk management encompasses the decision to approve or not approve a GM food for human consumption (Teng, 2008). Typically, this is the step where balance is sought between safety and 'nonsafety' considerations such as socioeconomic, cultural and political concerns that may be associated with GM food products. Additionally, although it happens rarely in practice, potential benefits associated with a GM food may also be applied to informing the decision. Transparency is critical at this point in the regulatory process as the extent to which safety and other concerns are taken into account, and how this is accomplished, should be a matter of public record.

6. APPROVED BIOTECH CROPS AND EVENTS

There are almost 30 species of biotech crops that have been developed and approved for commercial release since 1996. Annex is a list of biotech crops and events approved worldwide for import and also cultivation. It is first important to understand the definition of an 'event'. According to AgBiosafety, University of Nebraska (http://agbiosafety.unl.edu/glossary.htm), an event is the insertion of a particular transgene into a specific location on a chromosome. The term 'event' is often used to differentiate genetically engineered crop varieties. In other words, for the same trait, there might be a number of events as the transgene might have been inserted in different locations on the plant's chromosome. Each event requires individual application and approval. While this table displays a long list of approved biotech crops and events, not all of them are available in the market in these countries or are commercially cultivated.

The countries listed in Annex are a mix of countries that cultivate biotech crops and those that only approve the crops to be imported for FFP purposes. It is interesting to note that countries that have not approved the commercial cultivation of biotech crops such as Taiwan, South Korea, Singapore, European Union, Japan and New Zealand approve a large number of biotech crops as these countries are heavily dependent on these crops for their feed and many of them end up as ingredients in hundreds of food products.

There is also another major reason for developers of these crops to seek approval for crops that are not in the market yet (not cultivated yet) or that is not of importance to a country. This is to avoid low-level presence (LLP) in a country's food and feed supply. LLP is unapproved GM material and can occur in a number of circumstances (Stein & Rodríguez-Cerezo, 2010):

1. There can be 'asynchronous approval' (AA), i.e., at least one cultivating country has already authorized a GM crop while other (importing) countries have not.
2. There can be 'isolated foreign approval' (IFA, or 'asymmetric approval'), i.e., a cultivating country has authorized a GM crop, but its developer does not seek approval in (potential or unattractive) importing countries.
3. There can be LLP of research events, i.e., a country has authorized the cultivation of a GM crop in field trials only, but due to accidental admixture, traces end up in the commercial crop supply.

To avoid rejection of consignments that are contaminated with unapproved events, companies take preemptive measures by obtaining approvals for events that have possibilities of getting mixed with approved events.

There are over 60 countries that import GM crops—Argentina, Australia, Austria, Bangladesh, Belgium, Bolivia, Brazil, Bulgaria, Burkina Faso, Canada, Chile, China, Colombia, Costa Rica, Croatia, Cyprus, Czech Republic, Denmark, Egypt, Estonia, Finland, France, Germany, Greece, Honduras, Hungary, Indonesia, India, Iran, Ireland, Italy, Japan, Latvia, Lithuania, Luxembourg, Malaysia, Malta, Netherlands, Mexico, Myanmar, New Zealand, Norway, Pakistan, Panama, Philippines, Poland, Portugal, Romania, Russia, Slovakia, Slovenia, South Africa, South Korea, Spain, Sweden, Taiwan, Thailand, Turkey, United Kingdom, United States and Uruguay (GMOAnswers, https://gmoanswers.com/ask/what-countries-currently-have-regulations-gmos).

7. THE GLOBAL CULTIVATION STATUS OF BIOTECH CROPS

The year 2016 marked the first year into the second decade of the global commercialization of biotech crops. Biotech crops were first

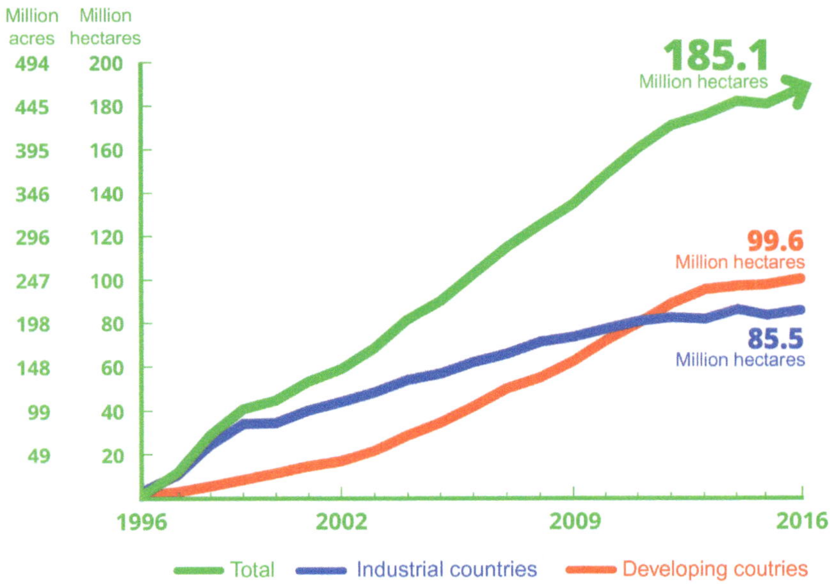

Fig. 2 Global area of biotech crops (1996–2016). *Source: ISAAA. (2016). Global status of commercialized biotech/GM crops: 2016. ISAAA Brief No. 52. ISAAA, Ithaca, NY.*

cultivated in 1996 on 1.7 million ha and in 2016 the area of biotech crops stood at 185.1 million, a 110-fold increase (ISAAA, 2016). Fig. 2 shows the global area of biotech crops, where 18 million farmers in 26 countries grow these crops.

A significant reflection from this figure is that the adoption of biotech crops in the developing countries has been higher than in developed countries for four consecutive years, starting from 2010. This has quashed the critics of biotech crops who in early years claimed that these crops will only benefit large-scale farmers in developed countries. Currently, 11 biotech crops are grown commercially, namely corn, cotton, soybean, canola, sugar beet, papaya, poplar, eggplant, alfalfa, squash and potato. Fig. 3 shows the countries that grow biotech crops, their area and the crops.

The accumulative economic benefits to farmers from 1996 to 2015 were estimated at US$167.7 billion and 90% of the beneficiaries were small, resource-poor farmers in developing countries. Besides economic benefits, these crops have also contributed to sustainable food production, food security and mitigation of climate change. From 1996 to 2015, crop biotechnology reduced the spraying of crop protection products by 619 million kg,

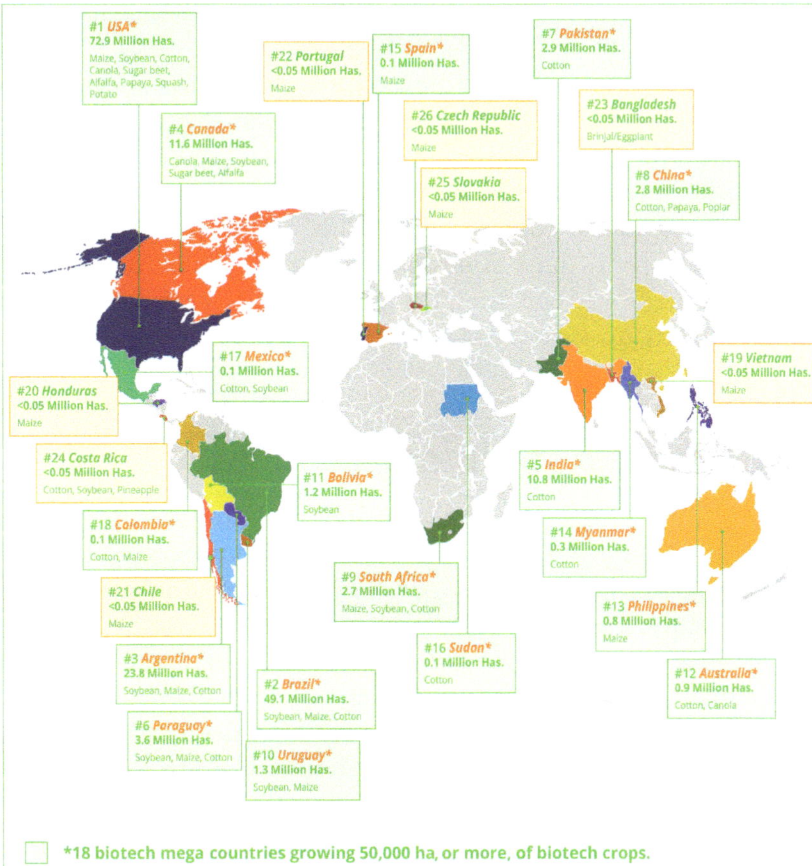

Fig. 3 Countries that grow biotech crops, the area and crops (James, 2015). *Source: ISAAA. (2016). Global status of commercialized biotech/GM crops: 2016. ISAAA Brief No. 52. ISAAA, Ithaca, NY.*

a global reduction of 8.1%. These crops have also saved 26.7 billion kg of carbon dioxide from being emitted into the atmosphere in 2015, which is the equivalent of adding 11.9 million cars to the roads (Brookes & Barfoot, 2017).

The forecast for new biotech crops looks promising and public research is expected to take a fair share of what will be commercially cultivated, with Asia and Africa taking the limelight. There are over 85 potential new products that are already being field tested globally for a variety of traits that includes multiple modes of resistance to weeds, pests and diseases, resistance to nematodes and other biotic/abiotic stresses associated with climate change (James, 2015). Golden rice is being field tested in the Philippines and Bangladesh. More propoor crops such as fortified bananas, pest resistant cowpea and WEMA (Water Efficient Maize for Africa) biotech maize are on track for release in Africa in 2017/18. Biotech corn is expected to contribute to a significant increase in biotech crop area where there is a global potential of at least an additional 100 million ha. This will largely come from Asia (60 million ha), with China alone contributing 35 million ha and another 35 million ha in African countries. Up to 10 countries are expected to grow B.t. cotton, each growing 100,000 ha (ISAAA, 2016).

8. LABELLING STATUS OF GMO FOODS

Labelling of GMO foods is a contentious issue that has created polarized debate for long and also has given rise to heterogeneity in policies around the world. A number of approaches have been proposed ranging from mandatory labelling to voluntary, and a third-party base (https://www.biofortified.org/2013/08/what-is-kosher/) like 'Kosher' which is not governed and audited by the government. In the United States, it is estimated that 70%–80% of foods contain ingredients that have been genetically modified (Lowe, 2009) and this might be true for foods consumed in many other parts of the world. While GM foods have been consumed for the last 2 decades without any health hazards, advocates for labelling are arguing that the safety of these foods is still questionable and that their risks are not understood fully. However, countries are under pressure from consumer right groups and critics of biotechnology to impose mandatory labelling.

The Codex Committee on Food Labelling, hosted and chaired by Canada, examines international food labelling issues; drafts labelling provisions that are applicable to all foods; and endorses labelling provisions prepared by Codex Committees charged with drafting standards, codes of practice and guidelines (MacKenzie, 2000). Member countries are encouraged to incorporate it into any relevant domestic rules and legislation but under WTO SPS Agreement, member countries retain the right to unilaterally impose more stringent food safety regulations deemed necessary to ensure domestic consumer protection, provided the different standards are scientifically justifiable and otherwise consistent with WTO SPS rules

(MacKenzie, 2000). While Codex may try to reach international consensus, biotechnology perception among the public and policymakers differ from one country to another, making harmonization in this area difficult.

8.1 The Heterogeneous Labelling Situation Around the Globe

The polarized debate on labelling GMO foods around the globe is resulting in a heterogeneous labelling regime with different standards and threshold. Harmonization of GMO foods labelling is not practical worldwide as countries devise their labelling laws according to their political, trade and societal needs.

For a long time, the US Food and Drug Administration stood by its position that labelling of GMO-containing foods is generally not required, as the agency does not have any evidence to show that GMO foods present any different or greater safety concern than foods developed by traditional plant breeding. However, due to pressure from the critics of GMO foods, the US Congress passed the legislation in July 2016 that requires food packages to carry a text label, a symbol or an electronic code readable by smartphone that indicates whether the food contains genetically modified organisms, or GMOs. Products containing highly refined oil and sweeteners like canola oil or high-fructose corn syrup are exempted as they contain no genetic material that would identify them as coming from a genetically modified source, which is what the law requires.

The United States became the 65th country that requires mandatory labelling of GMOs foods. The others are Russia, all of Europe and a dozen Asian countries (GMO FAQ, Genetic Literacy Project, http://gmo.geneticliteracyproject.org/FAQ/how-are-gmos-labeled-around-the-world/).

Table 4 lists the labelling regime and requirement for countries around the world. The information was gathered from numerous sources. While the European countries set their threshold between 0.9% and 1%, the threshold for Asian countries is between 3% and 5%.

According to the Center of Food Safety, these are the countries that require labelling of GMO Foods: Australia, Austria, Belarus, Belgium, Bolivia, Bosnia and Herzegovina, Brazil, Bulgaria, Cameroon, China, Croatia, Cyprus, Czech Republic, Denmark, Ecuador, El Salvador, Estonia, Ethiopia, Finland, France, Germany, Greece, Hungary, Iceland, India, Indonesia, Ireland, Italy, Japan, Jordan, Kazakhstan, Kenya, Latvia, Lithuania, Luxembourg,

Table 4 Labelling Regime of GMO Foods Around the World

Countries		**Type of Labelling and Requirement**
Europe (non-EU members)		
Bosnia and Herzegovina	Norway	Mandatory labelling with threshold of 0.9%–1% GMO content
Croatia	Switzerland	
United Kingdom		
Montenegro	Macedonia	No labelling law
Albania		
Serbia		Official ban on import and cultivation of GM crops/foods
Europe (EU members)		
Ireland	Portugal	Mandatory labelling with threshold of 0.9%–1% GMO content
Spain	France	
Belgium	Luxemburg	
Netherlands	Belgium	
Italy	Malta	
Germany	Denmark	
Greece	Germany	
Czech	Austria	
Slovakia	Slovenia	
Hungary	Romania	
Poland	Bulgaria	
Latvia	Estonia	
Lithunia	Finland	
Sweden		
Russia		Mandatory labelling with threshold of 0.9%–1% GMO content
Asia Pacific Oceania		
Australia	New Zealand	Mandatory labelling with threshold of 0.9%–1% GMO content

Table 4 Labelling Regime of GMO Foods Around the World—cont'd

Countries		**Type of Labelling and Requirement**
China	Malaysia	Mandatory labelling with threshold higher than 1% GMO content
South Korea	Japan	
Indonesia	Sri Lanka	
India	Thailand	Mandatory labelling with undefined threshold
Vietnam	Singapore	
Philippines	Mongolia	No labelling law
Nepal	Bangladesh	
Bhutan	Myanmar	
Laos	Pakistan	
Cambodia	Brunei	
Central Asia		
Kazakhstan		Mandatory labelling with threshold of 0.9%–1% GMO content
Uzbekistan	Kyrgyzstan	No labelling law
Azerbaijan	Turkmenistan	
Tajikistan	Afghanistan	
Middle East		
Iran	Iraq	No labelling law
Kuwait	Lebanon	
Syria	Qatar	
Bahrain	UAE	
Oman	Yemen	
Saudi Arabia		Mandatory labelling with threshold of 0.9%–1% GMO content
Jordan		Mandatory labelling with undefined threshold
African continent		
Morocco	Mauritania	No labelling law
Algeria	Libya	
Egypt	Niger	

Continued

Table 4 Labelling Regime of GMO Foods Around the World—cont'd

Countries		**Type of Labelling and Requirement**
Chad	Sudan	
Eritrea	Djibouti	
Somalia	Guinea	
Liberia	Ghana	
Burkina Faso	Cote d'Ivoire	
Togo	Nigeria	
South Sudan	Central African Republic	
Congo	DR Congo	
Gabon	Rwanda	
Uganda	Burundi	
Tanzania	Angola	
Malawi	Namibia	
Botswana	Zimbabwe	
Madagascar	Mozambique	
Swaziland	Lesotho	
Kenya	South Africa	Threshold of 1% or higher
Tunisia	Mali	Mandatory labelling with undefined threshold
Senegal	Cameroon	
Ethiopia		
Benin	Zambia	Official ban on import and cultivation of GM crops/foods
Americas		
Canada		No labelling law. Mandatory labelling only if there is a health concern, e.g., presence of allergen or compositional change
United States		Mandatory labelling law passed by Senate in 2016. Highly refined products are exempted

Table 4 Labelling Regime of GMO Foods Around the World—cont'd

Countries		Type of Labelling and Requirement
Brazil		Mandatory labelling with threshold of 1% or higher
Mexico	Belize	No labelling law
Guatemala	Honduras	
Nicaragua	Costa Rica	
Panama	Cuba	
Columbia	Venezuela	
Guyana	Paraguay	
Chile	Argentina	
Uruguay		
Ecuador	Peru	Mandatory labelling with undefined threshold
Bolivia		

Malaysia, Mali, Malta, Mauritius, Netherlands, New Zealand, Norway, Peru, Poland, Portugal, Romania, Russia, Saudi Arabia, Senegal, Slovakia, Slovenia, South Africa, South Korea, Spain, Sri Lanka, Sweden, Switzerland, Taiwan, Thailand, Tunisia, Turkey, Ukraine, United Kingdom, United States and Vietnam (https://www.centerforfoodsafety.org/issues/976/ge-food-labeling/international-labeling-laws).

The number of countries with labelling laws might be on the increase but in many countries the law is only partially enforced due to its practicality. Enforcement, detection and monitoring are challenges that are difficult to overcome due to the intensive human, technical and infrastructure resources required.

8.2 Consequences of Labelling

Labelling GMO foods imposes a number of consequences not only to food producers but also to consumers, as follows:

(i) *The need to substitute or eliminate ingredients*: Companies that claim to be GMO free need to alter their recipe and source for alternative ingredients. This is a Herculean task as it requires ensuring the entire supply chain is GMO free, which includes farmers, handlers, processors and distributors. Ben & Jerry's, a company that manufactures ice cream, frozen yogurt and sorbet took about 3 years just to remove GMOs from ingredients like cookie dough and caramel (Wall Street Journal, 2016). Swapping or eliminating ingredients may alter the taste of the product leading to consumer dissatisfaction.

(ii) *Product price increases*: As companies shift to producing non-GMO foods, it is becoming evident that sourcing for non-GMO ingredients comes with a cost, although the supporters of labelling deny it. Ben & Jerry's move to source for non-GMO ingredients resulted in new products averaging 11% higher in price (Wall Street Journal, 2016), although initially the company made a statement saying it would not inquire additional costs. The major cost will be segregating non-GMO product at the multiple stages of the supply chain which is a tedious and complicated process. This will also force small companies out of business and result in these companies losing segments of their markets.

(iii) *The loss of nutrients*: Swapping of ingredients or eliminating them results in loss of nutrients as it is difficult to source for non-GMOs with equivalent nutrients. The riboflavin (vitamin B2) in the new non-GMO Grape Nuts and original Cheerios from Post Foods is reduced to 2% from 25% of daily value after the company started using non-GMO ingredients (Food Navigator USA, 2014). This may result in consumers paying more for non-GMO foods with lesser nutritive value.

(iv) *Confusion among consumers*: There is an extensive body of scientific research on the safety of GMO foods conducted by independent scientists around the globe (Nicolia, Manzo, Veronesi, & Rosellini, 2013) and hundreds of peer-reviewed publications on feeding studies conducted on a wide variety of species (Flachowsky, Schafft, & Meyer, 2012). All studies show no detrimental effects of consuming GMO foods. Labelling these foods may cause confusion and stigmatization among the public and give an impression that GMO foods may have health hazards. In some cases, the same ingredient might be regulated by different agencies and one might need to be labelled while the other is exempted. In the United States, plain spaghetti is required to be labelled, while meatball spaghetti is exempted as meat-containing products are regulated by another agency. This is an indication that labelling does not adequately provide information to help in consumer choice, an intention lobbied by supporters of labelling.

(v) *Reduced investment in GM technology*: Introduction of policies that require labelling and any other obstacles to the evolution of GMOs may result in reduced investment in this technology, and consequently research and innovation (Zilberman, 2012). A similar effect has been seen in EU where publications, innovations and investment in GMOs were growing throughout the 1990s but peaked just before the ban was implemented in 1999 on GMOs (Graff et al., 2009). This might result in technological stagnation and reduced competitiveness in the farming sector of the affected countries.

(vi) *Effective enforcement is not possible*: In many countries, while the labelling law may look very impressive, its enforcement is difficult or impossible to achieve. Highly processed foods, such as oil and flour do not contain detectable DNA, which makes it very difficult to determine their status. Enforcement will then rely on monitoring the entire supply chain and documentation, making the process extremely expensive, impractical or impossible. It also requires extensive capacity building programmes for enforcement officer at various ports and laboratories that are well equipped for GMO testing.

(vii) *Trade disruption*: Companies might face trade disruptions where they have to opt out certain market areas with stringent labelling laws to avoid penalties and legal suits. Some companies might even have to change their base. This is already happening in the United States where small companies are contemplating to pull out from Vermont due to newly introduced labelling requirements (Mangan, 2016).

These are some of the known consequences of labelling GMO foods. A number of unforeseen consequences are also likely to arise when labelling is fully implemented.

9. GMO DATABASES

The public, regulators, industry and other stakeholders often seek information on approved GMOs around the world. There is no one site to acquire this information and in Table 5 a list of credible databases is provided that offer useful and current information on GMOs, their approval and countries involved. Some databases even provide novel traits acquired from conventional breeding techniques.

Table 5 The Databases That Maintain Information on GMOs and Related Activities

Owner	Link (URL)	Content
National Biological Impact Assessment Program (NBIAP) under USDA	http://www.isb.vt.edu/search-release-data.aspx	Development, testing and regulatory review of biotech crops, animals and microorganisms within the United States and abroad
Centre for Environmental Risk Assessment (CERA)	http://cera-gmc.org/GMCropDatabase	Plants produced using recombinant-DNA technologies (e.g. genetically engineered or transgenic plants), and plants with novel traits that may have been produced using traditional methods
BVL—the Federal Office of Consumer Protection and Food Safety (Berlin, DE) and RIKILT Wageningen UR (Wageningen, NL)	http://www.euginius.eu/euginius/pages/home.jsf	Presence, detection and identification of GMOs in the European Union and worldwide
ISAAA	http://www.isaaa.org/gmapprovaldatabase/countrylist/default.asp	Biotech crop approvals according to events worldwide
European Commission	http://gmo-crl.jrc.ec.europa.eu/gmomethods/	EU Database of Reference Methods for GMO Analysis

European Commission	https://ec.europa.eu/food/plant/gmo/eu_register_en	EU register of GM food and feed (Regulation (EC) 1829/2003) and the GM products withdrawn from the market
OECD	http://www2.oecd.org/biotech/	Products derived from the use of modern biotechnology, products with novel traits from the use of conventional breeding or mutagenesis that have been approved for commercial application in at least one country
Cartagena Protocol on Biosafety	https://bch.cbd.int/	An interactive portal where parties and observers share information on GMOs, approvals and regulations

10. CONCLUDING REMARKS

In this paper, we have traced the events and rationale behind the regulation of biotech crops and the labelling of food derived from these crops, i.e., 'GMO food'. We have also attempted to provide the legal and scientific basis for some of these regulations, since not all the regulations have a scientific basis and are strongly influenced by public attitudes and political demands. The evidence shows that worldwide, there are systems already in place to ensure safe research and testing of biotechnology crops. Some of these systems have regulations which are among the most stringent to evaluate food and environmental safety and have even resulted in large costs to move biotech crops from 'the lab to field'. A key question that has to be asked is how publicly developed biotech crops will be able to meet the same demands of proof of safety as has been imposed on the predominantly private-sector developed biotech crops of the first 20 years of commercialization.

While the scientific community and a majority of mainstream development organizations have endorsed the application of biotech crops to address current and anticipated problems associated with food security, there exists a small but vocal movement to stop or delay the use of this technology. The 20 years of planting biotech crops and an almost equal period of using these for human consumption, as animal feed and to produce fibre, have clearly demonstrated their safety. Yet an irrational fear promoted by groups with a general antitechnology and specific antibiotechnology stance, have delayed the application of this technology to benefit the many poor farmers and consumers worldwide. In some respects, these 'anti' groups have gained success against mainstream science and development organizations as evidenced by the number of countries which have mandatory labelling. But as pointed out in this paper, often having regulations on paper does not result in effective implementation due to the enormity of the challenge to ensure labelling is accurate. Conversely, the delay in sharing biotechnology applications has meant that the large numbers of people continue to suffer from diseases and hunger, both of which could have been partly or even fully ameliorated with the increased supply of extra, enriched food produced using biotech crops.

Going forward, the challenges posed by climate change on food production, the unequal nutrition of populations without adequate, diverse food and the potential food price increases caused by a mismatch of supply (production) and demand (consumption), are all considerations which argue for more widespread use by scientists and farmers of biotechnology. A rational and practical approach to safeguard the use of biotechnology is accepted by all in order to protect consumers AND concurrently allow farmers to grow crops efficiently and maintain their livelihoods. Food labelling regulations must be science-based, so must the regulations governing research, development and growing of biotech crops. While the consumers' right to choose what he/she eats, and the farmers' right to choose what he/she wants to grow is both worthy reasons for regulations, governments must not sacrifice one for the other and make expedient decisions that satisfy immediate criticism but result in longer term effects on their people.

ANNEX

Countries Which Have Approved Biotech Crops For Cultivation, Food, Feed and Processing (FFP)

Country	Total Number of Events Approved in the Country	Crops	No. of Biotech Crop Events Approved
Argentina	62	Cotton	7
		Corn	43
		Soybean	12
Australia	122	Alfalfa	3
		Argentine Canola	21
		Carnation	12
		Cotton	26
		Corn	27
		Potato	11
		Rice	1
		Rose	1
		Soybean	17
		Sugar beet	2
		Wheat	1

Continued

Countries Which Have Approved Biotech Crops For Cultivation, Food, Feed and Processing (FFP)—cont'd

Country	Total Number of Events Approved in the Country	Crops	No. of Biotech Crop Events Approved
Bangladesh	1	Eggplant	1
Bolivia	1	Soybean	1
Brazil	50	Bean (*Phaseolus vulgaris*)	1
		Cotton	15
		Eucalyptus	1
		Corn	38
		Soybean	11
Burkina Faso	1	Cotton	1
Canada	171	Alfalfa	3
		Apple	2
		Argentine Canola	18
		Cotton	25
		Flax	1
		Corn	64
		Papaya	1
		Polish Canola	4
		Potato	24
		Rice	1
		Soybean	21
		Squash	1
		Sugar beet	2
		Tomato	4

Countries Which Have Approved Biotech Crops For Cultivation, Food, Feed and Processing (FFP)—cont'd

Country	Total Number of Events Approved in the Country	Crops	No. of Biotech Crop Events Approved
Chile	3	Argentine Canola	1
		Cotton	1
		Corn	1
China	63	Argentine Canola	12
		Cotton	11
		Corn	17
		Papaya	1
		Petunia	1
		Poplar	2
		Rice	2
		Soybean	12
		Sugar beet	1
		Sweet pepper	1
		Tomato	3
Columbia	94	Carnation	8
		Cotton	14
		Flax	1
		Corn	49
		Rice	2
		Rose	2
		Soybean	16
		Sugar beet	1
		Wheat	1
Costa Rica	15	Cotton	13
		Soybean	2
Cuba	1	Corn	1
Egypt	1	Corn	1

Continued

Countries Which Have Approved Biotech Crops For Cultivation, Food, Feed and Processing (FFP)—cont'd

Country	Total Number of Events Approved in the Country	Crops	No. of Biotech Crop Events Approved
European Union	95	Argentine Canola	12
		Carnation	7
		Cotton	11
		Corn	48
		Potato	1
		Soybean	15
		Sugar beet	1
Honduras	8	Corn	7
		Rice	1
India	11	Cotton	6
		Soybean	5
Indonesia	18	Corn	9
		Soybean	6
		Sugar cane	3
Iran	1	Rice	1
Japan	310	Alfalfa	5
		Argentine Canola	20
		Carnation	6
		Cotton	37
		Corn	198
		Papaya	1
		Potato	8
		Rice	1
		Rose	2
		Soybean	29
		Sugar beet	3

Countries Which Have Approved Biotech Crops For Cultivation, Food, Feed and Processing (FFP)—cont'd

Country	Total Number of Events Approved in the Country	Crops	No. of Biotech Crop Events Approved
Malaysia	34	Argentine Canola	1
		Carnation	8
		Cotton	4
		Corn	14
		Soybean	7
Mexico	158	Alfalfa	5
		Argentine Canola	13
		Cotton	30
		Corn	68
		Potato	13
		Rice	1
		Soybean	22
		Sugar beet	1
		Tomato	5
Myanmar	1	Cotton	1
New Zealand	97	Alfalfa	3
		Argentine Canola	14
		Cotton	21
		Corn	27
		Potato	11
		Rice	1
		Soybean	17
		Sugar beet	2
		Wheat	1

Continued

Countries Which Have Approved Biotech Crops For Cultivation, Food, Feed and Processing (FFP)—cont'd

Country	Total Number of Events Approved in the Country	Crops	No. of Biotech Crop Events Approved
Norway	11	Carnation	11
Pakistan	2	Cotton	2
Panama	1	Corn	1
Paraguay	20	Cotton	3
		Corn	14
		Soybean	3
Philippines	88	Alfalfa	2
		Argentine Canola	2
		Cotton	8
		Corn	58
		Potato	8
		Rice	1
		Soybean	14
		Sugar beet	1
Russian Federation	22	Corn	12
		Potato	2
		Rice	1
		Soybean	7
		Sugar beet	1
Singapore	26	Alfalfa	2
		Argentine Canola	1
		Cotton	4
		Corn	10
		Soybean	8
		Sugar beet	1

Countries Which Have Approved Biotech Crops For Cultivation, Food, Feed and Processing (FFP)—cont'd

Country	Total Number of Events Approved in the Country	Crops	No. of Biotech Crop Events Approved
South Africa	70	Argentine Canola	5
		Cotton	10
		Corn	42
		Rice	1
		Soybean	12
South Korea	158	Alfalfa	5
		Argentine Canola	14
		Cotton	29
		Corn	75
		Potato	9
		Soybean	25
		Sugar beet	1
Sudan	1	Cotton	1
Switzerland	4	Corn	3
		Soybean	1
Taiwan	124	Argentine Canola	6
		Cotton	22
		Corn	71
		Soybean	24
		Sugar beet	1
Thailand	15	Corn	12
		Soybean	3
Turkey	32	Corn	25
		Soybean	7

Continued

Countries Which Have Approved Biotech Crops For Cultivation, Food, Feed and Processing (FFP)—cont'd

Country	Total Number of Events Approved in the Country	Crops	No. of Biotech Crop Events Approved
United States	195	Alfalfa	3
		Apple	3
		Argentine Canola	20
		Chicory	3
		Cotton	28
		Creeping bent grass	1
		Flax	1
		Corn	43
		Melon	2
		Papaya	3
		Plum	1
		Potato	43
		Rice	3
		Rose	2
		Soybean	24
		Squash	2
		Sugar beet	3
		Tobacco	1
		Tomato	8
		Wheat	1
Uruguay	17	Corn	10
		Soybean	7
Vietnam	22	Corn	14
		Soybean	8

Source: International Service for the Acquisition of Agribiotech Applications http://www.isaaa.org/gmapprovaldatabase/countrylist/default.asp (accessed 9 June 2017).

REFERENCES

Asia Food Journal. (2011). *Codex abandons definition and labelling work on GMOs*. https://www.asiafoodjournal.com/2011/06/codex-abandons-definition-and-labelling-work-on-gmos/. Accessed 6 December 2017.

Asian Productivity Organization. (2007). *Business potential of agricultural biotechnology products* (p. 197).Tokyo: Asian Productivity Organization. ISBN: 92-833-7054-6.

Brookes, G., & Barfoot, P. (2017). *GM crops: Global socio-economic and environmental impacts 1996–2015*. www.pgeconomics.co.uk. Accessed 6 December 2017.

Cartagena Protocol on Biosafety to the Conventional on Biological Diversity. (2000). Secretariat of the Conventional on Biological Diversity, Montreal, Canada.

Convention on Biological Diversity. (1992). United Nations.

Crow, J. F. (1998). 90 years ago: The beginning of hybrid maize. *Genetics*, *148*(3), 923–928.

FAO. (2009). *Building biosafety capacities*. FAO's experience and outlook.

Flachowsky, G., Schafft, H., & Meyer, U. (2012). Animal feeding studies for nutritional and safety assessments of feeds from genetically modified plants: A review. *Journal fur Verbraucherschutz und Lebelnsmittelsicherheit—Journal of Consumer Protection and Food Safety*, 7, 179–194.

Food Navigator USA. (2014). *Goodbye GMOs (and vitamin B12, D, A, and Riboflavin): New non-GMO grape nuts and original Cheerios are 'certainly less nutritious'*. claims professor, https://www.foodnavigator-usa.com/Article/2014/02/20/Non-GMO-Grape-Nuts-Cheerios-less-nutritious-due-to-fewer-vitamins. Accessed 6 December 2017.

Gould, F. (1998). Sustainability of transgenic insecticidal cultivars: Integrating pest genetics and ecology. *Annual Review of Entomology*, *43*, 701–726.

Graff, G., et al. (2009). The contraction of Agbiotech product quality innovation. *Nature Biotechnology*, *27*(8), 702–704.

Gupta, K., Karihaloo, J. L., & Khetarpal, R. K. (2014). *Biosafety regulations for GM crops in Asia-Pacific*. Bangkok: Asia-Pacific Consortium on Agricultural Biotechnology, New Delhi and Asia-Pacific Association of Agricultural Research Institutions. i–xii + 1–160.

Halle, M. (2008). *GMOs and the WTO rules*. International Institute for Sustainable Development. *www.trade.ecoaccord.org/s2008/6.ppt*. Accessed 6 December 2017.

ISAAA. (2016). *Global status of commercialized biotech/GM crops: 2016*. ISAAA, Brief No. 52Ithaca, NY: ISAAA.

James, C. (2015). *Global status of commercialized biotech/GM crops: 2015*. ISAAA, Brief No. 51Ithaca, NY: ISAAA.

Jindal, S. K., & Sharma, M. C. (2010). *Biotechnology in animal health and production*. New India Publishing Agency.

Josling, T. (2015). A review of WTO rules and GMO trades. *BioResources*, *9*(3), 4–7.

Levidow, L., Carr, S., & Wield, D. (2000). Genetically modified crops in the European Union: Regulatory conflicts as precautionary opportunities. *Journal of Risk Research*, *3*(3), 189–208. https://doi.org/10.1080/13669870050043062.

Lowe, R. (2009). *GMA praises introduction of national food labelling bill*. Grocery Manufacturers Association. http://www.gmaonline.org/news-events/newsroom/gma-praises-introduction-of-national-food-labeling-bill/.

MacKenzie, A. A. (2000). The process of developing labelling standards for GM foods in the codex alimentarius. *AgBioForum*, *3*(4), 203–208.

Mackenzie, R., Burhenne-Guilmin, F., La Viña, A. G. M., Werksman, J. D., in cooperation with, Ascencio, A., et al. (2003). *An explanatory guide to the cartagena protocol on biosafety*. Gland, Switzerland and Cambridge, UK: IUCN, xvi +295pp.

Mangan, M. (2016). *Six consequences for consumers of labelling GMOs. Genetic Literary Project*. https://geneticliteracyproject.org/2016/07/05/six-consequences-consumers-labeling-gmos/. Accessed 6 December 2017.

McDougall, P. (2011). *The cost and time involved in the discovery, development and authorisation of a new plant biotechnology derived trait.* Crop Life International.

Nicolia, A., Manzo, A., Veronesi, F., & Rosellini, D. (2013). An overview of the last 10 years of genetically engineered crop safety research. *Critical Review Biotechnology, 34*, 77–88. https://doi.org/10.3109/07388551.2013.823595.

OECD. (2000). *Genetically modified foods.* Widening the Debate on Health and Safety.

Pertry, I., Sabbadini, S., Goormachtig, S., Lokkoo, Y., Gheysen, G., Burssens, S., et al. (2013). Biosafety capacity building: Experiences and challenges from a distance learning approach. *New Biotechnology, 31*(1), 64–68. https://doi.org/10.1016/j.nbt.2013.08.008.

Pew Initiative on Food and Biotechnology. (2001). *Guide to U.S. regulation of agricultural biotechnology products.* Pew Initiative on Food and Biotechnology.

Plant Breeding and Genetics Newsletter. (2015). *World wide success in mutation breeding for food security—Achievement and outstanding achievement awards.* Joint FAO/IAEA Programme. No. 34 (p. 24).

Pocket K. No. 13. (2006). *Conventional breeding.* ISAAA. https://www.isaaa.org/resources/publications/pocketk/document/Doc-Pocket%20K13.pdf. Accessed 6 December 2017.

Stein, A. J., & Rodríguez-Cerezo, E. (2010). Low-level presence of new GM crops: An issue on the rise for countries where they lack approval. *AgBioForum, 13*(2), 173–182. http://www.agbioforum.org/v13n2/v13n2a08-cerezo.htm. Accessed 6 December 2017.

Tagliabue, G. (2016). Nature as a totem, "Genetically Modified Organisms" as a contemporary taboo. *North American Journal of Psychology, 18*(2), 283–294.

Teng, P. S. (2006). Agriculture—Where does Asia stand in accepting agribiotechnology crops? *Asia Pacific Biotech News, 10*(15), 838–845.

Teng, P. S. (2008). *BioScience entrepreneurship—Creating value from biology* (p. 335). World Scientific.

Thomas, J., & Fuchs, R. (2002). *Biotechnology and safety assessment.* Academic Press. ISBN: 9780126887211. 400 p.

UNEP. (2006). *A comparative analysis of experiences and lessons from the UNEP-GEF biosafety projects.* UNEP GEF.

UNEP-GEF. (2012). *National biosafety frameworks.* https://www.unenvironment.org/search/node?keys=National+Biosafety+Frameworks. Accessed 6 December 2017.

Wall Street Journal. (2016). *GMO labelling law roils food companies.* https://www.wsj.com/articles/gmo-labelling-law-roils-food-companies-1458510332. Accessed 6 December 2017.

Zilberman, D. (2012). The logic and consequences of labelling GMOs. *ARE Update, 15*(5), 5–8. University of California Giannini Foundation of Agricultural Economics.

FURTHER READING

International Service for the Acquisition of Agribiotech Applications. http://www.isaaa.org/gmapprovaldatabase/countrylist/default.asp. Accessed 6 December 2017.

CHAPTER FOUR

Regulating Safety of Novel Food and Genetically Modified Crops

Andrew Bartholomaeus[1]
Faculty of Health, University of Canberra, Canberra, ACT, Australia
Diamantina Institute, Faculty of Medicine, University of Queensland, Brisbane, QLD, Australia
[1]Corresponding author: e-mail address: a.bartholomaeus@uq.edu.au

Contents

Abstract

The principle of proportionality, embodying concepts of fairness, equity, and consistency, is fundamental to human rights, national and international law, and subordinate regulation. This principle, in theory, provides some limits on the potential unintended consequences that may result from disproportionate regulatory burdens distorting individual and corporate behaviour, the consequences of which may exceed the real or imagined harms the original regulations were intended to prevent. Current regulatory burdens applied in a number of jurisdictions on recombinant DNA technology and the new biotechnologies, however, as opposed to other less precise mechanisms of gene alteration in common use, are applied discriminately, are disproportionate to the known (lack of) plausible food safety risks, are ignorant of the broader knowledge of natural plant genome plasticity, and are consequently ethically highly questionable at best. Although major corporations developing GM crops are arguably beneficiaries of the reduced competition resulting from disproportionate regulatory burdens and their associated costs, this comes at the substantial detriment both to the respective jurisdictions and to developing economies seeking to improve the welfare of disadvantaged

Advances in Botanical Research, Volume 86
ISSN 0065-2296
https://doi.org/10.1016/bs.abr.2017.11.003

communities through the use of advanced plant breeding technologies. Disproportionate regulation of GMOs is consequently risk generating rather than risk mitigating and is contrary to the intent of the precautionary principle. The key principles underlying rational, ethical, risk proportionate regulation of new plant varieties developed by any technique, conventional or otherwise, are discussed.

1. INTRODUCTION

The prosperity of the developed world and the emerging economies, and the capacity of the earth to support a population headed for over 10 billion, is entirely dependent on the application of sophisticated technologies, mostly developed over only the last century or two. Ironically, as Carl Sagan once noted "We have designed our civilization based on science and technology, and at the same time arranged things so that almost no-one understands anything at all about science and technology—this is a clear prescription for disaster". Agricultural technologies play a central, if not pivotal, role in sustaining both the expanding population and their current and future prosperity, yet a profound ignorance of the importance of these technologies at all levels of society has led to regulatory regimes that limit or prevent their application at a time of increasing need for their benefits.

The modern world is facing a combination of daunting developments that will undoubtedly challenge the technological capacity of our species and the complacency of those currently privileged to live in the comparative wealth and luxury afforded by the developed world. The world's population in 2015 was 7.3 billion and is projected to approach 10 billion by 2050 (United Nations, 2015) an increase of 2.7 billion over a 35-year period, but even a decade ago 1 in 7 people, or 845 million, were going to bed hungry each night (von Braun, 2007). Although distribution and waste were, and remain, critical determinants of food security in many regions, yield and especially yield in the region of consumption are equally as critical and will become increasingly so as the frequency of severe drought increases with climate change (IPCC, 2007). Concurrently with population growth, the disposable income of the citizens of the world's most populous nations, China and India, are increasing considerably more rapidly than that of the citizens of the developed Nations. The growth of disposable income of Chinese households, for example, has grown by greater than 10% pa for the past 10 years (Strateford & Cowling, 2016), whereas in the European Union the rate of growth has averaged just under 2% over the same period

(OECD, 2015). To feed an additional 2.7 billion people requires an additional 8 billion meals a day or roughly 3 billion cubic metres of additional food every year. Without substantial, and sustainable, increases in crop yields the increasing wealth of the citizens of the most populous economically emergent countries will result in increased competition for, and therefore increased cost, and declining availability, of supply of available food. This will be most apparent in those Nations and regions which are demographically or geographically unable, or culturally or politically unwilling, to achieve self-sufficiency in food production. The proportion of household income required to provide food to a family in developed nations (10%) is a fraction of that required in India (36%) or China (34%) (Washington State University, 2008) and a fraction of the historical level (42% in 1900 USA) (US Department of Labor, 2006). As the prosperity of the emergent nations increases rapidly over the next decade, the privileged access to cheap food enjoyed by the citizens of the developed world will be severely challenged and likely associated with significant social disaffection and disruption. New and existing biotechnologies provide a critical mechanism for improving yields, providing greater food production sustainability, facilitating adaptation to climate change, and reducing requirements for agricultural inputs, but only if they are not obfuscated by regulatory barriers (Farre, Ramessar, Twyman, Capell, & Christou, 2010). Consequently, the application of the principle of risk proportionality and robust cost–benefit analysis to the establishment of regulatory regimes for these and future plant development technologies based on the best available scientific evidence is essential for the future stability, food security, and prosperity of the international community.

This chapter discusses the discordance between current regulatory requirements for biotechnology-derived crop varieties, with a focus on safety of consumption, and both the principles of ethical regulation and the moral and economic imperatives arising from the challenges of future food sustainability.

2. BASIC PRINCIPLES OF ETHICAL REGULATION

Regulation is not a morally neutral or costless process. All regulation imposes constraints, obligations, or liabilities on individuals and/or corporations and is therefore a restriction of their freedom. In an ideal regulatory environment, such imposts are strictly proportional to the risks or potential adverse outcomes associated with the activity being restricted, and the costs of implementation and compliance are proportionate to, and commensurate with, those associated with the risks being mitigated or obviated. The principles of proportionality and equality before the law have existed in the criminal code of various societies for thousands of years and are widely recognised as a foundation of National and International Law and subsidiary regulation (Cottier, 2012; Ferran, 2015; OECD, 1995, 2012).

Equally, regulatory risk should ideally not exceed the regulated risk. That is, the potential costs and broader immediate and longer term societal risks generated by the regulation should not be greater than the risk the regulation is intended to prevent. Unfortunately, we do not live in an ideal world and in practice regulation is conducted in a contentious and contested arena. In this environment evidence based and dispassionate consideration of the costs and benefits of regulation are prone to be obfuscated or misdirected by political, cultural, and ideological expediency based on selective, or mis-, interpretation/representation of the broader evidence base. Greenpeace, for example, in responding to a racketeering suit brought against them by a sustainable forestry firm in Canada has admitted in court that their public statements "do not hew to strict literalism or scientific precision", but rather should be seen as "hyperbole", "heated rhetoric", and "nonverifiable statements of subjective opinion" that should not be taken "literally" (Garneau, 2017). Such disregard for basic honesty by major contributors to public debate exploit the general inability of the majority of the public, politicians, and policy makers to grasp the technical complexity of most scientific issues, and skew the political discourse and resultant regulation. The regulatory environment of the EU, for example, has recently been argued to be "between Nonsense and Protectionism" (Masip et al., 2013; Tagliabue, 2017), despite the extensive EC funded GMO research programs consistently affirming the safety of GM crops (Directorate-General for Research and Innovation, 2010).

Additionally, most regulation by its nature is developed and administered by agencies constrained by statute to confine their considerations to the specific field or scope of activity for which they have been assigned authority. Although some lip service is paid to the need to weigh costs and benefits of regulatory burdens, in practice the scope of such analyses is limited to immediately quantifiable tangible costs and benefits and does not generally consider the costs of lost future opportunities or less immediately tangible benefits such as food security. This is particularly true where the immediate risk lies within one agencies responsibility but costs or regulatory risk lie in that of a different agency.

3. IDENTIFYING NOVELTY

In developing an appropriate, risk proportionate, regulatory regime for the products of any new technology, the focus is not the technology itself but the novelty introduced to the products the technology is used to generate.

In most cases, the perceptions of novelty during the early adoption of a new technology is erroneous, and largely due to a lack of knowledge on the natural preexisting occurrence of the perceived novelty. The development and early regulation of food irradiation provide a close parallel to the more recent experience with the introduction of biotechnology for food production, including most of the same errors of judgement apparent in the current regulatory excess for food derived from biotechnology.

The history of research on food irradiation to preserve wholesomeness dates back to the early 1900s (Bartholomaeus, Parrott, Bondy, & Walker, 2013). A key concern in the early phase of the introduction of food irradiation was the potential for unintended compositional consequences. Irradiation of food was known to produce a range of unintended compositional effects through the production and reaction of radiolytic products and through damage to micronutrients. Irradiation of food was first proposed in the 1890s with the first rodent bioassay conducted on an irradiated food in the 1920s (Giddings, 1992). Since that time, a large number of animal studies, including more than 30 lifetime studies, have been conducted to address the concern that irradiation may unintentionally alter food composition resulting in the generation of substances capable of causing adverse effects in humans, a concern closely analogous to that postulated for GM foods. Tens of thousands of experimental animals (mostly rats and mice but also dogs, hamsters, quail, primates, and chickens) were sacrificed over a number of decades in the misinformed and misguided belief that such studies would demonstrate the safety or otherwise irradiated food (Bartholomaeus et al., 2013; Elias, 1980). As the chemistry of processed and unprocessed food was better investigated and therefore understood, the products of food irradiation were found to mirror that produced by cooking, other processing, or occurring naturally. Overwhelmingly, the animal studies yielded no data that would call into question the validity of safety assessments based solely on compositional analysis and an understanding of the chemistry of untreated food. In a major review of the safety of irradiated food, the WHO (WHO, FAO, & IAEA, 1999) concluded that:

> *"...abundant and convincing data indicate that high-dose irradiated foods do not contain either measurable levels of induced radioactivity or significant levels of any radiolysis products distinct from those found in un-irradiated foods. The theoretical maximum levels that might be formed would be so low as to be of no toxicological consequence".*

This expert group further concluded that:

> *"the determination of wholesomeness for a representative food could be extrapolated to other foods of similar composition on the basis of available chemical data" (i.e. without toxicity testing) and*
> *"the committee ... recognised the value of chemical studies as a basis for evaluating the wholesomeness of irradiated foods".*

The perception of novelty of food irradiation in the minds of consumers, regulators, some scientists, and activist groups proved illusory and was founded in the absence of knowledge of the broader compositional characteristics of untreated and otherwise processed foods. The state of the art and science for biotechnologies used for food production are today at an analogous stage of understanding and, as discussed further in Section 4, confirm a lack of novelty arising from the technology itself—as opposed to the functionality introduced by the technology.

The EU, in repeating the misguided and ill-informed whole food feeding trials of irradiated food, conducted a major project to evaluate the role of 90 day, and longer, whole food feeding studies on GMOs. This project titled GMO Risk Assessment and Communication of Evidence (GRACE) (GRACE, 2015) yielded data consistent with previous criticisms of whole food studies (Bartholomaeus et al., 2013; Elias, 1980) and, with respect to MON810, for example, drew the facile and wholly predictable, and in general terms predicted (Bartholomaeus et al., 2013), conclusion that;

"... GRACE data did not provide any indication that the performance of 90-day feeding studies (following OECD or EFSA guidelines and current practice) with whole food/feed would provide additional information on the safety of maize MON810 when compared to the compositional comparison of the GM line and its closest conventional comparator in terms of an initial comparative safety assessment".

4. TESTING HYPOTHESES OF RISK

In the case of biotechnology, the postulated novelty/concerns regarding food safety tend to revolve around:

- De novo production of high potency protein toxins
- De novo generation of biochemical machinery to produce a novel toxic secondary metabolite unrelated to the parent plant or the source or function of the transgene itself
- Reactivation of dormant pathways giving rise to toxicity
- Altered nutritional composition
- Production of new allergens
- Systemic uptake of dietary DNA, RNAi, etc., by humans and animals and its incorporation into the cells of the consuming organism
- Coexpression of adjoining genes resulting in production of toxic genes

The speculated sources of risk novelty that have led to popular perceptions of food safety risk, identified above, can be rephrased into hypotheses testable against the now voluminous data on plant genome plasticity and compositional variability:

- How variable is the plant composition independently of genome plasticity,
- How common are toxins in nature and how variable are their levels in a given species/variety,
- What drives the variability within a species—phenotype or genotype,
- What is their purpose in nature,
- How common are gene insertions in nature,
  - What are the sources of the inserted genes
  - What are the consequences of those insertions
- How common are point mutations in nature,
  - What are the consequences
- Is there evidence for activation of dormant pathways from either SNPs or transposons,
  - Tissue specific vs genus/species dormany
- Is there evidence for de novo generation of toxins from mutations or insertions.

4.1 Genome Plasticity

In 2017, we know considerably more about the natural genome plasticity of food crops and the natural compositional variation of those crops independently of genome variation (Bartholomaeus et al., 2013). The natural variability of crop and all plants is quite substantial. More than 80% of a typical maize genome, for example, consists of transposable elements (Schnable et al., 2009) and a typical maize variety has more than 50 million identified single-nucleotide polymorphisms (Gore et al., 2009). So great is the variation that any protein of 300–400 amino acids taken at random from two lines of maize differed by 3.5 amino acids on average due to background rates of mutation (Tenaillon et al., 2001) and there can be greater genetic variability between two lines of conventional corn than between humans and chimpanzees (Tenaillon et al., 2001; Zhao et al., 2000).

Similar findings have been obtained with soybean (Tian et al., 2012), and in rice approximately one new gene insertion occurs per three plants per generation in field cropping conditions (Naito et al., 2006). Thus, gene mutation and random gene insertions are occurring routinely and frequently in conventional varieties in the field. Similarly, the movement and insertion of genes from mitochondria to the nucleus (Buchanan, Gruissem, & Jones, 2015; Lough et al., 2008) and from retroviruses to the nucleus (Harper, Hull, Lockhart, & Olszewski, 2002) are also known to occur naturally.

Even the now classical, and increasingly obsolescent, recombinant DNA techniques utilising *Agrobacterium tumefaciens* T-DNA has been shown to occur naturally, as discussed in more detail earlier in this volume (Otten, 2018). Kyndt et al. (2015) while assembling small interfering RNAs, or siRNAs, of sweet potato plants for metagenomic analysis, were surprised to find sequences homologous to T-DNA sequences from *Agrobacterium* spp., and were able to unambiguously demonstrate that "two different T-DNA regions are present in the cultivated sweet potato (*Ipomoea batatas* [L.] Lam.) genome and that these foreign genes are expressed at detectable levels in different tissues of the sweet potato plant". Thus, "conventional" sweet potato is naturally transgenic.

Despite the demonstrably extensive genome plasticity and naturally high rate of mutation and gene insertion in conventional crops such as soy or corn, there is not a single example of a line of either crop spontaneously developing toxicity either from conversion of a nontoxic protein to toxicity, or from the de novo generation of novel secondary metabolites. Similarly, no novel allergens have been produced. This outcome is entirely predictable from a knowledge of the requirements for protein toxicity and the implausibility of the de novo generation of the machinery necessary to develop new secondary metabolites. There are no known plausible additions, deletions, or substitutions of amino acids in endogenous crop proteins that would convert an otherwise nontoxic or nonallergenic protein into a protein with toxic or allergenic properties (Steiner et al., 2013). Consequently, the proposition that the *process* of insertion of a gene through recombinant DNA technology or modification of an existing gene through the newer biotechnologies might result in the de novo development of toxicity or allergenicity is highly implausible, and discordant with the increasingly voluminous evidence on, and understanding of, plant genetics.

4.2 Compositional Variability

Knowledge of the natural variability of crop composition has also expanded very greatly over the past 10 years, such that we now know that nutrient (Shewry et al., 2013), secondary metabolite and, where present, allergen levels can and do vary many fold due to variability in climate, soil, agronomic practices, and pest pressure. The variability of the natural, background level of allergens in conventional soy bean, for example, can vary by more than an order of magnitude (5- to 19-fold) simply due to agronomic and environmental variation at the point of cultivation (Geng et al., 2016). This level of variation very substantially exceeds that resulting from the processes involved in plant transformation (Harrigan et al., 2010). Similarly, the level of the psoralen contact toxins, which can produce photocontact dermatitis, in conventional celery can vary by more than 200-fold in response to infestation with *Sclerotinia sclerotiorum* (Bourgaud, Gravot, Milesi, & Gontier, 2001).

4.3 Dormant Pathways

Kessler, Taylor, Maryanski, Flamm, and Kahl (1992) postulated:

> *"Additionally, plants, like other organisms, have metabolic pathways that no longer function because of mutations that occurred during evolution. Products or intermediates of some of these pathways may include toxicants. In rare cases, such silent pathways may be activated by the introduction or rearrangement of regulatory elements, or by the inactivation of repressor genes by point mutations, insertional mutations, or chromosomal rearrangements".*

The authors offered no examples of such silent or dormant pathways being reactivated. Indeed, given the current knowledge of the extensive and frequent mutations that naturally occur in plants the proposition that a pathway dormant throughout the entire plant, and therefore performing no function essential to the viability of the plant, would retain sufficient integrity over more than a small number of generations to be capable of reactivation is untenable. The progressive degradation of the respective genes through accumulated mutations could not plausibly be reactivated through random mutations and no example of such reactivation is present in the literature. Conversely, many plants have pathways active in some tissues but inactive in others. In such cases, the integrity of the functional pathways is maintained and activation of that pathway in a tissue where it is normally silenced is a plausible outcome of both "conventional" and biotechnology techniques of plant breeding. Maize kernels, for example, were historically white like its ancestor teosinte due to inactive phytoene synthase in the kernels, although the pathway is active in the leaves. Selection for yellow endosperm from the early 20th century has produced the now ubiquitous yellow corn. The pivotal event for yellow endosperm is reactivation of the silenced phytoene synthase pathway in the endosperm (Palaisa, Morgante, Williams, & Rafalski, 2003).

Similarly, poisoning from bitter zucchini associated with elevated levels of cucurbitacin levels in the fruit has been reported (Herrington, 1983). These chemicals have evolved in cucurbitaceae as herbivore antifeedants and pest deterrents. The poisoning incident described by Herrington is likely to reflect either activation of the relevant pathway in the fruit or upregulation of a preexisting pathway in the plant that has been downregulated by domestication. In either case, cucurbitacin is an endogenous component of the crop and the poisoning is due to increased levels and/or altered distribution of a preexisting toxicant. The downregulation of cucurbitacin pathways during domestication of cucumbers has been described by Shang et al. (2014). This example illustrates the need for simple, targeted, analytical investigation of levels of endogenous secondary metabolites with toxicological potential as a routine component of all plant breeding, whether defined as "conventional" or biotechnology.

Thus, reactivation of dormant pathways is a plausible concern for both "conventional" and biotechnology-derived crops only where the pathway is active and therefore conserved in at least one tissue of the plant. Where such pathways are associated with the production of potentially toxic secondary metabolites, routine analytical investigation of the levels and distribution within edible portions of a crop are appropriate.

4.4 History of Safe Use

Crops developed through the use of biotechnology are currently grown on more than 175 million hectares in more than 30 countries. More than

18 million farmers choose to grow such crops due to a combination of higher yields, lower agricultural inputs, and improved sustainability of production (James, 2015). A large proportion of these crops, 70%–90%, are utilised as feed for food animal production in multiple countries including the United States, Europe, and East Asia. In the United States alone some 9 billion animals are produced for human food each year with 95% being raised on feed containing GM ingredients (Van Eenennaam & Young, 2014). Similarly, in Europe "Whilst less than 0.1% of the global acreage of GM crops is cultivated in Europe, more than 70% of EU animal protein feed requirements are imported as GM crop products" (Baulcombe, Dunwell, Jones, Pickett, & Puigdomenech, 2014). From both a toxicological and animal production perspective two of the most critical measures of animal health are food efficiency (weight gained per weight of feed consumed) and reproductive performance. The almost universal acceptance of the wholesomeness of GM feed for animal production by farmers is backed by an analysis of food animal health in the United States from 1983 (before the introduction of GM crops) through to 2011 (Van Eenennaam & Young, 2014) which identified no observable adverse impact of the move to GM animal feeds. The extraordinary number of animals raised on GM feed every year across the world arguably provides more robust proof of the food safety of such crops than any compositional or toxicology study could ever attain. Indeed, many of the current regulatory safety testing requirements have proved to be entirely unnecessary, scientifically unjustifiable, and/or redundant (Bartholomaeus, Batista, Burachik, & Parrott, 2015; Bartholomaeus et al., 2013; GRACE, 2015; Hoekenga, Srinivasan, Barry, & Bartholomaeus, 2013; McClain et al., 2015).

5. PLAUSIBLE RISKS OF PLANT MODIFICATION

Where then might risk come from during crop development? All plants naturally produce secondary metabolites many of which act as protectants or natural pesticides (Ames, Profet, & Gold, 1990). Even common food plants produce toxic metabolites in some tissues. Apples and other pome fruits, grapes, cassava, and a wide range of other food crops produce cyanogenic glycosides in their seeds or other tissues (Bolarinwa, Oke, Olaniyan, & Ajala, 2016). In some cases, the cyanide is not present in the commonly eaten portion (such as the apple flesh) or is not readily released from the consumed seed. In other cases, such as cassava and bamboo shoots, humans have learnt to process the food before consumption in such a way as to remove most of the toxic component. Similarly, potatoes naturally produce glycoalkaloids as pest deterrents and antifeedants (Friedman, 2006). The glycoalkaloids are important flavour components and provide pest protection, but the levels of these can be increased by poor handling, pest pressure, and agronomic factors (Najim, Hadi, Fazeli, Darzi, & Rahi, 2012; Sinden & Webb, 1974) resulting in human toxicity.

Overall, the key sources of food safety risk arising from "conventional" and biotechnology techniques of gene manipulation are the same and derive from the inherent characteristics of the plant being modified and the nature of the trait(s) being introduced. This observation is as true today as it was nearly 30 years ago when the National Research council reviewed the technology (National Research Council (NRC), 1989). Where the plant being modified naturally produces a toxin in some or all of its tissues, manipulation of the genome by any technique, has the potential to increase the levels of the toxin or alter its distribution. Relatively inexpensive, specifically targeted, analytical chemistry provides a means to control for this risk. The development of new varieties of potato or canola, for example, should, regardless of the technique being used, include an analysis for levels of potato glycoalkaloids and erucic acid, respectively. Similarly, where a genetic trait is introduced that results in the production of a protein or secondary metabolite that is new to the food supply, or will be present at substantially higher levels than previously, then the safety of the specific protein or metabolite produced should be assessed. Such assessments should follow a tiered approach proportionate to the degree of novelty and the properties of the new substances and include in silico and Threshold of Toxicological Concern approaches to exclude plausible hazards. A novel protein which is digestible and does not have structural homology with known allergens or toxins, for example, does not require extensive, and expensive, toxicological investigation.

6. RISKS OF RISK AVERSION

Perhaps the most perverse consequence of the current GM regulatory disconnect with well-established science demonstrating at most a negligible risk is illustrated by the example of high amylose wheat development by the Australian CSIRO. In 2005 the CSIRO was granted approval in Australia to conduct field trials of a high amylose wheat developed using RNAi gene silencing which was classified as genetically modified under Australian law (OGTR, 2005). However, the substantial regulatory overheads applied to commercialisation of GM crops (McDougall, 2011) combined with concerted disinformation campaigns by some NGOs, including the vandalising of a field trial site by Greenpeace in 2011 (Creagh, 2011), made it more cost effective to spend further years of research resources to reverse engineer the GM wheat using conventional methods that resulted in the same phenotype, rather than attempt to progress the GM wheat to commercialisation (Regina et al., 2015). Such resource wasteful strategies highlight the direct economic costs of risk disproportionate and technology discriminatory regulation.

Risk disproportionate regulation can also create broader adverse economic, equity, and health consequences that extend beyond the borders of the National regulatory authority. Thomas (2012) has observed that "focusing on the mitigation of low-probability risks with higher cost regulation that reflects the preferences of high-income households, effectively redistributes wealth from the poor to the middle class and the rich". This applies equally to the International as to the National community and Herring and Paarlberg (Herring & Paarlberg, 2016) have observed the "transnational diffusion of ideologies with attendant costs to poorer farmers and countries". The largely politically expedient regulatory burdens placed on biotechnology by the wealthy countries of the developed world, create a disproportionate burden on poorer nations wishing to utilise new technologies to improve the food security of the most vulnerable populations. Such outcomes fundamentally contravene the OECD guidance on improving the quality of government regulation which states that National regulatory policy should be: "paying particular attention to regulatory quality and transparency with respect to regulations that may have impacts on other countries" (OECD, 1995). The OECD guidance makes clear that "As individual governments take actions that affect other countries, they are expected to ensure that justification for their actions is empirical and transparent". To the extent that current regulations are discordant with the extensive body of evidence indicating that the technological processes used in biotechnology are in themselves no more inherently risky than those considered "conventional", that regulation is neither empirical in nature nor transparent in derivation and arguably highly unethical.

Biotechnology-derived crops have provided a substantial increase in the income of poorer farmers in developing nations (Brookes & Barfoot, 2013; Klümper & Qaim, 2014). However, even where a new GM crop is intended for cultivation only within the jurisdiction of a developing nation, the risk of adventitious presence in export products, asynchronous approval (where a crop is approved in the producing country but not the importing country) across trading partners and the consequent impact on food commodity trade necessitates similar regulatory burdens as for food crops intended for export. This otherwise redundant regulatory burden causes substantial slowing and impediment of the development and release of critical crop varieties, particularly those that are grown in small volume by remote communities where the cost of regulatory approval may preclude development. Some key subsistence commodities such as the potato are resistant to "conventional" methods of gene manipulation necessitating the use of biotechnology to achieve the desired traits such as late blight resistance (USAID, n.d.). The benefits of the development of pest resistant crops are not confined to increased yields from lower crop damage but include reduced pesticide requirements to the benefit of the health of both the farmer and the environment (Juma, 2015). Thus, nonscience-based regulatory affectations in wealthy countries can inhibit efforts to provide food security, improved prosperity and better health through decrease occupational exposure to pesticides, in the poorest communities of the world, and are not therefore ethically neutral.

The notably politicised demonization of biotechnology in the EU and some other developed regions also has the potential to undermine the food security of those regions. The logical and moral disconnect between an EU animal production industry dependent on imported GM feed for its viability and the concurrent institutionalised opposition to the cultivation of GM crops reflects the economic and food security necessity of access to cost effective animal feed. This economic dependency may, however, be impacted by the broader human food discourse underpinning the risk disproportionate regulatory environment. Currently, for example, there is no requirement to label food from animals produced using GM containing feed in the EU, but some countries in other regions have moved in that direction and this is likely to generate pressure from some quarters for the EU to follow, with consequent substantial potential economic and food security impacts. In Bolivia, for example, Regulation 002/2015 states: "Article 5: In the case of products of animal origin, the following additional text should be included: 'this product is derived from animals whose food contains genetically modified material'".

On multiple occasions, Nations in Africa have rejected food aid for their starving citizens because that offered was from GM crops. In 2002, for example, Zambia rejected GM food to aid 3 million of its citizens with President Levy Mwanawasa referring to the food as "poison" (BBC News, 2002) and in 2010, Zimbabwe refused GM maize as food aid after raising health and environmental concerns, despite 2.2 million of its citizens requiring emergency food aid due to crop failure following low seasonal rains (Mpofu, 2010). While it would be easy to lay the blame for such irrational decisions at the feet of the decision makers of the recipient countries, these attitudes reflect the, presumably, unintended impacts of risk disproportionate regulatory burdens placed on GM crops in many developed countries. The excessive and unjustifiable testing requirements of those regulations serve to reinforce the disingenuous exploitation of the general low level of science literacy of most of the population by NGOs such as Greenpeace.

Excessive, disproportionate regulatory regimes for negligible risk GM foods impact the capacity of food regulators to manage real risks with substantial real economic and human health impacts, such as food contamination with microorganisms or naturally occurring toxicants. The costs of such regulatory resource misallocation are most particularly severe in resource limited regulatory bodies of smaller and less developed nations. In the entire history GM food there is not one credible report of adverse consequences of consumption in either humans or animals. Conversely, there are substantial numbers of microbial or chemical contamination incidents every year, some of which have catastrophic human health outcomes such as the large Shiga toxin producing *Escherichia coli* outbreak in Germany in 2011 which was due to consumption of contaminated seed sprouts (Bucholz et al., 2011). A similar contamination issue occurred in Australia with Hepatitis A contaminated semidried tomatoes (Donnan et al., 2012) which resulted in a number of deaths, and multiple hospitalisations. Despite the obvious risk-based priorities and perpetually limited resources, many national regulators must allocate scarce resources to GM risk assessment even though no genuine risk has ever been identified. As the General Manager for Risk Assessment at Food Standards Australia New Zealand, for example, this author had almost as many staff dedicated to the assessment of no risk GM products as to the high-risk area of microbial contamination. As Australia and New Zealand are prosperous nations, resource allocation follows assigned regulatory function and such resource misallocation does not impact the resourcing of the higher risk areas. Many national regulators, however, do not enjoy such levels of resourcing, and regulatory activity around high-risk priorities are compromised by allocation of scarce resources to the regulation of no-risk GM crops. Here again, the regulatory affections of the rich can directly impact the food safety of the poorer nations.

7. TRANSITIONING TO A SCIENCE AND RISK-BASED REGULATORY MODEL

There is a growing recognition that disproportionate, discriminatory, and nonscience-based regulatory burdens imposed by some of the wealthiest nations are inhibiting the use of essential technologies to adapt to a combination of challenges that threaten the sustainability and prosperity of the worlds' population. Climate change, environmental degradation of arable land, a rapidly increasing population, and increased household spending power of some of the most populous nations combine to create arguable the single largest societal challenge the world has faced in at least the past century. The growing recognition of the magnitude of the challenge has led to increased scrutiny of existing regulatory burdens by a number of jurisdictions and authors, and consideration of more rational, evidence based and therefore morally defensible regulatory regimes (Bartholomaeus et al., 2015; Beker et al., 2016; Conko, Kershen, Miller, & Parrott, 2016). To achieve this outcome, policymakers need to abandon the current ethics and evidence ambivalent model of weighting equally opinion with evidence, and consider a more cost/benefit-based model that objectively weights the potential for regulation generated risk and adversity, with the postulated risks to be mitigated through regulation.

The current ambivalence towards objective evidence and potential regulation-derived risk is evident in the response of the US National Academy of Science (Gould, Amasino, Brossard, & Buell, 2017) to criticism of its report on experiences with GM Crops (The National Academies, 2016) by Giddings and Miller (2016) where US NAS authors dismiss criticism which in their view favours "an authoritarian role for science that we contend has proven counterproductive". But as Giddings and Miller note, biologist Don Kennedy (a member of the NAS panel) opined it is not excusable to "give up the difficult task of finding where the weight of scientific evidence lies, and instead attach equal weight to each side in an effort to approximate fairness". This position is increasingly pertinent given the manifest risks generated by regulations that seek to provide protection against nonexistent risk in a disproportionate and discriminatory manner.

Although the current regulations pertaining to genetically modified organisms are redundant and could arguably be removed without consequence, this course of action is unlikely to be politically tenable. As an intermediate step, the application of nondiscriminatory product-based regulatory regime that combines regulatory oversight with self-regulatory obligations, and that is agnostic on methodology used to achieve end use products, could yield substantially reduced regulatory obfuscation of crop development. Decades of experience have demonstrated that food safety risks associated with crop development are technology independent and derive from the characteristics of the crop being modified and the trait being introduced. Where an exogenous gene is introduced through recombinant DNA techniques the origin of the introduced trait may also present a source of risk and warrants consideration. In this context, there is scope to shift the regulatory focus onto the crop variety being modified and the trait being introduced rather than on an event by event assessment. Thus, crops that lack toxic proteins or toxic secondary metabolites or where the potentially toxic constituents are well characterised could be approved for modification generically with a requirement to exclude up regulation or altered distribution of any endogenous toxins known to be present in the crop. Most food crops would fall within this category. Separately, specific traits could be approved for introduction into food crops. Extensive experience with Bt toxin and herbicide resistance traits, for example, has comprehensibly demonstrated a lack of food safety risk associated with the introduction of those traits into multiple, genetically diverse, crops.

Similarly, development of crops enriched with specific nutrients, such as yellow rice, or modified to reduce endogenous toxins that present a health risk, such as cassava, clearly presents no plausible source of risk and does not require extensive demonstrations, or assessment, of risk beyond demonstration that the desired trait has indeed been achieved.

Such a pragmatic approach to reduce the burden of redundant regulation has been taken by some countries in regards to stacked events. In Australia, for example, where each event has been previously approved, the stacking of those events through conventional breeding does not require approval from the food regulator. Extending this approach to the approval of the gene construct itself rather than each individual event is a logical next step towards a rational science based, and minimalist, regulation of the existing and emergent biotechnologies.

8. CONCLUSIONS

Current regulatory burdens imposed discriminately on biotechnology developed crops generate greater risk than that they were intended to mitigate, and breach basic Human rights principles of equity and justice.

Postulated risks that were originally argued as supporting the current regulatory imposts have been comprehensively discredited by both experience and increased understanding of the underlying nature and consequences of normal plant genome plasticity and variability. The cost shifting of the consequences of the regulatory affectations of the wealthy developed nations, with the income to compensate for the costs of those affectations, to the poorest of nations and communities, is arguably one of the most significant ethical lapses of our time. Increasing prosperity in some of the most populous nations and an increased capacity to compete for the limited supply of high value food commodities, however, is likely to shift the consequences of regulatory affectations back to the societies that originally generated them, with significant potential societal impacts.

The challenges of climate change, substantially increasing world population, the shift in the world economic centre of gravity to Asia, and the overwhelming body of evidence for the safety of the broad range of plant development technologies indicate that a fundamental change to the regulatory regimes for "conventional" and biotechnology in food production is now urgently required. The new regulatory paradigm needs to embrace the principles of proportionality, consider the risks of regulation equally with the (lack of) risks being regulated and provide a balanced nondiscriminatory regulatory environment within which various technologies can be applied to the increasingly urgent objective of food security.

REFERENCES

Ames, B., Profet, M., & Gold, L. (1990). Dietary pesticides (99.99% all natural). *Proceedings of the National Academy of Sciences of the United States of America, 87*, 7777–7781.

Bartholomaeus, A., Batista, J., Burachik, M., & Parrott, W. (2015). Recommendations from the workshop on comparative approaches to safety assessment of GM Plant materials: A road toward harmonized criteria? *GM Crops & Food, 6*(2), 69–79.

Bartholomaeus, A., Parrott, W., Bondy, G., & Walker, K. (2013). The use of whole food animal studies in the safety assessment of genetically modified crops: Limitations and recommendations. *Critical Reviews in Toxicology, 43*(S2), 1–24.

Baulcombe, D., Dunwell, J., Jones, J., Pickett, J., & Puigdomenech, P. (2014). *GM science update. A report to the council for Science and Technology*. Retrieved from www.gov.uk/government/publications/genetic-modification-gm-technologies.

BBC News. (2002). *Famine-hit Zambia rejects GM food aid*. Retrieved from http://news.bbc.co.uk/2/hi/africa/2371675.stm.

Beker, M., Boari, P., Burachik, M., Cuadrado, V., Junco, M., Lede, S., et al. (2016). Development of a construct based risk assessment framework for genetic engineered crops. *Transgenic Research, 25*(5), 597–607.

Bolarinwa, I., Oke, M., Olaniyan, S., & Ajala, A. (2016). Chapter 8: A review of cyanogenic glycosides in edible plants. In S. S. Larramendy (Ed.), *Toxicology—New aspects to this scientific conundrum* (p. 179) (INTECH).

Bourgaud, F., Gravot, A., Milesi, S., & Gontier, E. (2001). Production of plant secondary metabolites: A historical perspective. *Plant Science*, *161*, 839–851.

Brookes, G., & Barfoot, P. (2013). The global income and production effects of genetically modified (GM) crops 1996–2011. *GM Crops and Food*, *4*(1), 74–83.

Buchanan, B., Gruissem, W., & Jones, R. (Eds.), (2015). *Biochemistry & molecular biology of plants* (2nd ed.). Chichester: Wiley.

Bucholz, U., Bernard, H., Werber, D., Böhmer, M., Remschmidt, C., et al. (2011). German outbreak of *Escherichia coli* O104:H4 associated with sprouts. *The New England Journal of Medicine*, *365*(19), 1763–1770.

Conko, G., Kershen, D., Miller, H., & Parrott, W. (2016). A risk based approach to the regulation of genetically engineered organisms. *Nature Biotechnology*, *34*, 493–503.

Cottier, T. E.-A.-G. (2012). *The principle of proportionality in international law*. Swiss National Centre of Competence in Research. Retrieved from http://www.nccr-trade.org/publication/the-principle-of-proportionality-in-international-law/.

Creagh, S. (2011). *Scientists appalled by whippersnipper attack on CSIRO GM wheat trial*. The Conversation. Retrieved from http://theconversation.com/scientists-appalled-by-whippersnipper-attack-on-csiro-gm-wheat-trial-2334.

Directorate-General for Research and Innovation. (2010). *A decade of EU-Funded GMO Research 2001–2010*. Luxembourg: European Commission. Retrieved from https://publications.europa.eu/en/publication-detail/-/publication/d1be9ff9-f3fa-4f3c-86a5-beb0882e0e65/language-en.

Donnan, E., Fielding, J., Gregory, J., Lalor, K., Rowe, S., Goldsmith, P., et al. (2012). A multistate outbreak of hepatitis A associated with semidried tomatoes in Australia 2009. *Clinical Infectious Diseases*, *54*(6), 775–781.

Elias, P. (1980). The wholesomeness of irradiated food. *Ecotoxicology and Environmental Safety*, *4*, 172–183.

Farre, G., Ramessar, K., Twyman, R., Capell, T., & Christou, P. (2010). The humanitarian impact of plant biotechnology: Recent breakthroughs vs bottlenecks for adoption. *Current Opinion in Plant Biology*, *13*, 219–225.

Ferran, E. (2015). *Principle of proportionality. In Presentation to EBA workshop "the application of the principle of proportionality in the context of institutional and regulatory reforms"*: European Banking Authority. Retrieved from https://www.eba.europa.eu/documents/10180/1044289/Session+1.+Proportionality+vs+simplicity+-+Prof+Eilis+Ferran.pdf.

Friedman, M. (2006). Potato glycoalkaloids and metabolites: Roles in the plant and in the diet. *Journal of Agricultural and Food Chemistry*, *54*(23), 8655–8681.

Garneau, R. (2017). A company unfairly attacked by the environmental group has sued it. *National Review* Retrieved from http://www.nationalreview.com/article/445373/greenpeace-environmental-groups-sued-resolute-forest-products-ontario-quebec.

Geng, T., Stojsin, D., Liu, K., Schaalje, B., Postin, C., Ward, J., et al. (2016). Natural variability of allergen levels in conventional soybeans; assessing variation across north and South America from five production years. *Journal of Agricultural and Food Chemistry*, *65*, 463–472.

Giddings, G. (1992). Chemical safety of irradiated food. In J. Finley, S. Robinson, & D. Armstrong (Eds.), *Food safety assessment* (pp. 332–345). Washington: American Chemical Society.

Giddings, V., & Miller, H. (2016). US National Academies Report misses the mark. *Nature Biotechnology*, *34*(December), 1226–1228.

Gore, M., Chia, J., Elshire, R., Sun, Q., Ersoz, E., Hurwitz, B., et al. (2009). A first generation haplotype of maize. *Science*, *326*(5956), 1115–1117.

Gould, F., Amasino, R., Brossard, D., & Buell, C. (2017). Elevating the conversation about GE crops. *Nature Biotechnology*, *35*(4), 302–304.

GRACE. (2015). Conclusions and recommendations on animal feeding trials and alternative approaches and on the use of systematic reviews on evidence maps for GMO impact assessment. In *GMO risk assessment and communication of evidence*. Retrieved from http://www.grace-fp7.eu/sites/default/files/GRACE_Conclusions%20&Recommendations.pdf.

Harper, G., Hull, R., Lockhart, B., & Olszewski, N. (2002). Viral sequences integrated into plant genomes. *Annual Review of Phytopathology*, *40*, 119–136.

Harrigan, G. G., Lundry, D., Drury, S., Berman, K., Riordan, S., Nemeth, M. A., et al. (2010). Natural variation in crop composition and the impact of transgenesis. *Nature Biotechnology*, *28*, 402–404.

Herring, R., & Paarlberg, R. (2016). The political economy of biotechnology. *Annual Review of Resource Economics*, *8*, 397–416.

Herrington, M. (1983). Intense bitterness in commercial zucchini. *Cucurbit Genetics Cooperative Report*, *6*, 75–76.

Hoekenga, O., Srinivasan, J., Barry, G., & Bartholomaeus, A. (2013). Compositional analysis of genetically modified (GM) crops: Key issues and future needs. *Journal of Agricultural and Food Chemistry*, *61*, 8248–8253.

IPCC. (2007). IPCC Fourth Assessment Report: Climate Change 2007. In *Intergovernmental panel on climate change*. Retrieved from https://www.ipcc.ch/publications_and_data/ar4/wg2/en/ch3s3-4-3.html.

James, C. (2015). *20th Anniversary (1996 to 2015) of the Global Commercialization of Biotech Crops and Biotech Crop Highlights in 2015*. Ithaca: International Service for the Aquisition of Agri-Biotech Applications (ISAAA).

Juma, C. (2015). Chapter 3: Agricultural innovation systems. In *The new harvest, agricultural innovation in Africa* (2nd ed.). New York: Oxford University Press.

Kessler, D., Taylor, M., Maryanski, J., Flamm, E., & Kahl, L. (1992). The safety of foods developed by biotechnology. *Science*, *256*(5065), 1747–1749.

Klümper, W., & Qaim, M. (2014). A meta-analysis of the impacts of genetically modified crops. *PLoS One*, *9*(11), e111629.

Kyndt, T., Quispe, D., Zhai, H., Jarret, R., Ghislain, M., Liu, Q., et al. (2015). The genome of cultivated sweet potato contains agrobacterium T-DNAs with expressed genes: An example of a naturally transgenic food crop. *Proceedings of the National Academy of Sciences of the United States of America*, *112*(8), 5844–5849.

Lough, A., Roark, L., Kato, A., Ream, T., Lamb, J., et al. (2008). Mitochondrial DNA transfer to the nucleus generates extensive insertion site variation in maize. *Genetics*, *178*(1), 47–55.

Masip, G., Sabalza, M., Perez-Massot, E., Banakar, R., Cebrian, D., Twyman, R., et al. (2013). Paradoxical EU agricultural policies on genetically engineered crops. *Trends in Plant Science*, *18*(6), 312–324.

McClain, S., Jones, W., He, X., Ladics, G., Bartholomaeus, A., Raybould, A., et al. (2015). Agricultural biotechnology safety assessment. *Zhonghua Yu Fang Yi Xue Za Zhi*, *49*(1), 78–87.

McDougall, P. (2011). *The cost and time involved in the discovery, development and authorisation of a new plant biotechnology derived trait*. Retrieved from https://croplife.org/wp-content/uploads/pdf_files/Getting-a-Biotech-Crop-to-Market-Phillips-McDougall-Study.pdf.

Mpofu, T. (2010). *Starving Zimbabwe rejects GM maize*. The National World. Retrieved from http://www.thenational.ae/news/world/africa/starving-zimbabwe-rejects-gm-maize.

Naito, k., Cho, E., Yang, G., Campbell, M., Okumoto, Y., Tanisaka, T., et al. (2006). Dramatic amplification of a rice transposable element during recent domestication. *Proceedings of the National Academy of Sciences of the United States of America*, *103*(47), 17620–17625.

Najim, A., Hadi, M., Fazeli, F., Darzi, M., & Rahi, A. (2012). Effect of integrated management of nitrogen fertilizer and cattle manure on the leaf chlorophyll, yield, and tuber glycoalkaloids of Agria potato. *Communications in Soil Science and Plant Analysis*, *43*(6), 912–923.

National Research Council (NRC). (1989). *Field Testing Genetically Modified Organisms: Framework for Decisions*. Washington, DC, USA: National Academy Press.

OECD. (1995). *Recommendation of the council of the OECD on improving the quality of government regulation*. Paris: OECD. Retrieved from http://www.oecd.org/officialdocuments/publicdisplaydocumentpdf/?doclanguage=en&cote=OCDE/GD(95)95.

OECD. (2012). *Recommendation of the council on regulatory policy and governance*. Paris: OECD. Retrieved from http://www.oecd.org/gov/regulatory-policy/49990817.pdf.

OECD. (2015). *National Accounts at a glance 2015*. OECD Publishing. Retrieved from http://www.oecd-ilibrary.org/docserver/download/3015071e.pdf?expires=1492573255&id=id&accname=guest&checksum=8A2A31393575F497998809BF3376AC46.

OGTR. (2005). *Application for licence for dealings involving an intentional release into the environment*. Canberra: Risk Assessment and Risk Management Plan, Office of the Gene Technology Regulator, Department of Health and Ageing. Retrieved from http://www.ogtr.gov.au/internet/ogtr/publishing.nsf/content/dir054-3/$FILE/dir054.1finalrarmp.pdf.

Otten, L. (2018). How Agrobacterium, a natural genetic engineer, became a tool for modern agriculture. *Advances in Botanical Research, 86*.

Palaisa, K., Morgante, M., Williams, M., & Rafalski, A. (2003). Contrasting effects of selection on sequence diversity and linkage disequilibrium at two phytoene synthase loci. *Plant Cell, 15*(8), 1795–1806.

Regina, A., Berbezy, P., Kosar-Hashemi, B., Li, S., Cmiel, M., Larroque, O., et al. (2015). A genetic strategy generating wheat with very high amylose content. *Plant Biotechnology Journal, 13*, 1276–1286.

Schnable, P., Ware, D., Fulton, R., Stein, J., Wei, F., & Paternack, S. (2009). The B73 maize genome: Complexity, diversity, and dynamics. *Science, 326*(5956), 1112–1115.

Shang, Y., Ma, Y., Zhou, Y., Zhang, H., Duan, L., Chen, H., et al. (2014). Biosynthesis, regulation, and domestication of bitterness in cucumber. *Science, 346*, 1084–1088.

Shewry, P., Hawkesford, M., Piironen, V., Lampi, A., Gebruers, K., Boros, D., et al. (2013). Natural variation in grain composition of wheat and related cereals. *Journal of Agricultural and Food Chemistry, 61*, 8295–8303.

Sinden, S., & Webb, R. (1974). *Effect of environment on glycoalkaloid content of six potato varieties at 39 locations*. United States Department of Agriculture.

Steiner, H., Halpin, C., Jez, J., Kough, J., Parrott, W., Underhill, L., et al. (2013). Evaluating the potential for adverse interactions within genetically engineered breeding stacks. *Plant Physiology, 161*, 1587–1594.

Strateford, K., & Cowling, A. (2016). Chinese household income, Consumption and savings. *RBA Bulletin, (September Quarter)*. Retrieved from https://www.rba.gov.au/publications/bulletin/2016/sep/pdf/rba-bulletin-2016-09-chinese-household-income-consumption-and-savings.pdf.

Tagliabue, G. (2017). The EU legislation on "GMOs" between nonsense and protectionism: An ongoing Schumpeterian chain of public choices. *GM Crops & Food, 8*, 35–51.

Tenaillon, M., Sawkins, M., Long, A., Gaut, R., Doebley, J., & Gaut, B. (2001). Patterns of DNA sequence polymorphism along chromosome 1 of maize (Zea mays ssp. mays L.). *Proceedings of the National Academy of Sciences of the United States of America, 98*, 9161–9166.

The National Academies. (2016). *Genetically engineered crops: Experiences and prospects*. National Academies Press.

Thomas, D. (2012). *Regressive effects of regulation*. Meractus Center—George Mason University. Retrieved from https://www.mercatus.org/publication/regressive-effects-regulation.

Tian, Z., Zhao, M., She, M., Du, J., Cannon, S., Liu, X., et al. (2012). Genome-wide characterization of nonreference transposons reveals evolutionary propensities of transposons in soybean. *The Plant Cell, 24*, 1–15.

United Nations. (2015). *World population prospects. The 2015 revision, key findings and advance tables*. Department of Economic and Social Affairs, Population Division. Retrieved from https://esa.un.org/unpd/wpp/Publications/Files/Key_Findings_WPP_2015.pdf.

US Department of Labor. (2006). *100 years of U.S. consumer spending. US Department of Labour.* Retrieved from https://www.bls.gov/opub/uscs/report991.pdf.
USAID. (n.d.). Disease resistant potatoes: South & South East Asia. U.S. Agency for International Development. Retrieved from http://pdf.usaid.gov/pdf_docs/Pdacp869.pdf
Van Eenennaam, A., & Young, A. (2014). Prevalence and impacts of genetically engineered feedstuffs on livestock populations. *Journal of Animal Science, 92*, 4255–4278.
von Braun, J. A. (2007). One in seven people goes to bed hungry. In T. A. Joachim von Braun (Ed.), *The challenge of hunger 2007: Global hunger index* (pp. 4–5). BONN: International Food Policy Research Institute.
Washington State University. (2008). Annual income spent on food. Retrieved from http://wsm.wsu.edu/researcher/2011fall_hunger_foodcost.php
WHO, FAO, & IAEA. (1999). High-dose irradiation: Wholesomeness of food irradiated with doses above 10kGy. In *Report of a Joint FAO/IAEA/WHO Study Group.* Geneva, Switzerland: World Health Organization.
Zhao, Z., Jin, L., Fu, Y., Ramsey, M., Jenkins, T., Leskinen, E., et al. (2000). Worldwide DNA sequence variation in a 10-kilobase noncoding region on human chromosome 22. *Proceedings of the National Academy of Sciences of the United States of America, 97*(21), 11354–11358.

FURTHER READING

Weber, N., Halpin, C., Hannah, L., & Jez, J. (2012). Crop genome plasticity and its relevance to food and feed safety of genetically engineered breeding stacks. *Plant Physiology, 160*, 1842–1853.

CHAPTER FIVE

Assessing the Environmental Safety of Transgenic Plants: Honey Bees as a Case Study

Agnès Ricroch*,†,1, Serife Akkoyunlu†, Jacqueline Martin-Laffon‡, Marcel Kuntz‡

*University of Paris-Sud, Collège d'Etudes Interdisciplinaires, Sceaux, France
†AgroParisTech, Paris, France
‡Univ. Grenoble Alpes, CNRS, CEA, INRA, Laboratoire de Physiologie Végétale & Cellulaire, Grenoble, France
[1]Corresponding author: e-mail address: agnes.ricroch@u-psud.fr

Contents

Abstract

Bees play an important role in the pollination of a wide range of plants and are likely to encounter genetically engineered crops (often termed "genetically modified" or "GM") during their foraging period, especially insect-resistant crops since these crops have been cultivated worldwide. Thus, it is important to assess potential impacts of these crops on the nontarget organism honey bee (*Apis mellifera* L.), the most important pollinator species worldwide. In the present study, we gathered all scientific data related to the effects of insect-resistant GM crops (mostly corn and cotton, and also oilseed rape, rice, soybean, and wheat) on honey bees. Assessments included feeding honey bees with purified insecticidal toxins or transgenic pollen collected from GM crops producing such toxins, namely protease inhibitors (PIs), Cry or VIP toxins from *Bacillus thuringiensis*. RNAi-producing and herbicide-tolerant crops were also included. A total of 64 peer-reviewed studies have been published between 1994 and 2017. We also compiled

Advances in Botanical Research, Volume 86
ISSN 0065-2296
https://doi.org/10.1016/bs.abr.2017.11.004

18 studies submitted to and examined by the US EPA between 1993 and 2002. Our analyses converge to the conclusion that the studied Cry proteins, RNAi or herbicide-tolerance proteins do not negatively affect the survival of honey bees and have no potential sublethal effect in controlled laboratory conditions or in field/semifield trials. The risk of PI will mainly depend on their concentration in pollen and need to be assessed case by case.

1. INTRODUCTION

Cultivation of genetically engineered (or "genetically modified" as legally termed in some countries; GM) crops has increased markedly since the first introduction of the transgenic technology in 1996, representing 185 millions of hectares (about 12.3% of all cropland worldwide) in 2016 (www.isaaa.org). Of the 14 GM traits approved for commercial production, the most widely planted GM crops are those expressing insecticidal toxins targeted at specific insect pests or those tolerant to one or more herbicides. Insect-resistant crops or so-called Bt maize, cotton, and soybean have been GM to express one or more insecticidal proteins from the natural occurring soil bacterium *Bacillus thuringiensis* (Bt). These Cry (δ-endotoxins found in crystals formed at the sporal stage) or VIP (vegetative insecticidal proteins) toxins bind to specific epithelial receptors in the gut wall of insect larvae, causing feeding cessation, cell lysis, and pore formation that eventually result in gut paralysis and death (Roh, Choi, Li, Jin, & Je, 2007). The specific binding of the particular Bt protein exerts a selective toxic effect on a specific group of target insects or taxonomic order, such as Lepidoptera (moths) or Coleoptera (beetles) in the case of current Bt crops (Bergé & Ricroch, 2011).

Risk assessment of the environmental effects of GM crops is part of the regulatory process for commercial approval. In this context, guidelines for evaluating the potential risks for nontarget arthropods and improvements in methodologies have been discussed (see reference list A in accompanying online Supplementary Data File (S) available on https://doi.org/10.1016/bs.abr.2017.11.004). Regarding official risk assessment agencies, EFSA guidance documents and discussion related to EFSA's view on unintended effects on nontarget organisms can also be cited (see reference list B in S available on https://doi.org/10.1016/bs.abr.2017.11.004). A number of recent reviews (see reference list C in S available on https://doi.org/10.1016/bs.abr.2017.11.004) dealing with Bt crop effects on nontarget arthropods have been published. In addition, recent metaanalyses are also available (see reference list D in S available on https://doi.org/10.1016/bs.abr.2017.11.004). In addition to the above-mentioned (marketed) traits, RNAi strategies against pests have become a burgeoning field of research (see reference list E in S available on https://doi.org/10.1016/bs.abr.2017.11.004). Their potential impact on nontarget organisms has fuelled much discussion (see reference list F in S available on https://doi.org/10.1016/bs.abr.2017.11.004). Other transgenic proteins expressed in GM plants for controlling insect pests include protease inhibitors (PIs) (Hilder, Gatehouse, Sheerman, Barker, & Boulter, 1987; Ryan, 1990). When ingested by insects, specific proteases can inhibit the digestive proteolytic enzymes in the insect gut, resulting in starvation and eventual death. However, none have been commercially planted to date with the exception of a PI trait in China (see below). A number of articles have also discussed the compatibility of Bt crops with integrated pest management. Some recent references, related to Bt cotton are listed here (see reference list G in S available on https://doi.org/10.1016/bs.abr.2017.11.004).

Honey bees, *Apis mellifera* L. (Hymenoptera: Apidae), play a major role in ecosystems, providing valuable pollination services for a diverse range of economically important fruits, vegetables, and agronomic crops. Beekeepers have always experienced colony losses during the winter due to starvation or weak colonies in the fall. Losses up to 15% are usual but have been sustainable in the beekeeping industry. However, unusual high colony losses, exceeding 30%, have been reported worldwide, since 1998 in Europe and since 2006 in the United States. This is known as colony collapse disorder (CDD), a die-off phenomenon associated with the sudden disappearance of worker bees from a colony, with immature bees and the queen often remaining (Van Engelsdorp, Cox-Foster, Frazier, Ostiguy, & Hayes, 2006). While the incidence of CDD symptomatic cases has declined, the continued high rate of colony losses is still a major concern. Bee scientists now agree that colony losses are the result of a multifactorial syndrome involving the interaction of many stress factors, including pathogens and parasites (particularly Varroa mites and viruses), pesticides, poor nutrition, management stress, and loss of foraging landscape (Berenbaum, 2014; Liao, Wu, & Berenbaum, 2017). While there is mounting evidence that parasitic mites and their interaction with viral diseases represent the major driver of colony stress, many beekeepers, and environmental groups still fiercely oppose GM crops which they view as a threat for honey bee health, but also for their own economic existence (Jany & Höfer, 2015).

Formulations of bioinsecticides containing *B. thuringiensis* proteins and spores have been safely used since 1961 and have not been known to harm bees (see D'Urso et al., 2017, and references therein). Therefore, Bt crops producing the same or similar active compounds would not be expected to harm bees either. However, unlike the proteins produced by the bacterium, which require solubilization and proteolytic degradation to activate the toxin, insecticidal proteins in Bt crops are produced constantly in a truncated, active form, sometimes with further modifications (Deist, Rausch, Fernandez-Luna, Adang, & Bonning, 2014).

Herbicide-tolerant crops have been genetically engineered with plant/bacterial genes encoding specific enzymes that either detoxify the herbicide or provide a metabolic backup pathway to tolerate the herbicide. The enzymes produced by these transgenes have no insecticidal activity and as expected, there have been no reports of direct adverse effect on insects, particularly bees. However, the relative efficacies of herbicide regimes may affect field biodiversity (see Hawes et al., 2003).

Despite 2 decades of commercial use without demonstrated adverse environmental effects, there is still much debate and concern about the possible nontarget effects of GM crops on bees and other beneficial insects. Clearly, bees are likely to be exposed to the pollen and nectar of GM crops which can contain varying concentrations of the expressed proteins. Canola, including GM varieties, serves as a major floral resource for bees, while maize and soybean (whose pollens are spread by wind) do not require insects for pollination. However, in case of large-scale monoculture, GM maize and soybean may become a major forage plant for honey bees during periods when more preferred floral sources are lacking.

Although previous literature reviews (see Duan, Marvier, Huesing, Dively, & Huang, 2008; Malone, 2004; Malone & Pham-Delegue, 2001; Marvier, McCreedy, Regetz, & Kareiva, 2007; O'Callaghan, Glare, Burgess, & Malone, 2005) did not provide evidence that cultivated GM crops have a significant adverse effect on nontarget plant-feeding insects and beneficial species, such adverse effects are still investigated by researchers. In this study, we have exhaustively screened the scientific literature on the assessment of the impact of insecticidal compounds (PIs, Cry, and VIP proteins from *B. thuringiensis*, RNAi) or proteins conferring tolerance to herbicides on nontarget organisms, with a specific focus on honey bees. The examined studies using either purified proteins or pollen (since potential risks due to products of the transgene will most likely be due to pollen feeding) in laboratory force-feeding studies, or that performed under field or semifield conditions.

2. METHODS

2.1 Searching

For the search of scientific publications on the assessment of the lethal and sublethal toxicity of insecticidal proteins on honey bees, we built up the following equation request for screening online databases of the Web of Science (all databases), Scopus, Pubmed, and Science Direct (research published until September 13, 2017).

Query 1: (honeybee* OR bee OR bees OR apis) AND (transgen* OR gmo OR « genetic* modif*» OR «genetically engineered» OR « GM plant* » OR cry OR bt OR mon810 OR «mon 810») AND (collapse* OR CCD OR disappear* OR dwindle OR «may disease» OR mortality OR death OR lethal OR survival OR abandon* OR «colony loss*» OR longevity OR decline OR «adverse effect*» OR killing).

Query 2: (honeybee* OR bee OR bees OR apis) NEAR/10 (transgen* OR gmo OR « genetic* modif*» OR «genetically engineered» OR « GM plant* » OR cry OR bt OR mon810 OR «mon 810»).

Final equation request: Query 1 OR Query 2.

Note: The above equation requests match with the Web of Science search rules and have been adapted to the other databases according to their own rules (proximity operators, adjacent terms searching, truncations, ...).

2.2 Analyses of Studies

The selected publications were sorted manually and classed according to their experimental conditions (laboratory conditions, field, or semifield trials) and the material tested (purified proteins or pollen). The resulting 64 scientific studies published in peer-reviewed journals are listed in Table 1. For each study, we examined the characteristics of the toxin and its origin, the dose and duration of exposure, response variables, and concluding remarks of the authors.

3. RESULTS

3.1 Study Types, Publications by Year and by Country

3.1.1 A Brief Historical Perspective

The first available studies, from 1993 onward, were produced by the plant biotech industry as part of the regulatory assessment of Bt plants (we focused on those examined by the US Environment Protection Agency, EPA,

Table 1 References (64 Studies)

Reference #	Article
1	Belzunces L. P., Lenfant C., Di Pasquale S., & Colin M.-E. (1994) In vivo and in vitro effects of wheat germ agglutinin and Bowman–Birk soybean trypsin inhibitor, two potential transgene products, on midgut esterase and protease activities from *Apis mellifera*. *Comparative Biochemistry and Physiology Part B: Comparative Biochemistry, 109,* 63–69. http://www.sciencedirect.com/science/article/pii/0305049194901422
2	Ahl G. P., Warren G., White J., Privalle L., Fearing P., & Vlachos D. (1995) Community impact: Effects on microorganisms or insects. *Mitteilungen aus der Biologischen Bundesanstalt für Land- und Forstwirtschaft 309,* 50–53 (unclear whether peer-reviewed)
3	Picard-Nizou A. L., Pham-Delègue M. H., Kerguelen V., Douault P., Marilleau R., Olsen L., et al. (1995) Foraging behaviour of honey bees (Apis mellifera L.) on transgenic oilseed rape (*Brassica napus* L. var. oleifera). *Transgenic Research, 4,* 270–276. http://link.springer.com/article/10.1007/BF01969121
4	Sims S. R. (1995) Bacillus thuringiensis var. kurstaki [Cry1A(c)] protein expressed in transgenic cotton: Effects on beneficial and other non-target insects. *Southwest Entomologist, 20,* 493–500. http://sswe.tamu.edu/articles/PDF/SWE_V20_N4_P493-500.pdf
5	Arpaia S. (1996) Ecological impact of Bt-transgenic plants: 1. Assessing possible effects of CryIIIB toxin on honey bee (Apis mellifera L.) colonies. *Journal of Genetics and Breeding, 50,* 315–319
6	Burgess E. P. J., Malone L. A., & Christeller J. T. (1996) Effects of two proteinase inhibitors on the digestive enzymes and survival of honey bees (Apis mellifera). *Journal of Insect Physiology, 42,* 823–828. http://www.sciencedirect.com/science/article/pii/0022191096000455
7	Picard-Nizou A. L., Grison R., Olsen L., Pioche C., Arnold G., & Pham-Delegue M. H. (1997) Impact of proteins used in plant genetic engineering: Toxicity and behavioral study in the honeybee. *Journal of Economic Entomology, 90,* 1710–1716. https://academic.oup.com/jee/article-abstract/90/6/1710/2216799/Impact-of-Proteins-Used-in-Plant-Genetic
8	Sims S. R. (1997) Host activity spectrum of the CryIIA Bacillus thuringiensis subsp. kurstaki protein: Effects on lepidoptera, diptera, and non-target arthropods. *Southwest Entomologist, 22,* 395–404. http://sswe.tamu.edu/PDF/SWE_V22_N4_P395-404.pdf

Table 1 References (64 Studies)—cont'd

Reference #	Article
9	Girard C., Picard-Nizou A. L., Grallien E., Zaccomer B., Jouanin L., & Pham-Delegue M. H. (1998) Effects of proteinase inhibitor ingestion on survival, learning abilities and digestive proteinases of the honeybee. *Transgenic Research, 7,* 239–246. http://link.springer.com/article/10.1023%2FA%3A1008813910751
10	Malone L. A., Burgess E. P. J., Christeller J. T., & Gatehouse H. S. (1998) In vivo responses of honey bee midgut proteases to two protease inhibitors from potato. *Journal of Insect Physiology, 44,* 141–147. http://www.sciencedirect.com/science/article/pii/S0022191097000966
11	Malone L. A., Burgess E. P. J., & Stefanovic D. (1999) Effects of a Bacillus thuringiensis toxin, two Bacillus thuringiensis biopesticide formulations, and a soybean trypsin inhibitor on honey bee (Apis mellifera L.) survival and food consumption. *Apidologie, 30,* 465–473. http://www.apidologie.org/articles/apido/abs/1999/05/Apidologie_0044-8435_1999_30_6_ART0001/Apidologie_0044-8435_1999_30_6_ART0001.html
12	Pham-Delegue M. H., Girard C., Le Metayer M., Picard-Nizou A. L., Hennequet C., Pons O., et al. (2000) Long-term effects of soybean protease inhibitors on digestive enzymes, survival and learning abilities of honeybees. *Entomologia Experimentalis Et Applicata, 95,* 21–29. http://onlinelibrary.wiley.com/doi/10.1046/j.1570-7458.2000.00637.x/abstract
13	Schur A., Tornier I., & Neumann C. (2000) Assessment of potential side effects of Bt-corn to the honey bee, Apis mellifera (Hymenoptera, Apidae) under field conditions. *Apidologie, 31,* 616–618
14	Malone L. A., Burgess E. P. J., Gatehouse H. S., Voisey C. R., Tregidga E. L., & Philip B. A. (2001) Effects of ingestion of a Bacillus thuringiensis toxin and a trypsin inhibitor on honey bee flight activity and longevity. *Apidologie, 32,* 57–68. http://www.apidologie.org/articles/apido/abs/2001/01/m1101/m1101.html
15	Brodsgaard H. F., Brodsgaard C. J., Hansen H., & Lovei G. L. (2003) Environmental risk assessment of transgene products using honey bee (Apis mellifera) larvae. *Apidologie, 34,* 139–145. http://www.apidologie.org/articles/apido/abs/2003/02/M3203/M3203.html
16	Hanley A. V., Huang Z. Y., & Pett W. L. (2003) Effects of dietary transgenic Bt corn pollen on larvae of Apis mellifera and Galleria mellonella. *Journal of Apicultural Research, 42,* 77–81. http://www.tandfonline.com/doi/abs/10.1080/00218839.2003.11101097

Continued

Table 1 References (64 Studies)—cont'd

Reference #	Article
17	Haughton J., Champion G. T., Hawes C., Heard M. S., Brooks D. R., Bohan D. A., et al. (2003) Invertebrate responses to the management of genetically modified herbicide-tolerant and conventional spring crops. II. Within-field epigeal and aerial arthropods. *Philosophical transactions of the Royal Society of London. Series B, 358,* 1863–1877
18	Pierre J., Marsault D., Genecque E., Renard M., Champolivier J., & Pham-Delegue M. H. (2003) Effects of herbicide-tolerant transgenic oilseed rape genotypes on honey bees and other pollinating insects under field conditions. *Entomologia Experimentalis Et Applicata, 108,* 159–168. http://onlinelibrary.wiley.com/doi/10.1046/j.1570-7458.2003.00081.x/abstract
19	Roy D. B., Bohan D. A., Haughton A. J., Hill M. O., Osborne J. L., Clark S. J., et al. (2003) Invertebrates and vegetation of field margins adjacent to crops subject to contrasting herbicide regimes in the Farm Scale Evaluations of genetically modified herbicide-tolerant crops. *Philosophical Transactions of the Royal Society of London, Series B, 358,* 1879–1898
20	Huang Z. Y., Hanley A. V., Pett W. L., Langenberger M., & Duan J. J. (2004) Field and semifield evaluation of impacts of transgenic canola pollen on survival and development of worker honey bees. *Journal of Economic Entomology, 97,* 1517–1523. https://www.ncbi.nlm.nih.gov/pubmed/15568338
21	Malone L. A., Todd J. H., Burgess E. P. J., & Christeller J. T. (2004) Development of hypopharyngeal glands in adult honey bees fed with a Bt toxin, a biotin-binding protein and a protease inhibitor. *Apidologie, 35,* 655–664. http://www.apidologie.org/articles/apido/abs/2004/07/M4045/M4045.html
22	Babendreier D., Kalberer N. M., Romeis J., Fluri P., Mulligan E., & Bigler F. (2005) Influence of Bt-transgenic pollen, Bt-toxin and protease inhibitor (SBTI) ingestion on development of the hypopharyngeal glands in honeybees. *Apidologie, 36,* 585–594. http://www.apidologie.org/articles/apido/abs/2005/04/M5407/M5407.html
23	Bailey J., Scott-Dupree C., Harris R., Tolman J., & Harris B. (2005) Contact and oral toxicity to honey bees (Apis mellifera) of agents registered for use for sweet corn insect control in Ontario, Canada. *Apidologie, 36,* 623–633. http://www.apidologie.org/articles/apido/abs/2005/04/M5406/M5406.html

Table 1 References (64 Studies)—cont'd

Reference #	Article
24	Dechaume-Moncharmont F. X., Azzouz H., Pons O., & Pham-Delegue M. H. (2005) Soybean proteinase inhibitor and the foraging strategy of free flying honeybees. *Apidologie, 36,* 421–430. http://www.apidologie.org/articles/apido/abs/2005/03/M4089/M4089.html
25	Liu B., Xu C. G., Yan F. M., & Gong R. Z. (2005) The impacts of the pollen of insect-resistant transgenic cotton on honeybees. *Biodiversity and Conservation, 14,* 3487–3496. http://link.springer.com/article/10.1007/s10531-004-0824-7
26	Ramirez-Romero R., Chaufaux J., & Pham-Delegue M. H. (2005) Effects of Cry1Ab protoxin, deltamethrin and imidacloprid on the foraging activity and the learning performances of the honeybee Apis mellifera, a comparative approach. *Apidologie, 36,* 601–611. http://www.apidologie.org/articles/apido/abs/2005/04/M4097/M4097.html
27	Sagili R. R., Pankiw T., & Zhu-Salzman K. (2005) Effects of soybean trypsin inhibitor on hypopharyngeal gland protein content, total midgut protease activity and survival of the honey bee (Apis mellifera L.). *Journal of Insect Physiology, 51,* 953–957. http://www.sciencedirect.com/science/article/pii/S0022191005000892
28	Babendreier D., Joller D., Romeis J., Bigler F., & Widmer F. (2007) Bacterial community structures in honeybee intestines and their response to two insecticidal proteins. *FEMS Microbiology Ecology, 59,* 600–610. https://www.ncbi.nlm.nih.gov/pubmed/17381517
29	Lehrman A. (2007) Does pea lectin expressed transgenically in oilseed rape (Brassica napus) influence honey bee (Apis mellifera) larvae? *Environmental Biosafety Research, 6,* 271–278. https://www.ncbi.nlm.nih.gov/pubmed/18289502
30	Mohr K. I., & Tebbe C. C. (2007) Field study results on the probability and risk of a horizontal gene transfer from transgenic herbicide-resistant oilseed rape pollen to gut bacteria of bees. *Applied Microbiology and Biotechnology, 75,* 573–582. http://link.springer.com/article/10.1007/s00253-007-0846-7
31	Rose R., Dively G. P., & Pettis J. (2007) Effects of Bt corn pollen on honey bees: Emphasis on protocol development. *Apidologie, 38,* 368–377. http://link.springer.com/article/10.1051/apido:2007022
32	Duan J. J., Marvier M., Huesing J., Dively G., & Huang Z. Y. (2008) A meta-analysis of effects of Bt crops on honey bees (Hymenoptera: Apidae). *PLoS One, 3,* e1415. http://journals.plos.org/plosone/article/asset?id=10.1371/journal.pone.0001415.PDF

Continued

Table 1 References (64 Studies)—cont'd

Reference #	Article
33	Porcar M., Gomez F., Gruppe A., Gomez-Pajuelo A., Segura I., & Schroder R. (2008) Hymenopteran specificity of Bacillus thuringiensis strain PS86Q3. *Biological Control, 45,* 427–432
34	Ramirez-Romero R., Desneux N., Decourtye A., Chaffiol A., & Pham-Delegue M. H. (2008) Does CrylAb protein affect learning performances of the honey bee Apis mellifera L. (Hymenoptera, Apidae)? *Ecotoxicology and Environmental Safety, 70,* 327–333
35	Lipinski Z., Farjan M., Zoltowska K., & Polaczek B. (2008) Effects of dietary transgenic Bacillus thuringiensis maize pollen on hive worker honeybees. *Polish Journal of Environmental Studies, 17,* 957–961. http://www.pjoes.com/abstracts/2008/Vol17/No06/14.html
36	Liu B. A., Shu C., Xue K., Zhou K. X., Li X. G., Liu D. D., et al. (2009) The oral toxicity of the transgenic Bt plus CpTI cotton pollen to honeybees (Apis mellifera). *Ecotoxicology and Environmental Safety, 72,* 1163–1169. http://www.sciencedirect.com/science/article/pii/S014765130900027X
37	Han P., Niu C. Y., Lei C. L., Cui J. J., & Desneux N. (2010) Quantification of toxins in a Cry1Ac + CpTI cotton cultivar and its potential effects on the honey bee Apis mellifera L. *Ecotoxicology, 19,* 1452–1459. https://www.ncbi.nlm.nih.gov/pubmed/20700762
38	Han P., Niu C. Y., Lei C. L., Cui J. J., & Desneux N. (2010) Use of an innovative T-tube maze assay and the proboscis extension response assay to assess sublethal effects of GM products and pesticides on learning capacity of the honey bee Apis mellifera L. *Ecotoxicology, 19,* 1612–1619. https://www.ncbi.nlm.nih.gov/pubmed/20872243
39	Hendriksma H. P., Hartel S., & Steffan-Dewenter I. (2011) Testing pollen of single and stacked insect-resistant Bt-maize on in vitro reared honey bee larvae. *PLoS One, 6,* e28174. http://journals.plos.org/plosone/article?id=10.1371/journal.pone.0028174
40	Lima M. A. P., Pires C. S. S., Guedes R. N. C., Nakasu E. Y. T., Lara M. S., Fontes E. M. G., et al. (2011) Does Cry1Ac Bt-toxin impair development of worker larvae of Africanized honey bee? *Journal of Applied Entomology, 135,* 415–422. http://onlinelibrary.wiley.com/doi/10.1111/j.1439-0418.2010.01573.x/abstract
41	Naik A., Viraktamath S., Vastrad A. S., Udikeri S. S., & Megeri S. N. (2011) Pollinator fauna and foraging activity of honey bees in two genotypes of Bt cotton. *Karnataka Journal of Agricultural Sciences, 24,* 400–403

Table 1 References (64 Studies)—cont'd

Reference #	Article
42	Raybould A., & Vachos D. (2011) Non-target organism effects tests on Vip3A and their application to the ecological risk assessment for cultivation of MIR162 maize. *Transgenic Research, 20,* 599–611. doi: https://doi.org/10.1007/s11248-010-9442-1. Epub Sept. 14, 2010. https://www.ncbi.nlm.nih.gov/pubmed/20839052
43	Dai P. L., Zhou W., Zhang J., Jiang W. Y., Wang Q., Cui H. J., et al. (2012) The effects of Bt Cry1Ah toxin on worker honeybees (Apis mellifera ligustica and *Apis cerana* cerana). *Apidologie, 43,* 384–391. http://link.springer.com/article/10.1007/s13592-011-0103-z
44	Dai P. L., Zhou W., Zhang J., Cui H. J., Wang Q., Jiang W. Y., et al. (2012) Field assessment of Bt cry1Ah corn pollen on the survival, development and behavior of Apis mellifera ligustica. *Ecotoxicology and Environmental Safety, 79,* 232–237. https://www.ncbi.nlm.nih.gov/pubmed/22364780
45	Grabowski M., & Dabrowski Z. T. (2012) Evaluation of the impact of the toxic protein Cry1Ab expressed by the genetically modified cultivar MON810 on honey bee (Apis mellifera L.) behaviour. *Medycyna Weterynaryjna, 68,* 630–633. http://medycynawet.edu.pl/index.php/archives/231/4215-summary-med-weter-68-10-630-633-2012
46	Han P., Niu C. Y., Biondi A., & Desneux N. (2012) Does transgenic Cry1Ac + CpTI cotton pollen affect hypopharyngeal gland development and midgut proteolytic enzyme activity in the honey bee Apis mellifera L. (Hymenoptera, Apidae)? *Ecotoxicology, 21,* 2214–2221. https://www.ncbi.nlm.nih.gov/pubmed/22868904
47	Hendriksma H. P., Hartel S., Babendreier D., von der Ohe W., & Steffan-Dewenter I. (2012) Effects of multiple Bt proteins and GNA lectin on in vitro-reared honey bee larvae. *Apidologie, 43,* 549–560. http://link.springer.com/article/10.1007/s13592-012-0123-3
48	Hendriksma H. P., Kuting M., Hartel S., Nather A., Dohrmann A. B., Steffan-Dewenter I., et al. (2013) Effect of stacked insecticidal Cry proteins from maize pollen on nurse bees (Apis mellifera carnica) and their gut bacteria. *PLoS One, 8,* e59589. http://journals.plos.org/plosone/article?id=10.1371/journal.pone.0059589
49	Geng L. L., Cui H. J., Dai P. L., Lang Z. H., Shu C. L., Zhou T., et al. (2013) The influence of Bt-transgenic maize pollen on the bacterial diversity in the midgut of Apis mellifera ligustica. *Apidologie, 44,* 198–208. http://link.springer.com/article/10.1007/s13592-012-0171-8

Continued

Table 1 References (64 Studies)—cont'd

Reference #	Article
50	Niu L., Ma Y., Mannakkara A., Zhao Y., Ma W., Lei C., et al. (2013) Impact of single and stacked insect-resistant bt-cotton on the honey bee and silkworm. *PLoS One, 8,* e72988. http://journals.plos.org/plosone/article?id=10.1371/journal.pone.0072988
51	Burns A., & Raybould A. (2014) Nontarget organism effects tests on eCry3.1Ab and their application to the ecological risk assessment for cultivation of event 5307 maize. *Transgenic Research, 23,* 985–994. https://pdfs.semanticscholar.org/e3db/bf5b7fa9a43aaa1bed8c9219e5843a7fb4fb.pdf
52	Pise V., & Viraktamath S. (2015) Comparative studies on the pollinator fauna and foraging activity of honey bees on Bt and non-Bt cotton hybrids. *Karnataka. Journal of Agricultural Sciences, 28,* 41–43
53	Wang Y. Y., Li Y. H., Huang Z. Y., Chen X. P., Romeis J., Dai P. L., et al. (2015) Toxicological, biochemical, and histopathological analyses demonstrating that Cry1C and Cry2A are not toxic to larvae of the honeybee, Apis mellifera. *Journal of Agricultural and Food Chemistry, 63,* 6126–6132. http://pubs.acs.org/doi/abs/10.1021/acs.jafc.5b01662
54	Tan J. G., Levine S. L., Bachman P. M., Jensen P. D., Mueller G. M., Uffman J. P., et al. (2016) No impact of DvSnf7 RNA on honey bee (Apis mellifera L.) adults and larvae in dietary feeding tests. *Environmental Toxicology and Chemistry, 35,* 287–294. https://www.ncbi.nlm.nih.gov/pmc/articles/PMC4744748/
55	Dai P. L., Jia H. R., Geng L. L., & Diao Q. Y. (2016) Bt toxin Cry1Ie causes no negative effects on survival, pollen consumption, or olfactory learning in worker honey bees (Hymenoptera: Apidae). *Journal of Economic Entomology, 109,* 1028–1033. https://academic.oup.com/jee/article-abstract/109/3/1028/2648796/Bt-Toxin-Cry1Ie-Causes-No-Negative-Effects-on
56	Jia H. R., Geng L. L., Li Y. H., Wang Q., Diao Q. Y., Zhou T., et al. (2016) The effects of Bt Cry1le toxin on bacterial diversity in the midgut of Apis mellifera ligustica (Hymenoptera: Apidae). *Scientific Reports, 6,* 24664–24664. https://www.ncbi.nlm.nih.gov/pmc/articles/PMC4835784/
57	Renzi M. T., Amichot M., Pauron D., Tchamitchian S., Brunet J. L., Kretzschmar A., et al. (2016) Chronic toxicity and physiological changes induced in the honey bee by the exposure to fipronil and Bacillus thuringiensis spores alone or combined. *Ecotoxicology and Environmental Safety, 127,* 205–213. http://www.sciencedirect.com/science/article/pii/S0147651316300288

Table 1 References (64 Studies)—cont'd

Reference #	Article
58	Steijven K., Steffan-Dewenter I., & Hartel S. (2016) Testing dose-dependent effects of stacked Bt maize pollen on in vitro-reared honey bee larvae. *Apidologie, 47,* 216–226. http://link.springer.com/article/10.1007/s13592-015-0392-8
59	Masood M., Everett C. P., Chan S. Y., & Snow J. W. (2016) Negligible uptake and transfer of diet-derived pollen microRNAs in adult honey bees. *RNA Biology, 13,* 109–118
60	Velez A. M., Jurzenski J., Matz N., Zhou X., Wang H., Ellis M., et al. (2016) Developing an in vivo toxicity assay for RNAi risk assessment in honey bees, Apis mellifera L. *Chemosphere, 144,* 1083–1090
61	Bachman P. M., Ahmad A., Ahrens J. E., Akbar W., Baum J. A., Brown S., et al. (2017) Characterization of the activity spectrum of MON 88702 and the plant-incorporated protectant Cry51Aa2.834-16. *PLoS One, 12,* e0169409
62	Niu L., Ma W., Lei C., Jurat-Fuentes J. L., & Chen L. (2017) Herbicide and insect resistant Bt cotton pollen assessment finds no detrimental effects on adult honey bees. *Environmental Pollution, 230,* 479–485
63	Ullah I., Asif M., Ranjha M. H., Iftikhar R., Ullah M., Khan N. M., et al. (2017) Biosafety risk assessment approaches for insect-resistant genetically modified crops. *Advancements in Life Sciences, 4,* 39–46
64	Wang Y. Y., Dai P. L., Chen X. P., Romeis J., Shi J. R., Peng Y. F., et al. (2017) Ingestion of Bt rice pollen does not reduce the survival or hypopharyngeal gland development of *Apis mellifera* adults. *Environmental Toxicology and Chemistry, 36,* 1243–1248. http://onlinelibrary.wiley.com/doi/10.1002/etc.3647/abstract

#: Reference numbering in this Table is used in following Tables.

between 1993 and 2005) and tested the effects of purified Cry proteins on adult and larval honey bees (see online Supplementary Data File (S) available on https://doi.org/10.1016/bs.abr.2017.11.004). Scientific publications describing laboratory studies using PIs such as trypsin inhibitors (BBI) or cysteine PIs were first published in the 1990s. Some studies also included proteins such as chitinase or lectins. Since 1994, scientific publications assessed Cry proteins and some VIPs, using either purified protein or pollen (with some using both pollen and purified protein). A few studies (starting in 2015) also examined the effect of RNA interference (RNAi) plants on honey bee adults and larvae. When studies used pollen, it was collected from corn and sweet corn, cotton, oilseed rape, and soybean only. The first field trial was published in Babendreier, Kalberer, Romeis, Fluri, and Bigler (2004) (see reference #22 in Table 1). A total of 64 publications have steadily accumulated over these last 20 years (Fig. 1).

3.1.2 Proportion of Studies Under Laboratory and/or Field/Semifield Conditions

In risk assessment, laboratory studies are often conducted prior to field or semifield trials for practical reasons. From 1994 to 2017, 52 laboratory

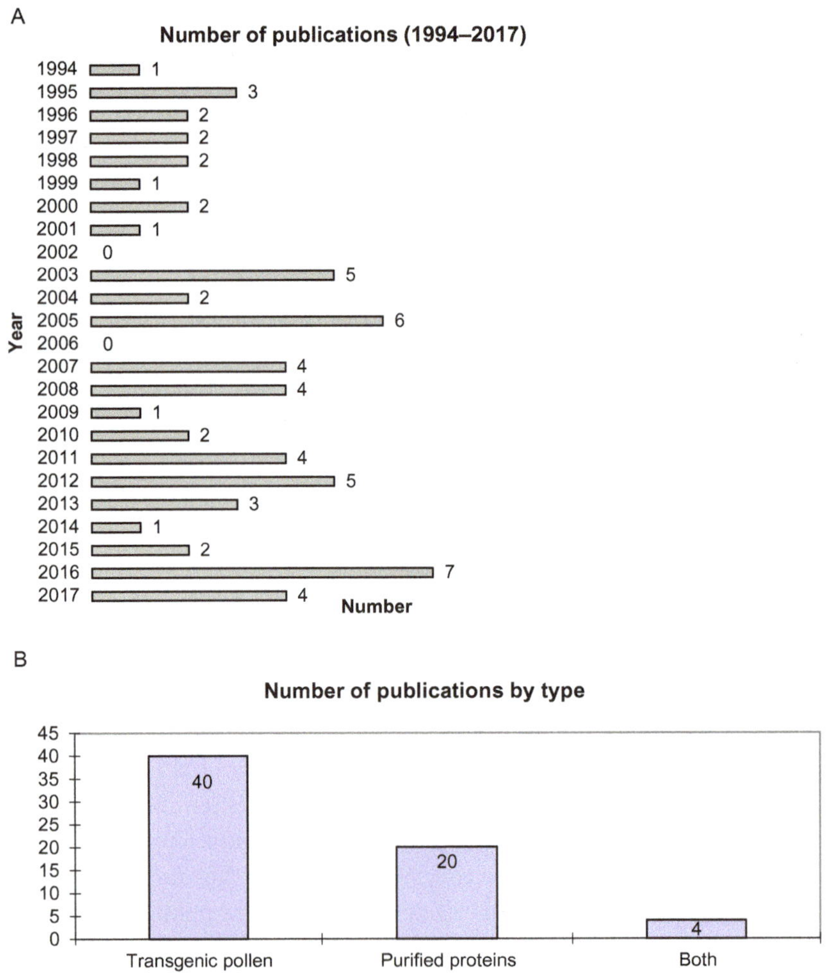

Fig. 1 Number of studies per year (1994–2017) (panel A) and by type (panel B) under laboratory conditions, using transgenic pollen or purified proteins or both, or field conditions.

studies and 12 field or semifield trials were published in the peer-reviewed literature (see panel 2a in S available on https://doi.org/10.1016/bs.abr.2017.11.004); 20 studies used purified proteins, 40 used GM pollen, and 4 publications used both (Fig. 1B). In addition, all of the first 18 studies provided to the US EPA (see panel 1 in S available on https://doi.org/10.1016/bs.abr.2017.11.004) were laboratory studies (two did not mention conditions but are likely to be laboratory studies) and used purified proteins.

3.1.3 Publications per Country

China, USA, followed by Europe (France and Germany) were the most implicated countries in such research published in peer-reviewed journals (Fig. 2). The distribution of these scientific publications per country and per year is shown in panel 3 in S (available on https://doi.org/10.1016/bs.abr.2017.11.004). France was the first country to conduct such studies (publications starting from 1994). However, no GM crops have been cultivated since 2008 in France (Kuntz, Davison, & Ricroch, 2013) and nor in Germany since 2009 (Katzek, 2014), in both cases for political reasons. China performed studies on two major staple GM crops (cotton and maize). This country is growing mainly GM cotton and smaller areas of virus-resistant papaya and Bt poplar (isaaa.org). GM maize is not authorized for cultivation in China, but there are indications that there may be illegal GM maize cultivation in some areas. It can be noted that China seems not to be affected as much as Europe and the United States by CDD. Indeed, losses in honey bees remain low in the surveys conducted between 2010 and 2013 in China (Liu et al., 2016).

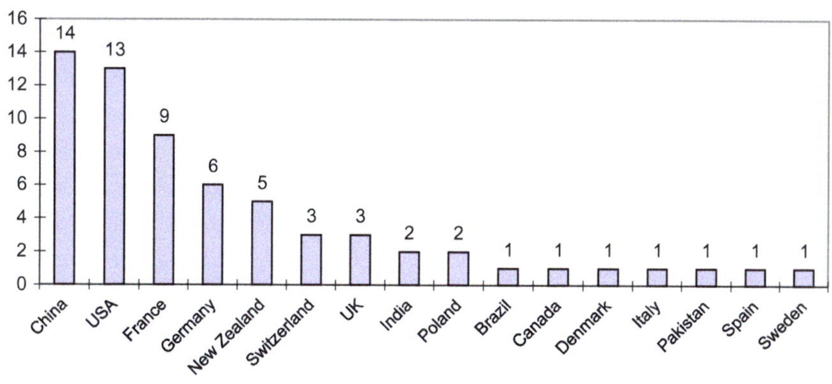

Fig. 2 Scientific publications (64) in peer-reviewed journals per country (1994–2017).

3.2 Results of Laboratory Studies

Pollen is the sole source of protein for honey bees and is foraged on various plants (Haydak, 1970). Thus, honey bees can be exposed to transgenic proteins through pollen consumption. This is not significantly the case with nectar, a secretion mainly composed of carbohydrates and some free amino acids. Therefore, laboratory studies used pollen collected from plants or purified toxins mixed in honey bee diet at different doses. In some experiments, the insecticidal solution was microinjected directly into the bee abdomen. Many studies used newly emerged adults, since in the life cycle of the honey bee, protein consumption is most marked at this stage. Other studies used larval stages since larvae also ingest pollen, but their food is mainly composed of glandular secretions from nurse bees. Tests included acute toxicity, survival (mortality), weight, development time, hypopharyngeal gland development, eclosion rate, midgut proteolytic enzyme activity, and behavioural parameters such as flight activity, food consumption, foraging behaviour, and olfactory learning abilities (for example, tested by the conditioned proboscis extension assay). Imidacloprid, a systemic insecticide belonging to a class of the neonicotinoids, or deltamethrin, a member of the class of synthetic pyrethroids, or other compounds were used in some studies as a positive control for sublethal effects.

3.2.1 Effects of PIs

PIs are proteins interfering with the digestive process by inhibiting proteolytic enzyme activities (Table 2; see panel 4a in S available on https://doi.org/10.1016/bs.abr.2017.11.004) and also regulatory processes. It is long known that PI can provide defensive proteins against insects and microorganisms (Ryan, 1990). Serine PIs and cysteine PIs have been tested, as well as chitinases and lectins. We collected 22 publications (under laboratory conditions; summarized in Table 2) which examined their effects on bees.

Some PIs seem to negatively affect honey bees at certain doses in feeding trials. This is consistently the case for the soybean trypsin inhibitor (SBTI) at the highest dose tested. This is not surprising since it is known that some PIs are not fully specific against pest insects and therefore are not the best available insecticidal compounds for biotechnological projects. This explains why they have not been widely used, unlike Cry proteins. It can be noted however that low doses of SBTI were shown not to have any negative effect. Transgenic cotton lines producing both Cry1Ac and cowpea trypsin inhibitor (CpTI) have been developed in China (see below). Obviously the risks

Table 2 Main Characteristics of 22 Publications Under Laboratory Conditions Using Protease Inhibitors (PIs)

Reference #	Active Compounds	Exposed Stage	Response Variable	Exposure Time	Concentration in Pollen Food or in Syrup (Sugar Solution)	Conclusions of Studies and Authors' Citations ("…")
64	SBTI (soybean trypsin inhibitor), included in study of Cry2A and Cry1C (see Table 3)	Adults (newly emerged)	Survival (mortality) Hypopharyngeal gland acini development	22 days	SBTI at 1 or 10 mg/g sucrose solution	Bee survival and hypopharyngeal gland development reduced already at 1 mg/g (but not with Cry2A not Cry1C)
24	BBI (Bowman–Birk trypsin inhibor)	Adults	Foraging strategy: mean number of visits of flower per hour, mean time spent on flower, and interval between consecutive visits	10 days	100 μg/mL sucrose solution	No statistically significant differences
27	SBTI	Adults	Hypopharyngeal gland protein content Midgut proteolytic enzyme activity Survival	30 days	Three different concentrations (0.1%, 0.5%, and 1%, w:w) (SBTI dissolved in sugar solution mixed with powdered pollen)	No significant differences between control, 0.1% and 0.5% for biochemical endpoints. Reduced values at 1% SBTI Lowest survival at 1% SBTI, followed by 0.5% and 0.1%. "Nurse bees fed a pollen diet containing at least 1% SBTI would be poor producers of larval food, potentially threatening colony growth and maintenance"

Continued

Table 2 Main Characteristics of 22 Publications Under Laboratory Conditions Using Protease Inhibitors (PIs)—cont'd

Reference #	Active Compounds	Exposed Stage	Response Variable	Exposure Time	Concentration in Pollen Food or in Syrup (Sugar Solution)	Conclusions of Studies and Authors' Citations ("…")
15	SBTI	Larvae	Survival (mortality) Juvenile Development Adult body mass (weight)	Up to 22 days	0.1% and 1.0% SBTI (w:w of total protein): considered as realistic low and high expression levels	1.0% SBTI "created significant additional larval mortality, slowed juvenile development and significantly decreased adult body mass"
14	BPTI (bovine pancreatic trypsin inhibitor), included in study of Cry 1Ba (see Table 3)	Adults (newly emerged)	Flight activity Longevity	7 days	2.5 mg/g BPTI in pollen food (equivalent to 0.25% of total soluble protein)	BPTI-fed bees began to fly and also died about 3 days sooner than control bees. Periods during which flights took place not affected. No differences for Cry1Ba
12	SBTI BBI	Adults	Survival Learning abilities (foraging behaviour) Proteolytic activities	15 days	1, 0.1, and 0.01 mg/mL	"Diets containing PIs at 1 mg/mL significantly increased the probability of death at a given time." Learning performance affected for BBI at 1 mg/mL, no convincing effects at lower concentrations. "A dose lower than 0.01 mg/mL of PI in the diet would be harmless"

11	SBTI, included in study of Cry1Ba (see Table 3)	Adults	Survival Food consumption	7 days	10, 5, and 0.5 mg/g 10, 2.5, and 0.25 mg/g in pollen food	"Adult bees unlikely to be harmed by transgenic plants expressing Cry1Ba or SBTI, or by Bt biopesticides that are used as recommended"
9	BBI, Cystatin, and OCI (oryzacystatin)	Adults (14–15-day-old)	Short and long-term mortality Olfactory learning abilities (PER) Midgut proteinase activity	24, 48, and 96 h for mortality. 3 days	330 μg/mL in sugar solution for 24 h Learning: 1, 5, 10 μg/mL (added to reward)	No short-term mortality. Learning performance unaffected. "Long-term ingestion of BBI or OCI did not disrupt total digestive proteolytic activity, but ingestion of BBI induced a new proteinase form"
10	POT-1 and POT-2 (potato protease inhibitor)	Adults	Midgut protease activities Survival	3 and 8 days	0.2% or 0.01% (w:v) in sugar solution 1% or 0.2% (w:w) in pollen food	"Survival at 0.2% in pollen or 0.01% in syrup did not differ from the controls"
6	BPTI SBTI	Adults	Midgut protease activities Survival	Over 60 days for survival Protease: 8 days and up to 75% mortality	Survival: 10, 5, 1, 0.1, and 0.01 mg/mL in sugar solution Protease: 10, 3, and 1 mg/mL	Reduced longevity at three highest concentrations. Some reduction of protease activity, especially at the highest concentrations. "Low levels of the inhibitors studied were not toxic to bees"

Continued

Table 2 Main Characteristics of 22 Publications Under Laboratory Conditions Using Protease Inhibitors (PIs)—cont'd

Reference #	Active Compounds	Exposed Stage	Response Variable	Exposure Time	Concentration in Pollen Food or in Syrup (Sugar Solution)	Conclusions of Studies and Authors' Citations ("…")
3	Chitinase produced in transgenic oilseed rape	Adults	Indoor flight room and outdoor cage in which five transgenic flowers were introduced. Foraging behaviour (number of flowers visited per bee, time spent on the plant, or number of nectar collection trials per bee)	Recorded over 45 min	Chitinase not detected in nectar or pollen	No effects on bee foraging behaviour due to the genetic modification, but differences between two lines with different genetic backgrounds
7	CpTI (cowpea trypsin inhibitor) Chitinase and glucanase (against fungi)	Adults	Acute toxicity Olfactory learning performances: -PER -flight room with artificial flowers (food source linked to odour), only CpTI tested	Mortality recorded at 24 and 48 h	Ingestion: 11 μg per bee (oral) over 24 h; Injection: 0.5 μg per bee 1, 5, and 10 μg/mL in sugar solution (reward) 10 μg/mL in sugar solution (food source)	Three proteins tested were nontoxic at concentrations tested. Chitinase did not affect learning performances, while glucanase and CpTI did, especially at high concentration

1	WGA (wheat germ agglutinin) BBI	Adults	Midgut protease activities in vivo and in vitro	3.5 days	1, 0.1, 0.01, and 0.001 mg/g in sugar solution	No acute toxicity observed. In vivo, BBI caused a decrease in the amount of trypsin activity but no effect on esterase activity. WGA at 1 mg/mL elicited a decrease in trypsin activity but no effect on esterase activity. Protease activity inhibition observed in vitro
29	PSL (*Pisum sativum* lectin) produced in oilseed rape	Larvae	Larval mortality, weight at LS stage, development time to LS and pupal stages	6 days for mortality, plus until LS and pupal stages	1.5% dried pollen in diet (w/w), expressing PSL up to 1.2% of total soluble protein in the pollen, which is 0.0012% PSL in diet (w/w)	"At the levels studied, no effect on bee larval survival, weight, or development time compared to non-transgenic pollen"
53	SBTI, included in study of Cry1C, Cry2A (see Table 3)	Larvae	Larval development time Mature larvae weight Eclosion rate	7 days	Stock solution of SBTI diluted and mixed into artificial diet to concentrations of 0, 0.1, 1, and 10 mg/g FW of diet	"Pupal and adult development and larval weight not significantly affected by SBTI at 0.1 mg/g. Survival was reduced and development was delayed when larvae were fed SBTI (1 mg/g diet). No larva reached the pupal stage at the highest SBTI concentration of 10 mg/g diet"

Continued

Table 2 Main Characteristics of 22 Publications Under Laboratory Conditions Using Protease Inhibitors (PIs)—cont'd

Reference #	Active Compounds	Exposed Stage	Response Variable	Exposure Time	Concentration in Pollen Food or in Syrup (Sugar Solution)	Conclusions of Studies and Authors' Citations ("…")
31	SBTI, included in study of Cry1Ab (see Table 3)	Adults (newly emerged)	Intestinal bacterial communities	10 days	SBTI (0.1% or 1%, w/v) in sugar solution mixed to pollen	High concentration of SBTI had an effect on bacterial communities in honeybee intestines and also increases bee mortality
34	SBTI, included in study of Cry1Ab (see Table 3)	Adults (newly emerged)	Survival Hypopharyngeal gland development	10 days	SBTI low concentration and SBTI high concentration: a 1% (w/v) and a 0.1% (w/v) sucrose solution Based on observations of nectar and pollen consumption of honey bees, this is equivalent to a toxin concentration in the pollen of about 7.5% and 0.75%	Feeding honeybees the protease inhibitor SBTI for the first 10 days of adult life negatively affects the development of hypopharyngeal glands at the concentrations tested (0.1% and 1%)
15	GNA lectin, included in study of Cry1A.105, Cry2Ab2, Cry3Bb1, and CP4 EPSPS (see Table 3)	Larvae (L2)	Survival Prepupal weight	Single dose of 10 μL	Each protein stock solution was serially diluted 10 × for an exponential concentration gradient	As a positive control, GNA lectin produced significant increase in larval mortality over the concentration gradient

27	CpTI, included in study of Cry1Ac (see Table 3)	Adults (newly emerged)	Survival Sublethal effect on feeding behaviour	7 days	Diets contained water, honey, and pollen at rate of 1:2:7 (weight) considered as realistic pollen consumption. Pollen collected in July (highest concentration of Cry1Ac protein: 300 ± 4.52 ng/g pollen) Sublethal control contained imidacloprid at 48 ng/g	"CCRI41 cotton carried no risk on survival of honey bees" in a worst case scenario. As for imidacloprid, "bees consumed significantly less CCRI41 cotton pollen than in the control group exposed to conventional pollen." "Differential consumption of the three food types not function of time." See text
26	CpTI, included in study of Cry1Ac (see Table 3)	Adults (newly emerged)	Visual learning (T-tube maze) Olfactory learning (PER)	7 days exposure and tests the subsequent days	As in 34	"Oral exposure to CCRI41 pollen did not affect learning capacities of honey bees in both tests"
25	CpTI, included in study of Cry1Ac (see Table 3)	Adults (newly emerged)	Oral acute toxicity Long-term toxicity (longevity) Superoxide dismutase (SOD) activity	Oral acute toxicity: one dose of treatment followed by sugar diet Long term: treatment renewed each day	Oral acute toxicity: pollen doses were 0.4, 0.2, and 0.1 g/mL in sugar solution, respectively. Mortality recorded at 24, 48, and 72 h Long-term toxicity evaluation: pollen dose of 0.2 g/mL SOD: after 48 h of acute toxicity experiment	"No acute oral toxic effect on worker bees. No significant differences observed in the superoxide dismutase activity or in the longevity of worker bees"

Continued

Table 2 Main Characteristics of 22 Publications Under Laboratory Conditions Using Protease Inhibitors (PIs)—cont'd

Reference #	Active Compounds	Exposed Stage	Response Variable	Exposure Time	Concentration in Pollen Food or in Syrup (Sugar Solution)	Conclusions of Studies and Authors' Citations ("...")
21	Avidin and BPTI, included in study of Cry1Ba (see Table 3)	Adults (newly emerged)	Survival Food consumption Development of hypopharyngeal glands (acinus diameter, gland mass, and protein content)	10 days treatment, survival measured up to 30 days, other parameters up to 10 days	Pollen food with 2.5 mg/g BPTI (equivalent to pollen expressing 1.25% (of total protein) BPTI, 11.2 μM avidin, or 0.3% Cry1Ba protein)	"None of the treatments significantly affected bee survival, or the mean diameters of gland acini, gland mass or protein content" Transgenic plants are insecticidal with minimum leaf expression levels of 3 μM avidin and 0.5% of total protein for SBTI

PER: conditioned proboscis extension reflex assay.
#: Reference numbering is taken from Table 1.

for honeybees of PI produced by transgenic plants will mainly depend on their concentration in pollen and need to be assessed case by case.

3.2.2 Effects of Bt Proteins, RNA Interference, and Proteins Conferring Herbicide Tolerance

3.2.2.1 Assessment of Cry and VIP Insecticidal Proteins

Purified Cry or VIP proteins or pollen containing these proteins have been tested for their effects on honey bees. These proteins are: Cry1Ab, Cry1Ac, Cry1A.105, Cry1Ah, Cry1Ba, Cry1C, Cry1F, Cry1Ie, Cry2A, Cry2Ab, Cry2Ab2, Cry3Bb1, Cry5, Cry51Aa2, and VIP3A. One study examined a spider venom toxin. Panel 2b in S (available on https://doi.org/10.1016/bs.abr.2017.11.004) lists the peer-reviewed studies for each one. Studies also include chimeric proteins: Cry1A.105 is a chimeric protein of domains from Cry1Ab, Cry1Ac1, and Cry1F (see https://bch.cbd.int/database/organisms/), and eCry3.1Ab is a chimera of Cry1Ab and modified Cry3A. Studies of Cry3A, Cry34Ab1/Cry35Ab1, and Cry2Ae (for which there were no peer-reviewed studies found) have been provided by the industry to regulatory bodies (for US EPA, see panel 2c in S available on https://doi.org/10.1016/bs.abr.2017.11.004). In most cases, these proteins are actually produced by a given transgenic species (see panel 4b in available on https://doi.org/10.1016/bs.abr.2017.11.004), but are not always on the market (see http://cera-gmc.org/GMCropDatabase).

The expression levels of these toxins are in the microgram range per gram (fresh weight) of pollen and sometimes less. Most studies used worst case scenario during experiments exposing honey bees at an adult or larval stage to purified proteins (doses superior to those they may actually face in nature). Experiences using pollen often used doses described as "realistic" by the authors, sometimes with increasing doses.

Examined parameters were those mentioned above for PIs. None of these individual studies reported an increased mortality of either larvae or adult bees, neither did a metaanalysis (Table 3). When positive controls were included in the experiments, lethal effects were observed with pollen of *Heliconia rostrata* (which is known to be toxic to bees) and in some experiments with high doses of synthetic insecticides, indicating that the performed bioassays were able to detect toxic effects. Few sublethal effects were identified using these insecticidal proteins (Table 2). These are discussed here.

Study #26 by Ramirez-Romero, Chaufaux, and Pham-Delegue (2005) exposed bee colonies to purified Cry1Ab protein (1000 ng/g). No higher mortality was observed during or after Cry1Ab treatment, compared to

Table 3 Main Characteristics of 38 Published Studies Using Insecticidal Proteins (Cry, VIP, or Spider Venom) or Protein Conferring Tolerance Against Glufosinate or Glyphosate Herbicide Under Controlled (Usually Laboratory) Conditions

Reference #	Cry Protein or PI or Protein Conferring Tolerance Against Herbicide	Protein Source and Transgenic Plant Event Name	Exposed Stage	Response Variable	Oral Chronic Exposure Period	Concentration	Conclusions of Studies and Authors' Citations ("…")
5	Cry3B	Purified protein (from recombinant *Escherichia coli*)	Larvae	Survival (mortality) Pupal dry weight	Fed to colonies over a 2-month period	In sugar syrup at concentrations of 0.066% or 0.332%	For both concentration: no effect on tested parameters
4	Cry1Ac and Cry1Ac (trypsinized)	Purified protein (from recombinant *E. coli*)	Larvae (1–3-day-old) Adults (3 days after emergence)	Survival	7 days	Cry1Ac: 100 μL/mL in water Cry1Ac(Tryp.): 50 μL/mL in water	No significant effect on survival. Toxin concentration more than "100 times the concentration of CryIAc protein found in the field as present in pollen and nectar of transgenic cotton"
8	Cry2A	Purified protein (from recombinant *E. coli*)	Larvae (1–3-day-old) Adults (3 days after emergence)	Survival	18 days (larvae), 21 days (adults)	50 μg/mL	No significant effect on survival

64	Cry1C Cry2A (and SBTI*)	Pollen from Cry1C-RICE (T1C-19) or Cry2A-RICE (T2A-1) or purified protein	Adults (newly emerged)	Survival Hypopharyngeal gland development	Pollen exp.: 7 days for survival, 10 days for gland development. Protein exp.: 4 weeks	Pollen mixed with honey and water at a ratio of 5:4:1 (w/w). or sucrose solution supplemented with Cry1C at 50 μg/g; Cry2A at 400 μg/g (concentration more than 10× greater than in pollen)	No effect on tested parameters for Cry1c or Cry2a, but effects for SBTI used as positive control
55	Cry1Ie	Purified protein	Adults (newly emerged worker bees)	Survival Pollen consumption Olfactory learning performances (PER)	18 days	20, 200, and 20,000 ng/mL in sugar solution	"Cry1Ie toxin carries no risk to survival, pollen consumption, or learning capabilities of young adult honey bees." As a positive control, midacloprid affected learning behaviour
56	Cry1Ie	Purified protein	Adults	Midgut bacteria diversity	15 or 30 days	20 ng/mL, 200 ng/mL, and 20 μg/mL in sugar solution	No significant differences

Continued

Table 3 Main Characteristics of 38 Published Studies Using Insecticidal Proteins (Cry, VIP, or Spider Venom) or Protein Conferring Tolerance Against Glufosinate or Glyphosate Herbicide Under Controlled (Usually Laboratory) Conditions—cont'd

Reference #	Cry Protein or PI or Protein Conferring Tolerance Against Herbicide	Protein Source and Transgenic Plant Event Name	Exposed Stage	Response Variable	Oral Chronic Exposure Period	Concentration	Conclusions of Studies and Authors' Citations ("…")
57	Cry	Spores of *Bacillus thuringiensis* (kurstaki strain)	Adults	Survival Feeding behaviour Enzyme activities	10 days exposure, measurement after 25 days (survival, feeding) or 10 and 20 days for enzymes	100 and 1000 μg/L	No toxicity revealed, including for fipronil. Bt spores (with or without Cry proteins) found to "differentially modulating enzyme activities," as well as fipronil
58	Cry1A.105 Cry2Ab2 Cry3Bb1 and CP4 EPSPS (glyphosate tolerance)	Maize pollen from stacked MON89034 (Cry1A.105 and Cry2Ab2), and MON88017 (Cry3Bb1; CP4 EPSPS)	Larvae (L3)	Pollen digestion Survival Developmental delay Prepupal body weight	4-day pollen feeding, measurements 6 h after feeding for digestion, after 11 days for survival and development	Pollen: 1, 2, 5, and 10 mg/larva (to test for potential dose-dependent effects)	No "significant dose-dependent effects on survival rates, pollen digestion, and developmental delay," whereas *Heliconia rostrata* pollen (positive control) negatively affected all parameters. "Trend towards reduced perpupal weight with higher amount of maize pollen," slightly more pronounced with staked Bt maize. See text

53	Cry1C Cry2A (and SBTI*)	Purified protein (from recombinant *E. coli*)	Larvae	Survival Development time Weight (larva and adult) Enzyme activities Histopathology	Larval stage Enz. + Hist.: 4 days	Cry1C: 50 μg/g Cry2A: 400 μg/g	"Cry1C and Cry2A have no acute toxicity to *A. mellifera* larvae at concentrations >10 × higher than those detected in pollen from Bt plants" Survival reduced and development delayed with SBTI (positive control)
43	Cry1Ah	Purified protein (*B. thuringiensis*)	Adults (newly emerged)	Survival Pollen consumption Hypopharyngeal gland mass	21 days	10 μg/mL, 10 ng/mL, and 1 ng/mL in sugar solution	No significant effects
46	Cry1Ac (and CpTI*)	Pollen from cotton CCRI41	Adults (newly emerged)	Survival Hypopharyngeal gland development Midgut proteolytic activity	7 days	Bees received a mixture of water, honey, pollen, or Imidacloprid, or pollen-containing 1% SBTI. Mean contents of Cry1Ac protein were 517.3 ± 7.32 ng/g and 651.0 ± 9.01 ng/g for August 6th (mid-stage bloom) and September 4th (late bloom), respectively	No measured parameter significantly affected, while imidacloprid (mortality, gland) and SBTI (gland, digestion) did show effects. Previously found "antifeedant effect" "not linked to decrease in midgut proteolytic activity" See text

Continued

Table 3 Main Characteristics of 38 Published Studies Using Insecticidal Proteins (Cry, VIP, or Spider Venom) or Protein Conferring Tolerance Against Glufosinate or Glyphosate Herbicide Under Controlled (Usually Laboratory) Conditions—cont'd

Reference #	Cry Protein or PI or Protein Conferring Tolerance Against Herbicide	Protein Source and Transgenic Plant Event Name	Exposed Stage	Response Variable	Oral Chronic Exposure Period	Concentration	Conclusions of Studies and Authors' Citations ("…")
47	Cry1A.105 Cry2Ab2 Cry3Bb1 CP4 EPSPS (glyphosate tolerance) (and GNA lectin*)	Purified protein	Larvae (L2)	Survival Prepupal weight	Single dose of 10 μL	Per 10 μL: 3.2 μg Cry1A.105, 0.124 μg Cry2Ab2.820, 3.0 μg Cry3Bb1, 6.4 μg CP4 EPSPS or 7.03 μg Cry1, Cry2, Cry3, and CP4 in the proportion as in Mon89034 × Mon88017 pollen	Individually or in combination, the tested proteins showed no adverse effects on bee larvae, even at an individual Bt toxin dose exceeding by 186 × (or 18 × for the combined proteins) a normal 2 mg Bt maize (Mon89034 × Mon88017). As a positive control, GNA lectin produced significant increase in larval mortality over the concentration gradient

39	Cry1Ab Cry1A.105 Cry2Ab2 Cry3Bb1 CP4 EPSPS (glyphosate tolerance)	MAIZE pollen from MON810 (Cry1Ab) and from stacked MON89034 (Cry1A.105 and Cry2Ab2) and MON88017 (Cry3Bb1; CP4 EPSPS)	Larvae (L3)	Larvae survival Prepupal weight	5–6 days	A "realistic" 2 mg dietary pollen dose at day 6 consumed over the remaining total exposure time (day 10–11)	Pollen of the tested Bt varieties does not harm the survival and development of in vitro reared larvae. As a positive control, pollen from *Heliconia rostrata* caused significant toxic effects
40	Cry1Ac	Purified protein	Larvae	Larvae survival Development time of larvae Development time of pupae	5 days	50 μg Cry1Ac diluted in water provided over 5 days in a total of 164 μL diet (as 4, 15, 25, 50, and 70 μL of diet, respectively, every day). Feeding on Bt cotton, larvae will consume 1.2 ng of Cry1Ac	"Cry1Ac ingestion did not affect survival, developmental time, and neither adult body mass nor size" at a dose far higher than expected exposure

Continued

Table 3 Main Characteristics of 38 Published Studies Using Insecticidal Proteins (Cry, VIP, or Spider Venom) or Protein Conferring Tolerance Against Glufosinate or Glyphosate Herbicide Under Controlled (Usually Laboratory) Conditions—cont'd

Reference #	Cry Protein or PI or Protein Conferring Tolerance Against Herbicide	Protein Source and Transgenic Plant Event Name	Exposed Stage	Response Variable	Oral Chronic Exposure Period	Concentration	Conclusions of Studies and Authors' Citations ("...")
37	Cry1Ac (and CpTI*)	Pollen from cotton CCRI41	Adults (newly emerged)	Survival Sublethal effect (feeding behaviour)	7 days	Diets contained water, honey, and pollen at rate of 1:2:7 (weight) considered as realistic pollen consumption. Pollen collected in July (highest concentration of Cry1Ac protein: 300 ± 4.52 ng/g pollen) Sublethal control contained **imidacloprid** at 48 ng/g	"CCRI41 cotton carried no risk on survival of honey bees" in a worst case scenario. As for imidacloprid, "bees consumed significantly less CCRI41 cotton pollen than control group exposed to conventional pollen." "Differential consumption of the three food types not function of time." See text
38	Cry1Ac (and CpTI*)	Pollen from cotton CCRI41	Adults (newly emerged)	Visual learning (T-tube maze) Olfactory learning (PER)	7 days exposure and tests the subsequent days	As in 34	"Oral exposure to CCRI41 pollen did not affect learning capacities of honey bees in both tests"

36	Cry1Ac (and CpTI*)	Pollen from cotton Zhong-41	Adults (newly emerged)	Oral acute toxicity Long-term toxicity (longevity) Superoxide dismutase (SOD) activity	Oral acute toxicity: one dose of treatment followed by sugar diet, Long term: treatment renewed each day	Oral acute toxicity: pollen doses were 0.4, 0.2, and 0.1 g/mL in sugar solution, respectively. Mortality recorded at 24, 48, and 72 h. Long-term toxicity evaluation: pollen dose of 0.2 g/mL SOD: after 48 h of acute toxicity experiment	"No acute oral toxic effect on worker bees. No significant differences observed in the superoxide dismutase activity or in the longevity of worker bees"
35	Cry3Bbl Cry1Ab	Pollen from maize MON863 (Cry3Bbl) X MON810 (Cry1Ab)	2-, 3-, 4-, and 5-day-old workers	Consumption rate Weight Protein content Antioxidant enzyme activities Total antioxidant status	5 days	Mixture of rapeseed honey and pollen in 3:1 ratio	"no significant differences in any of the [measured] parameters," but "experimental diets were different from real diets of honey bees in the field"
33	Cry5	Spore/crystal suspension of *B. thuringiensis* strain PS86Q3	Bee hives larvae	Number of larvae (open cells), pupae (sealed cells), and empty cells was recorded	11 days	0.4 or 1.2 mg/mL of spore/crystal suspension	"despite the hymenoptera-specific target spectrum of strain PS86Q3, the results reported here show that this strain is also safe for honeybees"

Continued

Table 3 Main Characteristics of 38 Published Studies Using Insecticidal Proteins (Cry, VIP, or Spider Venom) or Protein Conferring Tolerance Against Glufosinate or Glyphosate Herbicide Under Controlled (Usually Laboratory) Conditions—cont'd

Reference #	Cry Protein or PI or Protein Conferring Tolerance Against Herbicide	Protein Source and Transgenic Plant Event Name	Exposed Stage	Response Variable	Oral Chronic Exposure Period	Concentration	Conclusions of Studies and Authors' Citations ("…")
34	Cry1Ab	Purified protein (from recombinant *E. coli*)	Adults (2 day-old) 14–15-day-old for learning test	Survival during subchronic exposure Feeding behaviour Learning performance (PER)	12 days	3 and 5000 ng/g of Cry1Ab in sugar solution, similar to concentration in maize pollen from MON810 (2 ng/g) or event 176 (5000 ng/g)	No lethal effect. "disturbed learning performances" at the highest concentration See text
28	Cry1Ab (and SBTI*)	Pollen from maize MON810 (Cry1Ab) and purified proteins	Adults (newly emerged)	Intestinal bacterial communities	10 days	Pollen: food provided ad libitum (4 g of pollen and 25 mL of sugar solution); Cry1Ab concentration below quantification level. Purified Cry1Ab (0.0014%, w/v), Bt-toxin concentration at least 10 × higher than in pollen of Bt 176	"Neither Bt-maize pollen nor high concentrations of Cry1Ab significantly affected bacterial communities in honeybee intestines"

31	Cry1Ab	Pollen from maize Bt11 In lab and also field (see Table 3)	Adults (newly emerged)	Survival weight	35 days (lab)	Lab: pollen provided on days 0, 7, 14, as 5 g cakes (80% pollen and 20% honey, w/w)	"Laboratory feeding studies showed no effects on the weight and survival of honey bees feeding on Cry1Ab-expressing sweet corn pollen for 35 days"
22	Cry1Ab (and SBTI*)	Pollen from maize MON 810 (Cry1Ab) and purified protein	Adults (newly emerged)	Survival Hypopharyngeal gland development	10 days	Pollen: food provided ad libitum (4 g of pollen and 25 mL of sugar solution); Cry1Ab concentration below quantification level. Cry1Ab: 0.001% or 0.0014% (w/v) in sugar solution. Bt-toxin concentration at least 10× higher than in pollen of Bt 176	"Neither the Bt-pollen treatment nor the Bt toxin treatment affected hypopharyngeal gland development." Effects with SBTI (see Table 1)

Continued

Table 3 Main Characteristics of 38 Published Studies Using Insecticidal Proteins (Cry, VIP, or Spider Venom) or Protein Conferring Tolerance Against Glufosinate or Glyphosate Herbicide Under Controlled (Usually Laboratory) Conditions—cont'd

Reference #	Cry Protein or PI or Protein Conferring Tolerance Against Herbicide	Protein Source and Transgenic Plant Event Name	Exposed Stage	Response Variable	Oral Chronic Exposure Period	Concentration	Conclusions of Studies and Authors' Citations ("…")
25	Cry1Ac	Pollen from cotton SK-310	Adults (newly emerged)	Acute toxicity (Survival) Superoxide dismutase (SOD) activity Long-term survival	Mortality recorded at 24, 48, and 96 h for acute tox. And up to 46 days for long term. After 48 h of acute tox. exp. for SOD	Acute: 0.16, 0.08, and 0.04 g pollen in 0.4 mL of sucrose solution, consumed *ad libidum* Long term: sucrose solution containing 20% (g/mL) cotton pollen renewed every 1 or 2 days	"no direct adverse impacts on caged worker bees"
38	Cry1Ab	Purified protein (from recombinant *E. coli*)	Adults (forager bees) Larvae	Survival Syrup consumption Foraging activity and olfactory learning (artificial flower)	Measurements 4-day before treatment, 3-day recovery, 4-day treatment, 3-day recovery, 4-day posttreatment measurements	Cry1Ab protoxin, deltamethrin or imidacloprid at 1000 ng/g, 500 ng/kg, and 48 ng/kg, respectively, in sugar solution. For Cry1Ab, amount higher than in pollen of MON810 (~2 ng/g) and lower than in Event 176 (~5000 ng/g)	"Cry1Ab did not affect mortality, syrup consumption or learning capacities. However, foraging activity was reduced during and after the treatment" (see text) As positive controls, "Deltamethrin and imidacloprid both affected syrup consumption and foraging activity. Deltamethrin also induced a reduction in learning capacities"

21	Cry1Ba (and avidin and BPTI*)	Purified protein (from *Bacillus thuringiensis* Bt4412)	Adults (newly emerged)	Survival Food consumption Development of hypopharyngeal glands (acinus diameter, gland mass, and protein content)	10 days treatment, survival measured up to 30 days, and other parameters up to 10 days	Pollen food with 0.625 mg/g Cry1Ba or 0.174 mg/g avidin, or 2.5 mg/g BPTI (equivalent to pollen expressing 1.25% (of total protein) BPTI 11.2 μM avidin or 0.3% Cry1Ba protein)	"None of the treatments significantly affected bee survival, or the mean diameters of gland acini, gland mass or protein content"
16	Cry1Ab Cry1F	Pollen from maize Bt11 (Cry1Ab) or TC1507 (Cry1F)	Larvae (4–5 days old): Section of brood patches isolated in a colony for various treatments	Laval mortality Pupal mortality Pupal weight Haemolymph protein concentration of newly emerged adults	Single dose	Each larva fed 6 μL of treatment solution (containing 1.5 mg pollen)	"Transgenic Bt corn pollen, whether Cry1A (b) or Cry1F, has no significant detrimental effects on honey bees when fed to them at the immature stage." As a positive control, diazinon-treated pollen showed effects in all parameters measured
14	Cry1Ba (and BPTI*)	Purified protein (from *B. thuringiensis* Bt4412)	Adults (newly emerged)	Survival Flight activity	7 days	Pollen food with 625 μg/g Cry1Ba toxin or 2.5 mg/g BPTI (equivalent to 0.25% or 1% of total soluble protein, respectively)	"Cry1Ba-fed bees did not differ significantly from control bees in the timing of their first flight, the period during which flights took place or in estimated longevity"

Continued

Table 3 Main Characteristics of 38 Published Studies Using Insecticidal Proteins (Cry, VIP, or Spider Venom) or Protein Conferring Tolerance Against Glufosinate or Glyphosate Herbicide Under Controlled (Usually Laboratory) Conditions—cont'd

Reference #	Cry Protein or PI or Protein Conferring Tolerance Against Herbicide	Protein Source and Transgenic Plant Event Name	Exposed Stage	Response Variable	Oral Chronic Exposure Period	Concentration	Conclusions of Studies and Authors' Citations ("…")
11	Cry1Ba and biopesticide formulations (and SBTI*)	Purified protein (from *B. thuringiensis* Bt4412)	Adults (newly emerged)	Survival Food consumption	7 days	10, 2.5, and 0.25 mg/g of Cry1Ba or SBTI As control, Bt biopesticides Dipel 2X (1% and 0.25%) and Foray 48B (0.25%)	"Bee survival time unaffected, and food consumed at the same rate as control food for all treatments, except 1% Dipel, where both survival and food consumption were significantly reduced"
23	Cry1Ab	Bt biopesticide DIPEL, Sweet corn tassels and pollen (BC 0801)	Adults (newly emerged or foragers)	Survival: 1. Biopesticide contact assays (forager bees sprayed once with insecticides) 2. Tassel contact assays using foragers 3. Pollen oral assays using newly emerged bees	1. 24 h after exposure 2. Up to 4 days. 3. after 24 h	Various synthetic insecticides and Bt formulations tested at 4–5 concentrations ranging from 0.00008% to 1.0% solution	"Bt-sweet corn tassels had no impact on honey bee mortality. Pollen collected from insecticide field treated corn and fed to honey bees had no impact on mortality." "Compared to the chemicals studied, Bt was unique being the only control agent non-toxic by direct contact to honey bees"

49	Cry1Ah	Maize In lab and also field (see Table 3)	Lab: newly emerged adults Field: larvae and adults (worker bees)	Effects on the intestinal bacterial community	Larvae sampled after 0, 3, 6, and 9 days. Adults collected 0, 15, and 30 days after emergence	Lab: treatments were 160 mg/mL of transgenic pollen, 160 mg/mL of nontransgenic pollen, and 900 ng/mL of plasmid DNA in sugar solution	"No significant effects were observed in the midgut communities between larvae and adult honeybees fed transgenic cry1Ah-maize pollen and non-transgenic maize pollen in the field trials" Lab: "no significant differences in the midgut bacterial community composition among the five treatments." "The full-length cry1Ah gene from transgenic pollen or high-dose treatments did not transfer to indigenous bacteria in the midgut of *A. mellifera* ligustica" Field trials (Table 3) and lab experiments yielded identical conclusions
42	Vip3A (VIPAa20)	Purified protein (from recombinant *E. coli*)	Brood	Survival (adult emergence)	Single dose exposure, observation for 24 days	50, 200, and 500 μg/g diet in sugar solution	No statistically significant differences in survival

Continued

Table 3 Main Characteristics of 38 Published Studies Using Insecticidal Proteins (Cry, VIP, or Spider Venom) or Protein Conferring Tolerance Against Glufosinate or Glyphosate Herbicide Under Controlled (Usually Laboratory) Conditions—cont'd

Reference #	Cry Protein or PI or Protein Conferring Tolerance Against Herbicide	Protein Source and Transgenic Plant Event Name	Exposed Stage	Response Variable	Oral Chronic Exposure Period	Concentration	Conclusions of Studies and Authors' Citations ("…")
51	eCry3.1Ab	Purified protein (from recombinant *E. coli*)	Brood (field conditions)	Mortality (eggs)	Single dose exposure, observation for 24 days	50 μg/g sucrose solution	No statistically significant differences in survival
32	Various Cry protein	Purified proteins and pollen	Larvae and adults	Survival	Metaanalysis		"Bt Cry proteins used in GM crops for control of lepidopteran and coleopteran pests do not negatively affect the survival of either honey bee larvae or adults in laboratory" settings
61	Cry51Aa2.834_16	Purified proteins	Larvae	Survival Development time	Single dose exposure, observation for 12 days	2000 μg/mL diet solution (sucrose)	"These results demonstrate no observed adverse effects on the survival and development time of *A. mellifera* larvae after dietary exposure to Cry51Aa2.834_16 protein at 20 μg Cry51Aa2.834_16/ larvae"

62	Cry1Ac + Cry2Ab, Cry1Ac + EPSPS	Pollen from two Bt cotton varieties ZMSJ (producing Cry1Ac/Cry2Ab) and ZMKCKC (producing Cry1Ac/EPSPS)	Emerged adult worker bees (0 d)	Survival	7 days	(water: honey: pollen 1:2:7 by weight) without any sugar. On average, a worker consumes about 7–10 mg of cotton pollen per day	"Binding assays showed no Cry1Ac or Cry2Ab binding to midgut brush border membrane proteins from *A. mellifera* adults. Taken together, these results support minimal risk for potential negative effects on *A. mellifera* by exposure to Bt cotton"
63 see also semifield conditions	Spider venom toxin ω-HXTXHv1a or Cry2Ab2	Purified protein Pollen Also semifield conditions	Emerged adult worker bees	Survival and longevity of worker bees		Hvt (0, 40 μg/mL) or Cry2Ab2 (10 μg/mL) proteins in sucrose solution. Pollen in sucrose	Neither purified proteins nor pollen affected the survival or longevity of the bees

Some studies also included a protease inhibitor (* see Table 2).
PER: conditioned proboscis extension reflex assay.
#: Reference numbering is taken from Table 1.

the mortality rate before treatment. The same results were obtained for food consumption. The authors also measured the foraging activity of honey bees on an artificial flower device. The number of visits decreased by c.29% between the week before treatment and the week of treatment, and by a further c.23% between the week of treatment and the week after treatment. It should be noted that no untreated control was included in the experiment (only chemical insecticide treatments), therefore it is difficult to evaluate the biological meaning of the observed decreases. It should also be mentioned that the concentration used is c.500 higher than measured in pollen of Cry1Ab-producing maize MON810.

In a similar experimental design, study #34 by Ramirez-Romero, Desneux, Decourtye, Chaffiol, and Pham-Delegue (2008) fed adult bees with two concentrations of purified Cry1Ab protein (3 and 5000 ng/g). Survival was not affected. Feeding behaviour, namely the mean time to consume the syrup-based food was increased by c.30%, but only at the highest Cry1Ab concentration. The low concentration is realistic for maize MON810 pollen (the higher concentration corresponds to a pollen concentration in maize Bt176 which is no longer cultivated). This effect was no longer found when bees were fed syrup without Cry1Ab after consumption of the Cry1Ab syrup. The study also included a proboscis extension reflex (PER) assay to test the learning performances of honey bees. The assay associates an odour to the delivery of food and has a conditioning (learning) phase followed by an extinction phase (odour provided without food delivery). No statistically differences were observed during the learning phase nor during the first four extinction phase trials. The last extinction trial showed an effect for the highest Cry1Ab concentration: Cry1Ab-fed bees kept their memory of the conditioned learning slightly longer than bees fed control food, while bees fed imidacloprid as a positive control lost this memory faster. It can be noted that the values of the extinction trial at high Cry1Ab concentration do not reflect a continuous decrease in the response, as expected, but the last extinction trial shows a slight increase with respect to the preceding trial, which sheds doubt on their validity. It should be noted that in their study published in 2005, the authors also performed an olfactory learning assay (visits to scented flowers) which did not reveal any difference before treatment, during Cry1Ab treatment or after, "suggesting no deleterious effects of Cry1Ab on olfactory learning performances" (to cite the authors) at 1000 ng/g Cry1Ab.

Study #37 by Han, Niu, Lei, Cui, and Desneux (2010) measured the mean daily quantity of food consumed by emerged bees and concluded that they consumed significantly less CCRI41 cotton pollen than control pollen. However, their data are puzzling since the CCRI41-fed bees consumed already less after 1 day (consumption at day 0 was not shown) and their decrease in consumption declined at the same rate as for control bees over several days, but then almost reached the consumption of control bees at day 7. In a subsequent paper (Han, Niu, Biondi, & Desneux, 2012), in which they did not show an experiment confirming this "antifeedant effect," they concluded that it "is not linked to effects on their midgut proteolytic enzymes or on the development of their hypopharyngeal glands." If true, it is unclear whether this "antifeedant effect" is due to Cry1Ac or CpTI which are both produced by this cotton or to other factors.

Study #58 by Steijven, Steffan-Dewenter, and Hartel (2016) found no differences between control larvae and larvae fed pollen from a maize variety with stacked genes (Cry1A.105, Cry2Ab2, Cry3Bb1, and CP4 EPSPS) for pollen digestion, survival, and developmental delay. However, they interpreted their prepupal weight data as showing a dose-dependent response to the transgenic Bt pollen. However, while the negative effects of a control toxic pollen was obvious in this study, those of the transgenic pollen are far from evident, and their data could be interpreted as showing a slight effect only at the highest dose, which is higher than the expected pollen consumption by larvae. To the best of our knowledge, a repetition of this experiment has not been published.

In conclusion, it appears that the four laboratory studies (Han et al., 2010; Ramirez-Romero et al., 2005, 2008; Steijven et al., 2016) reporting sublethal effects used either nonrealistic concentrations of purified Cry proteins and/or would gain from demonstrating the reproducibility of their findings which are not fully convincing based on a critical examination of these published single experiments.

3.2.2.2 RNA Interference

Plant micro-RNAs can affect bee development. It has been shown that a higher concentration of miRNA in beebread than in royal jelly prevents larval differentiation into queens and induces development into worker bees (Zhu et al., 2017). On the other hand, no evidence was found that diet-derived miRNAs (via pollen) are efficiently transferred across epithelia of the adult honey bee under normal conditions (Masood, Everett, Chan, & Snow, 2016). Nevertheless, RNA interference is a promising new tool for pest control in agriculture. By targeting a specific gene in the genome of the insect, the RNA produced by the plant can ultimately lead to the death of the pest (Agrawal et al., 2003).

For example, MON 87411 maize expresses a double-stranded RNA (dsRNA) targeting the *Snf7* ortholog (*DvSnf7*) in Western corn rootworm (WCR, *Diabrotica virgifera virgifera*) and this RNAi strategy downregulates the targeted *DvSnf7* gene from WCR leading to mortality. In feeding experiments, Tan et al. (2016, see study #54) tested the impact of a dsRNA containing a fragment of the *DvSnf7* gene on adults and larvae of honey bees at higher exposure levels and found no adverse effects in either larval or adult honey bees at high exposure levels. In addition, the effect of dsRNA targeting the vATPase subunit A of the pest insect *D. virgifera virgifera* was tested in larval and adult honey bees after dietary exposure. Larval development, survival, adult eclosion, adult life span, and relative gene expression were not affected (Velez et al., 2016).

3.2.2.3 Proteins Conferring Herbicide Tolerance

As shown in Table 3, one study dealt with purified CP4 EPSPS protein (conferring tolerance to the herbicide glyphosate) fed to bee larvae. In this study, larvae survival and prepupal weight were not affected by the treatment. Four studies fed bees with maize or cotton pollen producing this enzyme together with CRY protein(s). For study #58, see discussion above. In study #50, survival, cumulative consumption, and total haemocyte count of adult bees were not affected. In study #39 larvae survival and prepupal weight were not affected. In study #62, detoxification response gene expression was not affected after pollen exposure.

3.3 Results of Field Studies

As detailed in Table 4, two studies (glufosinate-tolerant oilseed rape and insect-resistant MON810 maize producing Cry1Ab) placed hives between transgenic and control plots and established that these transgenic lines were equally attractive to foraging insects. The UK Farm-Scale Evaluations (www.scri.ac.uk/research/epi/agroecology/fieldandlandscape/fseofgmhtcrops) compared the effects on arthropod and weed densities of weed management of conventional or GM herbicide-tolerant crops grown on adjacent plots and also in field margins. The authors concluded that the observed effects were indirect and related to herbicide management, not to the genetic background of the cultivars. Other studies listed in Table 3 did not find any difference in foraging activity in Bt cotton.

Table 4 Publications (15) Carried Out in Field

Reference #	Conditions	Cry Protein or PI or Protein Conferring Tolerance Against Herbicide	Protein Source and Event Name	Exposed Stage	Response Variable	Experimental Design	Concentration	Results ("authors' citations")	Concluding Remarks ("authors' citations")
18	Field	Bar or pat (phosphinothricine amino transferase) conferring tolerance to glufosinate	Winter oilseed rape (*Brassica napus*) varieties	Adult bees and other taxa	Pollinators counted over 8 days at flowering period. Pollinator density (number of insects/1000 flowers). Foraging behaviour (flowers visited per min) and posture	1. Hive placed between transgenic and control plots 2. No hive added		"No indications of a possible pleiotropic effect of the pat transgene on nectar and pollen production and nectar composition." "No difference among pairs of genotypes in the distribution of taxa." "The analysis of insect density (insects/1000 flowers) showed no genotype effect." No effect on foraging posture	Study indicates that these transgenic lines are equally attractive to foraging insects

Continued

Table 4 Publications (15) Carried Out in Field—cont'd

Reference #	Conditions	Cry Protein or PI or Protein Conferring Tolerance Against Herbicide	Protein Source and Event Name	Exposed Stage	Response Variable	Experimental Design	Concentration	Results ("authors' citations")	Concluding Remarks ("authors' citations")
20	Field and semifield	Protein conferring tolerance to glyphosate	Oilseed rape canola (pollen)	Larvae and adults (worker bees) Larvae	Field: Larval survival, pupal weight, adult recovery, and haemolymph protein concentrations. Semifield: Larval survival, pupal survival, and weight, haemolymph concentration	Colonies moved to transgenic or nontransgenic fields (each at least 40 ha). Repeated 2 years Semifield: colonies brood patch of 4–5-day-old larvae selected and fed with pollen and then returned to their natal colonies	Semifield: Each larva was fed 6 μL of the designated treatment solution (containing 1.5 mg pollen)	Field: Colonies in transgenic fields showed significantly higher haemolymph protein in newly emerged bees, but difference not detected the other year. Semifield: "no significant differences in any of the parameters between larvae fed transgenic canola pollen and those fed non-transgenic pollen"	"Transgenic canola pollen does not have adverse effects on honey bee development and that the use of transgenic canola does not pose any threat to honey bees." NB: Year to year difference in endpoint measurement are commonly observed in impact on NTO studies

45	Field	CryIAb	Maize MON810 (DKC3421 variety)	Adults	Number of bees visiting the flowers of maize plots	Hive placed between transgenic and control plots at blossom. Repeated 2 years		"Research has not revealed any significant differences in the number of honey bees visiting the two maize cultivars that would be caused by the potential negative impact of the genetically modified maize MON810"	Study indicates that these transgenic lines are equally attractive to foraging insects
49	Field and also lab (see Table 3)	Cry1Ah	Maize	Field: larvae and adults (worker bees) Lab: newly emerged adults	Effects on the intestinal bacterial community	Field: Three colonies introduced in field plot. Hives contained predominantly open and capped brood and eggs less than 48 h old 15 days. Larvae sampled after 0, 3, 6, and 9 days. Adults collected 0, 15, and 30 days after emergence		"No significant effects were observed in the midgut communities between larvae and adult honeybees fed transgenic cry1Ah-maize pollen and non-transgenic maize pollen in the field trials"	Field trials and lab experiments (Table 2) yielded identical conclusions

Continued

Table 4 Publications (15) Carried Out in Field—cont'd

Reference #	Conditions	Cry Protein or PI or Protein Conferring Tolerance Against Herbicide	Protein Source and Event Name	Exposed Stage	Response Variable	Experimental Design	Concentration	Results ("authors' citations")	Concluding Remarks ("authors' citations")
48	Semifield (flight cage)	Cry1A.105 Cry2Ab2 Cry3Bb1 CP4 EPSPS	Maize MON89034 (Cry1A.105 and Cry2Ab2) x MON88017 (cry3Bb1 and CP4 EPSPS)	Nurse bees	Survival Body weight Rates of pollen digestion Gut bacteria	When 5%–10% of the maize anthers opened Two colonies place into a flight cage 10 days. Bees collected after 9 days		"Consumption of Bt maize pollen had no effect on their survival rate, body weight and rates of pollen digestion compared to the conventional maize varieties." Gut bacterial population sizes not affected "Cry proteins from natural sources, most likely *B. thuringiensis*, were detected in bees with no exposure to Bt maize"	Bees are naturally facing Cry proteins in their guts, which confirm they are not sensitive to them. "The natural occurrence of Cry proteins and the lack of detectable effects on nurse bees and their gut bacteria give no indication for harmful effects of this Bt maize on nurse honey bees"

44	Field and lab	Cry1Ah	Bt maize MON810 (Z31 variety)	Larvae and Adults	Duration of immature stage Worker survival Body weight of larvae and adults Hypopharyngeal gland weight Colony performance Foraging behaviour Olfactory learning abilities (PER, in lab)	Colonies moved to flight cage in field. Survival of adults measured over 25 days. Exposition for 15 days during the early adult stage before PER	10 μg/mL, 10 ng/mL, and 1 ng/mL Cry1Ah in sugar solution mixed with pollen food	"cry1Ah corn carries no risk for the survival, development, colony performance or behaviour of the honeybee *A. mellifera* ligustica"	Results confirm a laboratory study by same team #43
31	Field and also lab (see Table 3)	Cry1Ab	Pollen from sweet maize Bt11 hybrid (Attribute GSS0966)	Adults	Foraging activity Weight Colony performance	Hives placed in centre of maize plot prior to anthesis. Colonies allowed to forage for 28 days Each colony was provisioned with pollen cakes to increase the level of exposure	Pollen collected from plants within each respective plot and processed into pollen cakes. Cakes containing 10, 100, or 1000 μg of imidacloprid per kg assigned to positive control plots	Field: no significant effect on foraging activity, "bees from the imidacloprid-exposed hives were foraging at about the same level but 22% fewer bees were bringing back pollen" No adverse effects on bee weight, colony performance and brood development	Field: "only differences in stores of pollen and honey greater than 75% from the control could be detected with 80% power" "The most consistent endpoints were the number of foragers returning with pollen loads, pollen load weight, and foraging bee weight"

Continued

Table 4 Publications (15) Carried Out in Field—cont'd

Reference #	Conditions	Cry Protein or PI or Protein Conferring Tolerance Against Herbicide	Protein Source and Event Name	Exposed Stage	Response Variable	Experimental Design	Concentration	Results ("authors' citations")	Concluding Remarks ("authors' citations")
30	Field	Pat gene (tolerance to glufosinate)	Oilseed rape	Larvae	Horizontal gene transfer from larvae to gut bacteria	At beginning of oilseed rape growing season, hives of three bees species established next to field		At oilseed rape flowering period, larvae were harvested. 96 gut bacterial strains were isolated and characterized	"Resistant phenotypes were found in all phylogenetic groups. None of the resistant phenotypes carried the recombinant pat gene in its genome"
17	Field (UK Farm-Scale Evaluation, FSE)	Transgenic herbicide-tolerant crops compared to crops with conventional herbicide treatment	Oilseed rape Beet Maize	Foraging adults	Sampling all invertebrates	Adjacent field plots of HT and control varieties		More bees found in oilseed rape compared to beet and maize "Growing GMHT beet allowed more efficient weed control, which resulted in fewer mature and flowering weed plants in the GMHT treatments, a consequent reduction in the available nectar and pollen resources, and less total foraging by bees." No statistically significant differences for maize or oilseed rape	"The results for bees and butterflies relate to foraging preferences and might or might not translate into effects on population densities, depending on whether adoption leads to forage reductions over large areas"

19	Field (UK FSE)	As 17	As 17	As 17	As 17	As 17, but in field margins		"For all three crops, counts were low and variable and no differences were found in total density of all bees"	
41	Field	Cry1Ac (Bunny BG-I) Cry 1Ac + cry2Ab (Bunny BG-II)	Bt cotton Bunny BG-I and Bunny BG-II	Foraging adults	Different honey bee species visiting flowers identified and counted			No significant change in foraging activity	Study indicates that both transgenic lines are equally attractive to foraging insects. No control (non—Bt) included
52	Field	Cry 1Ac + Cry2Ab (Bunny BG-II)	Bt cotton (BG-II, NCS-145)	Foraging adults	Different honey bee species visiting flowers identified and counted			Pollinator fauna did not vary in Bt cotton compared to non-Bt	Study indicates that this transgenic line is equally attractive to foraging insects
2	Field and semifield	Cry1Ab	Maize Bt11	Adults and larvae	Field sampling of entomofauna. Semifield: larvae survival and adult emergence	Semifield: brood frames with young larvae treated in lab. With pollen and returned to hive	Semifield: 1 mg pollen per cell	Field: "no difference in the kind and number of insects" associated with Bt maize. Semifield: survival higher for Bt maize compared to control	Semifield results explained by "hive vigour or genetic variability"

Continued

Table 4 Publications (15) Carried Out in Field—cont'd

Reference #	Conditions	Cry Protein or PI or Protein Conferring Tolerance Against Herbicide	Protein Source and Event Name	Exposed Stage	Response Variable	Experimental Design	Concentration	Results ("authors' citations")	Concluding Remarks ("authors' citations")
13	Field and semifield	Cry1Ab	Maize Bt11	Bee workers	Mortality, flight intensity, behaviour of bees, brood development	Small bee colonies	Cages on the corn plots	No significant change in mortality, foraging activity; No adverse effects on brood development	"No effect recorded on bees and their brood due to exposure to Bt-corn"
63	Semifield assays See also lab experiments with purified proteins and pollen	Spider venom peptide ω-HXTX-Hv1a and Cry1Ac	Transgenic cotton expressing a spider venom peptide and Bt-cotton line IR-1524, (Mon531 event)	Bee workers	Development and survival		Cages		"Field experiments did not yield any meaningful data due to high rate of mortality in all treatments including the control"

PER: conditioned proboscis extension reflex assay.
#: Reference numbering is taken from Table 1.

Effects on the intestinal bacterial community were studied for a Cry1Ah maize and a maize producing Cry1A.105, Cry2Ab2, and Cry3Bb1 and tolerant to glyphosate. Neither observed significant effects due to the transgene(s). Interestingly, one study #48 (Hendriksma et al., 2013) detected the presence of Cry proteins in the guts of bees not exposed to Bt maize, confirming that bees (which are likely to face *B. thuringiensis* in nature) are not sensitive to these toxins. One study examined the probability of gene transfer from pollen of glufosinate-tolerant oilseed rape to gut bacteria in larvae of three bee species. While bacteria resistant to glufosinate were found in all tested phylogenetic groups, none carried the transgene in its genome.

Other studies measured lethal or sublethal endpoints using maize producing the above-mentioned Cry proteins and glyphosate-tolerance protein. None reported adverse effects. When laboratory studies were also performed, there was no contradiction between field and laboratory results. Rose, Dively, and Pettis (2007) (study #31) discussed the statistical validity of their study and experimental designs allowing to reaching a 50% difference in response with 80% statistical power. They found that for statistical power, the most consistent parameters for sublethal effect (namely foraging activity) were "number of foragers returning with pollen loads, pollen load weight, and forager weight." It should also be mentioned that differences in endpoint values were obtained in one study from 1 year to the next, a trend often observed when impacts on other nontarget organisms are studied.

4. DISCUSSION

Risk assessment of GM plants is performed case by case and by a comparative approach. Thus, studies listed here studied either purified proteins added to feed compared to a control feed not containing these proteins, or pollen from a transgenic line compared with pollen from a conventional line, preferably harvested from the same location at the same time. In field experiments, a transgenic plant plot is compared to a conventional one for its impact on bees. Thus, the impact on bees is estimated with respect to an impact considered as acceptable. In addition to such a negative control, some studies included a positive control (expected to produce lethal or sublethal effects) in order to show that the bioassays performed are able to detect adverse effects, when they exist. Risk is defined as hazard × exposure to hazard. Thus, in laboratory studies it is important to correlate the doses to which bees were exposed to the realistic doses that bees may encounter in a real life situation. Worst case scenarios and even fully unrealistic doses are sometimes used in laboratory studies in order to characterize dose with effects.

It is clear from the present compilation that an impressive amount of scientific work has been devoted to the potential impact on honey bees of transgenic lines or their novel proteins. These studies included larvae as well as adult bees. Apart from minor impacts detected in a limited number of laboratory studies, which would need to be reproduced before being convincing (see comments on these studies above in Section 3.2.2.1), the general conclusion corroborates the idea that, when expressed in transgenic plants, Bt proteins have kept their specificity for the targeted order of insects and that bees are not sensitive to them. Thus, this compilation supports the conclusion that the current generation of Bt crops are unlikely to have lethal or sublethal direct effects on honey bees. This conclusion is less obvious for PIs which appear less specific and their effects on bees would depend on their concentrations in pollen. To the best of our knowledge, such an inhibitor is only produced in a Chinese GM cotton (CpTI coproduced with Cry1Ac). Laboratory studies did not point to lethal or sublethal effects on bees for this cotton (with the exception of an "antifeedant effect," which has not been confirmed; see comments above).

In this reassuring context, and also taking into account that Bt crops do often lead to a reduction in chemical insecticide use (see its dramatic decline in the United States as a consequence of Bt maize adoption: http://www.sciencemag.org/site/special/pesticides/infographic.xhtml), it has to be concluded that the fierce opposition to Bt crops by some beekeepers (see http://www.cropgen.org./article_513.html; Jany & Höfer, 2015) is not guided by a scientific rationality. This is the case especially in Europe where many beekeepers are, rightly or wrongly, convinced that chemical insecticides are the main cause for CDD, while still opposing GM Bt maize cultivation with reduced field usage of chemical insecticides. On the other hand, it has to be said that because of a general deadlock in new GM crop authorizations, for political reasons, the only Bt event that has been authorized for cultivation in Europe is of the Cry1Ab type, to combat some lepidopteran pests which only represent a localized problem. Therefore, its cultivation would only reduce the use of some insecticides in some European locations with heavy infestation of these pests. No current GM event is currently able to substitute for the insecticides which are criticized most vigorously by beekeepers. Nevertheless, it is equally true that beekeepers have contributed to block new plant biotechnology approaches for solving European agricultural challenges (see Ricroch et al., 2016).

May other plant products (apart from pollen) affect honey bees? No laboratory study used nectar as a possible source of proteins that may impact honeybees. This is because most nectars are free of protein, with exceptions but in these cases they contain very low levels of proteins, and to the best of our knowledge this is not the case for species for which GM lines are currently marketed. This is also true for honeydew: phloem sap does contain proteins but it is unclear whether considerable amounts of intact plant proteins would persist after excretion by sap-sucking insects. To the best of our knowledge, no study has examined the case of plant gums and resins which are collected by bees to make propolis. In some species, these gums or resin may contain low levels of proteins. There is no evidence that this is the case for marketed GM species.

The studies performed by public research since the mid-1990s were guided by the reasonable idea that new technologies for the benefits of farmers should also bear less risks for the environment. However, it is worth noting that some researchers justified their work by the necessity "to investigate potential impacts of insecticidal *transgene* products" (see study #21) while actually studying an insecticidal protein (Cry1Ba) which still has not been marketed in GM crops, but which is produced by *B. thuringiensis* biopesticide formulations such as the widely used ones based on its *kurstaki* ecotype. Thus, the risk of these formulations does not seem to attract as much attention as the one of new biotechnological innovations. Considering the ongoing struggle over "GM" crops, it must be concluded that such a deep focus on risk of new biotechnologies has not contributed to reassurance of lay people as to the safety of the said technologies.

5. CONCLUSIONS

Assessment of the effects on nontarget organisms is one of the most important components of environmental risk assessment. All 64 studies examined conducted with purified toxins or transgenic pollen, under laboratory conditions or in fields or semifield trials, converged to the conclusion that current GM plants have no adverse effects on larvae or adult honey bees. Toxins produced by current Bt plants and/or herbicide-tolerance proteins have no lethal effects on survival and no demonstrated sublethal effects (studied as midgut bacteria population, foraging activity, or olfactory learning). Thus, there is no evidence that consumption of transgenic pollen from these crops is contributing to CDD. Plants producing PIs or RNAi will need case by case evaluation.

ACKNOWLEDGEMENT

We deeply thank Dr. Galen Dively and Dr. Florence Courtois for valuable comments.

REFERENCES

Agrawal, N., Dasaradhi, P. V., Mohmmed, A., Malhotra, P., Bhatnagar, R. K., & Mukherjee, S. K. (2003). RNA interference: Biology, mechanism, and applications. *Microbiology and Molecular Biology Reviews*, *67*, 657–685.

Babendreier, D., Kalberer, N., Romeis, J., Fluri, P., & Bigler, F. (2004). Pollen consumption in honey bee larvae: A step forward in the risk assessment of transgenic plants. *Apidologie*, *35*, 293–300.

Berenbaum, M. R. (2014). Bees in crisis: Colony collapse, honey laundering, and other problems bee-setting American apiculture. *Proceedings of the American Philosophical Society*, *158*, 229–247.

Bergé, J. B., & Ricroch, A. (2011). La gestion de la durabilité des PGM résistantes à certains insectes. In A. Ricroch, Y. Dattée, & M. Fellous (Eds.), *Biotechnologies végétales, environnement, alimentation, santé* (p. 272): Vuibert.

Deist, B. R., Rausch, M. A., Fernandez-Luna, M. T., Adang, M. J., & Bonning, B. C. (2014). Bt toxin modification for enhanced efficacy. *Toxins*, *6*, 3005–3027.

Duan, J. J., Marvier, M., Huesing, J., Dively, G., & Huang, Z. Y. (2008). A meta-analysis of effects of Bt crops on honey bees (Hymenoptera: Apidae). *PLoS One*, *3*, e1415.

D'Urso, V., Mazzeo, G., Vaccalluzzo, V., Sabella, G., Bucchieri, F., Viscuso, R., et al. (2017). Observations on midgut of Apis mellifera workers (Hymenoptera: Apoidea) under controlled acute exposures to a Bacillus thuringiensis-based biopesticide. *Apidologie*, *48*, 51–62.

Han, P., Niu, C. Y., Biondi, A., & Desneux, N. (2012). Does transgenic Cry1Ac + CpTI cotton pollen affect hypopharyngeal gland development and midgut proteolytic enzyme activity in the honey bee Apis mellifera L. (Hymenoptera, Apidae)? *Ecotoxicology*, *21*, 2214–2221.

Han, P., Niu, C. Y., Lei, C. L., Cui, J. J., & Desneux, N. (2010). Quantification of toxins in a Cry1Ac + CpTI cotton cultivar and its potential effects on the honey bee Apis mellifera L. *Ecotoxicology*, *19*, 1452–1459.

Hawes, C., Haughton, A. J., Osborne, J. L., Roy, D. B., Clark, S. J., Perry, J. N., et al. (2003). Responses of plants and invertebrate trophic groups to contrasting herbicide regimes in the farm scale evaluations of genetically modified herbicide-tolerant crops. *Philosophical Transactions of the Royal Society B—Biological Sciences*, *358*, 1899–1913.

Haydak, M. H. (1970). Honey bee nutrition. *Annual Review of Entomology*, *15*, 143–156.

Hendriksma, H. P., Kuting, M., Hartel, S., Nather, A., Dohrmann, A. B., Steffan-Dewenter, I., et al. (2013). Effect of stacked insecticidal cry proteins from maize pollen on nurse bees (Apis mellifera carnica) and their gut bacteria. *PLoS One*, *8*, e59589.

Hilder, V. A., Gatehouse, A. M. R., Sheerman, S. E., Barker, R. F., & Boulter, D. (1987). A novel mechanism of insect resistance engineered into tobacco. *Nature*, *300*, 160–163.

Jany, K. D., & Höfer, E. (2015). *The German beekeepers and modern plant biotechnology: Chronological overview of the fight of German beekeepers against modern plant biotechnology and genetic engineering to protect honey and other bee products: Bt toxins/Bt plants and bee health* (p. 184). Hamburg: Verlag Dr. Kovač.

Katzek, J. (2014). At the end of the day everything boils down to politics: The evolving of German policy toward GMO crops and the existing stagnation. *GM Crops & Food*, *5*, 178–182.

Kuntz, M., Davison, J., & Ricroch, A. (2013). What the French ban of Bt MON810 maize means for science-based risk assessment. *Nature Biotechnology*, *31*, 498–500.

Liao, L.-H., Wu, W.-Y., & Berenbaum, MR. (2017). Impacts of dietary phytochemicals in the presence and absence of pesticides on longevity of honey bees (*Apis mellifera*). In: Cook, S., & Evans, J. D. (Eds.), *Insects*, *8*(1), 22. https://doi.org/10.3390/insects8010022.

Liu, Z., Chen, C., Niu, Q., Qi, W., Yuan, C., Su, S., et al. (2016). Survey results of honey bee (Apis mellifera) colony losses in China (2010 – 2013). *Journal of Apicultural Research*, *55*, 29–37.

Malone, L. A. (2004). Potential effects of GM crops on honey bee health. *Bee World*, *85*, 29–36.

Malone, L. A., & Pham-Delegue, M.-H. (2001). Effects of transgene products on honey bees (Apis mellifera) and bumblebees (Bombus sp.). *Apidologie*, *32*, 287–304.

Marvier, M., McCreedy, C., Regetz, J., & Kareiva, P. (2007). A meta-analysis of effects of Bt cotton and maize on nontarget invertebrates. *Science*, *316*, 1475–1477.

Masood, M., Everett, C. P., Chan, S. Y., & Snow, J. W. (2016). Negligible uptake and transfer of diet-derived pollen microRNAs in adult honey bees. *RNA Biology*, *13*, 109–118.

O'Callaghan, M., Glare, T. R., Burgess, E. P. J., & Malone, L. A. (2005). Effects of plants genetically modified for insect resistance on nontarget organisms. *Annual Review of Entomology*, *50*, 271–292. Annual Reviews, Palo Alto.

Ramirez-Romero, R., Chaufaux, J., & Pham-Delegue, M. H. (2005). Effects of Cry1Ab protoxin, deltamethrin and imidacloprid on the foraging activity and the learning performances of the honeybee Apis mellifera, a comparative approach. *Apidologie*, *36*, 601–611.

Ramirez-Romero, R., Desneux, N., Decourtye, A., Chaffiol, A., & Pham-Delegue, M. H. (2008). Does CrylAb protein affect learning performances of the honey bee Apis mellifera L. (Hymenoptera, Apidae)? *Ecotoxicology and Environmental Safety*, *70*, 327–333.

Ricroch, A., Harwood, W., Svobodova, Z., Sagi, L., Hundleby, P., Badea, E. M., et al. (2016). Challenges facing European agriculture and possible biotechnological solutions. *Critical Reviews in Biotechnology*, *36*, 875–883.

Roh, J. Y., Choi, J. Y., Li, M. S., Jin, B. R., & Je, Y. H. (2007). Bacillus thuringiensis as a specific, safe, and effective tool for insect pest control. *Journal of Microbiology and Biotechnology*, *17*, 547–559.

Rose, R., Dively, G. P., & Pettis, J. (2007). Effects of Bt corn pollen on honey bees: Emphasis on protocol development. *Apidologie*, *38*, 368–377.

Ryan, C. A. (1990). Protease inhibitors in plants: Genes for improving defenses against insects and pathogens. *Annual Review of Phytopathology*, *28*, 425–449.

Steijven, K., Steffan-Dewenter, I., & Hartel, S. (2016). Testing dose-dependent effects of stacked Bt maize pollen on in vitro-reared honey bee larvae. *Apidologie*, *47*, 216–226.

Tan, J. G., Levine, S. L., Bachman, P. M., Jensen, P. D., Mueller, G. M., Uffman, J. P., et al. (2016). No impact of DvSnf7 RNA on honey bee (Apis mellifera L.) adults and larvae in dietary feeding tests. *Environmental Toxicology and Chemistry*, *35*, 287–294.

Van Engelsdorp, D., Cox-Foster, D., Frazier, M., Ostiguy, N., & Hayes, J. (2006). *Colony collapse disorder preliminary report*. Mid-Atlantic Apiculture Research and Extension Consortium (MAAREC)—CCD Working Group, p 22.

Velez, A. M., Jurzenski, J., Matz, N., Zhou, X. G., Wang, H. C., Ellis, M., et al. (2016). Developing an in vivo toxicity assay for RNAi risk assessment in honey bees, Apis mellifera L. *Chemosphere*, *144*, 1083–1090.

Zhu, K., Liu, M., Fu, Z., Zhou, Z., Kong, Y., Liang, H., et al. (2017). Plant microRNAs in larval food regulate honeybee caste development. *PLoS Genetics*, *13*, e1006946.

CHAPTER SIX

Genetic Engineering of Crop Plants: Colombia as a Case Study

Julián Mora-Oberlaender*, Adriana Castaño Hernández†, Silvio Alejandro López-Pazos‡, Alejandro Chaparro-Giraldo*,[1]

*Grupo de Ingeniería Genética de Plantas, Universidad Nacional de Colombia, Bogotá, Colombia
†GMO Biosafety and AgroBiotech Consultant, Bogotá, Colombia
‡Facultad de Ciencias, Universidad Antonio Nariño, Bogotá, Colombia
[1]Corresponding author: e-mail address: achaparrog@bt.unal.edu.co

Contents

Abstract

Colombia was one of the leading countries in the formulation and negotiation of the Cartagena Protocol on Biosafety, and as a megadiverse country has taken the challenge of developing technical and institutional capacities to ensure that applications of biotechnological developments do not pose risks to human and animal health or the environment. Since 2000 there are GM crops which have been approved for environmental release in Colombia. The first approved crop was blue carnation. To date, GM maize, cotton, flowers, and soybeans have been authorized for growing. The approval of foods derived from GM plants for human consumption in Colombia is a process that has been

Advances in Botanical Research, Volume 86
ISSN 0065-2296
https://doi.org/10.1016/bs.abr.2017.11.005

going on since 2003 and to date there are 104 genetically modified events approved for this purpose. Those events are present in maize, soybean, cotton, canola, sugar beet, rice and wheat. For their use as animal feed there are 59 approved events from six different crop species, mainly maize, soybean and cotton. Colombia has various research institutions, working on genetic transformation of crop plants of economic importance for the country, but to date none of them have reached the market. The adoption of agbiogenerics as an alternative approach by national research groups may reduce the time and cost associated and facilitate the eventual commercialization of a Colombian biotechnological crop. This presents a challenge that involves all stages of GM crop development.

1. INTRODUCTION

Colombia was one of the leading countries in the formulation and negotiation of the Cartagena Protocol on Biosafety (CPB) as well as the Nagoya-Kuala Lumpur Protocol on Liability and Redress Supplementary to the Cartagena Protocol. The Focal Point of the Cartagena Protocol designated for Colombia is the Ministry of Foreign Affairs and the Focal Point designated for the Biosafety Clearing-House (CIISB-BCH) is the Research Institute for Biological Resources Alexander von Humboldt (IAvH) (Hodson de Jaramillo & Castaño, 2013).

In addition to being a party to the Convention on Biological Diversity (CBD) and the Cartagena Protocol, the general regulatory framework on biotechnology and biosafety in Colombia includes adherence to various international and regional conventions and agreements on Intellectual Property (IP) rights and other related aspects such as trade, access to genetic resources and biodiversity such as the World Trade Organization (WTO), the Paris Agreement (industrial property), the General Agreement on Tariffs and Trade (GATT), the International Union for the Protection of New Varieties of Plants (UPOV), the Agreement on the Application of Sanitary and Phytosanitary Measures (SPS), the Agreement on Technical Barriers to Trade, the International Plant Protection Convention, the FAO-WHO *Codex Alimentarius*, the Aarhus Convention, and others. As a member of the Andean Community of Nations (CAN), it governs normative aspects related to biotechnology and related areas through agreements: Decisions 345 (UPOV), 391 (access to genetic resources), 486 (industrial property), 523 (Regional Biodiversity Strategy) (Hodson de Jaramillo, Castaño, & Uscátegui, 2012).

Based on data provided by the United Nation, from 2014 to 2030, the population of Colombia will grow from 46.3 to 52.7 million inhabitants.

Every 10 years, about 4 million Colombians will join in consuming goods and food in the country, which are an extraordinary factor and an opportunity for local agricultural producers. Different sources, including, but not limited to, International Monetary Fund, World Bank, and USDA, put Colombia on a clear trend of economic growth for the coming years due to its good macroeconomic fundamentals. Given this economic growth, substantial growth in local incomes is expected, with GDP per capita as an indicator of local income. From 2014, GDP per capita in Colombia is expected to grow from US $8.9 to US $12.7 thousand in 2020 (USDA-United States Department of Agriculture, 2014).

Genetically Modified (GM) cotton was introduced in Colombia in the 2003/04 planting season, accompanying the worldwide adoption of the technology. Since then, the generation of economic benefits from this technology reached US $134.1 million. Of this total, 63.2% (US $84.7 million) was achieved by Colombian cotton producers, in the form of a reduction in production costs (US $12.4 million) and an increase in productivity (US $72.3 millions). The remainder was accumulated by the technology breeders, 37.1% or US $49.4 million (Céleres, 2015).

The same study shows that, in the case of maize, the Colombian farmer who adopts transgenic technology earns US $28 per hectare using these technologies (considering events with stacked genes). In addition to the profit derived from a reduction in the cost of production, the Colombian rural producer derives economic benefits from increased productivity, which will result in higher sales, through the prices paid per tonne of maize. In the case of the breeder of transgenic technology (industry), the economic gain is calculated by the price difference between GM and conventional seed, which results in the achievement of economic benefits, which were stipulated, on average, in US $71.0 per hectare. Since the adoption of biotechnology in maize in Colombia in 2007/08, to the 2014/15 sowing season, rural producers have accumulated direct and indirect economic benefits. In the case of rural producers, the direct economic benefits accrued came to US $77.9 million, including cost reduction and the surplus of production generated with the technology.

Although Colombia still presents disparate conditions regarding the intensity and insertion of agricultural technologies, cases where local producers already use the most modern techniques of agricultural production show the productive potential of the country.

Colombian farmers perceive the direct and indirect advantages, both economic and socio-environmental, in the continuous use of genetic modification technologies, with a high degree of confidence and expectations in their use, recognizing the intrinsic value and the capacity that they have to transfer to the cotton, maize and soy production chains. Of these, the last one has not yet been adopted, despite the fact that it is already covered by the Colombian regulatory framework (Céleres, 2015).

Despite the active participation of the country in international regulatory policy formulation and the adoption of the technology by farmers, no commercial GM crops have been developed by Colombian institutions. There is, however, a significant body of research groups working in this area, and recent changes in the intellectual property landscape and available technologies may lead to more effective technology development by local groups, as may be the case with the emergence of agbiogenerics.

2. REGULATION FOR GENETICALLY MODIFIED ORGANISMS IN COLOMBIA

In January 2000 in the city of Montreal, and after more than 5 years of negotiations, the International Biosafety protocol, also called CPB and part of the CBD, was signed by 128 countries who were the parties to the CBD. Today there are 170 parties (https://bch.cbd.int/protocol/parties/), with the objective of: "...to contribute to ensuring an adequate level of protection in the field of the safe transfer, handling and use of living modified organisms resulting from modern biotechnology that may have adverse effects on the conservation and sustainable use of biological diversity, taking also into account risks to human health, and specifically focusing on transboundary movements" (Secretariat of the Convention on Biological Diversity, 2000).

In 1995, the Secretariat of the CBD included Colombia as part of the panel of experts meeting in Cairo to discuss whether or not a Biosafety Protocol was required, and to analyse which of the terms, Living Modified Organism (LMO) or Genetically Modified Organism (GMO), was the most convenient. They came to the conclusion of focusing on LMOs, since they considered that the term GMOs opened the door to regulate other products (vaccines, yeasts and some antibiotics using techniques of rDNA), and difficulties could arise in terms of international trade (Artunduaga et al., 1997). Not surprisingly, the Protocol bears the name Cartagena, after the city in Colombia where the Biosafety Protocol was originally scheduled to be concluded in February 1999. However, due to a number of outstanding issues, the Protocol was finalized and adopted in Montreal, Canada.

Later, between the years 2004 and 2010, during the negotiations of article 27 of the Cartagena Protocol, regarding liability and redress for damages resulting from the transboundary movements of LMOs, which would lead to the issuance of the Nagoya-Kuala Lumpur Supplementary Protocol on Liability and Redress to the CPB, Colombia exercised cochairing throughout these negotiations, with a significant recognition at the end of them.

Since then, and after its ratification in Colombia (Colombia, 2002, Law 740), the country has worked on its implementation, always seeking the joint participation of the different governmental actors, in accordance not only with the Protocol's directives, but in accordance with local environmental, agricultural, health and trade standards.

The country has sought the active participation in the different international negotiations in which the subject of Biotechnology and GMOs is approached. This is how the participation in the meetings of the Conference of the Parties of the Cartagena Protocol, leading the negotiation of article 27 on liability and redress, participation in the discussions in the framework of the *Codex Alimentarius* on methods of GMO detection, risk analysis, and labelling of foods obtained by biotechnology, among other topics under discussion by those intergovernmental body has been a key point.

Recognizing that Biosafety is a multisectoral responsibility, the Ministers of the Environment, Agriculture and Rural Development, Health, and International Trade, convened a task force made up of representatives of these four Ministries, the Institute for Colombian Agriculture (Instituto Colombiano Agropecuario—ICA) and the IAvH in June, 2001. This group has been developing a Plan of Action to promote a coordinated approach to Biosafety in Colombia. Central to the Plan of Action would be the development of a mechanism to facilitate an intersectorial approach to Biosafety and to policy formulation and decision-making at the national level (Silva, 2001).

Based on that, Colombia has a strong national biosafety system, which was strengthened between 2004 and 2008 with the implementation of the GEF-BM PROJECT-PO 77171 Capacity Building for Implementation of the Cartagena Protocol. The fundamental objective was to develop scientific, technical and institutional competences to comply the commitments of the Cartagena Protocol, including the assessment, management and monitoring of potential risks related with the use of GMOs on biodiversity and human and animal health (Hodson de Jaramillo & Castaño, 2013).

2.1 Biosafety Framework for Environmental Release, Food and Feed

The regulatory framework for biosafety in Colombia in relation to GM crops has two basic components. The first is composed of constitutional mandates and different legislative acts in relation to the environment, biodiversity, genetic resources and agricultural issues. Although none of these deals directly with biosafety issues, they constitute the legal framework for GMOs in Colombia in the areas of principles, general competences, definitions and concepts, risks, responsibilities, sanctions and prohibitions, export, import, marketing, and introduction of species, among others. The second component refers to the specific biosafety legislation developed, majorly in compliance with the Cartagena Protocol and the International Agreements of which Colombia is a country Party (Hodson de Jaramillo et al., 2012).

In Colombia the term GMO is understood as any living organism that possesses a new combination of genetic material, which has been obtained by the application of recombinant DNA technology, its developments or advances, as well as its parts, derivatives or products containing them, capable of reproducing or transmitting genetic information. This concept includes the term LMO as referred to in the CPB. In such a way in Colombia the legal term used is GMO, however, since the term LMO is included, it can be found in some regulations without distinction.

The first regulatory approximation to GMOs was through the ICA that issued two agreements (No. 0013/98 and 004/02) (ICA, 1998a, 2002) and two resolutions (No. 3492/1998 and 2935/01) (ICA, 1998b, 2001) that outline procedures for the introduction, research, production, release and marketing of LMOs for agricultural and livestock use. The agreements signed by the Minister of Agriculture as a head of the Board of Directors of ICA established the first National Technical Committee (CTN in Spanish), as an advisory body on the applications of LMOs for the above uses (Rosillo, 2007).

On this basis, the first commercial release, in 2000, of blue carnation for purposes of multiplication of cutflowers was authorized for the small-scale. There were also approvals for field trials of Monsanto's insect-resistant Bollgard® Cotton, confined research for insect-resistant *Bt* cassava and the beginning of experimental work on transformation of *Brachiaria* and *Stiloshantes* for the Centro Internacional de Agricultura Tropical (CIAT), coffee for Cenicafé and sugarcane for Cenicaña (Chaparro-Giraldo, 2015). The biosafety system and the procedure for the approval of GM crops for planting is the one with the longest creation and operation time compared to those related to health and environmental issues.

Following the ratification of the Cartagena Protocol in 2002, it was necessary to redefine the structure of the National Biosafety System and the measures to implement it. In December 2005, the Ministries of Agriculture, Social Protection (today Ministry of Health) and Environment issued Decree 4525 regulating Law 740 of 2002 (Colombia, 2005). This decree established the regulatory framework for GMOs in relation to: Authorization to develop activities with GMOs (understood as the prior informed consent (AFP) for the first movement of GMOs), risk assessment and management, the creation of National Biosafety Committees, monitoring and surveillance, education and research (Decree 4525, 2005; Hodson de Jaramillo & Castaño, 2013).

The regulatory system for GMOs is based on the precautionary approach and a case-by-case assessment during all the GM crop development process, from laboratory to field and market, and also taking into account the concepts of familiarity and substantial equivalence.

Three designated National Competent Authorities (NCAs) were established by Decree 4525. Each NCA has a Biosafety Technical Committee which is responsible for carrying out the risk assessment and risk analysis, based on which they issue a concept to the competent authority to recommend the authorization or not of a GMO for a specific use[a]. Based on this concept, the respective competent authority is the one that makes the decision to authorize or not the use of a GMO in Colombia through a resolution act (Castaño, 2015). Authorization decisions taken by each of the competent authorities are nonbinding, which means that there may be GMOs authorized for several different uses (cultivation, human consumption or animal consumption), for one use or none. The NCAs are:

1. The Ministry of Agriculture and Rural Development, through the ICA, is responsible for evaluating and authorizing GMOs exclusively for agriculture, livestock, fishery, commercial forest plantations and agro-industry. For purposes of assessment and authorization, the Ministry of Agriculture gives the competence to the ICA.

[a] **CTNBio** (for its name in Spanish): National Technical Committee on Biosafety of GMOs for agricultural, livestock, fishery, commercial forest plantations and agro-industry use.

CTNSalud (for its name in Spanish): National Technical Committee on Biosafety of GMOs for use in health and human consumption.

CTNAmbiente (for its name in Spanish): National Technical Committee on Biosafety of GMOs for environmental use.

2. The Ministry of Health and Social Protection is responsible for evaluating and authorizing GMOs whose destination is the use in human food or health.
3. The Ministry of Environment and Sustainable Development is responsible for evaluating and authorizing GMOs exclusively for environmental use.

2.2 Risk Assessment Process for Environmental Release

The risk assessment strategy seeks to implement appropriate methodologies and approaches to compare the GM crop with its non-GM counterparts. This comparison is the starting point of the evaluation that focuses on intentional or unintentional differences. To this end, the concept of familiarity of the OECD (1993a, 1993b) and WHO/FAO (2000) has been developed for the environmental safety assessment of GMOs.

With the issuance of Decree 4525 of 2005, ICA promulgated Resolution 946 (ICA, 2006), which establishes the process for applications for GMOs for agricultural, livestock, agro-industrial and forest use. As general evaluation criteria, the ICA is guided by the following considerations: case-by-case and step-by-step studies; comparative approach (conventional vs GM crops); sequential and successive evaluation; evaluation during the entire plant development period; and environmental assessments. Biosafety studies carried out include:

1. Agronomic behaviour of the modified variety in different sowing regions in the country.
2. Evaluation of weed control and crop damage (if applicable).
3. Costs for management and crop yields.
4. Evaluation of wild species in the GM planting areas.
5. Determination of the effects of the new proteins expressed on nontarget pests (if applicable).
6. Establishment of planting distances between GM crops and conventional crops, by determining the transport of pollen from the GM material to be planted.

In addition, a detailed molecular review of the transformation event is performed, based on information from the donor organism, the transformation vector, the receiving organism and the description of the GMO.

The GM planting areas in Colombia respond to a Management Plan that is approved by the ICA. Permanent verification of its compliance during each of the sowing cycles is made by the ICA.

Evaluations made for more than 15 years by the ICA, have made it possible to establish, for the case of maize, the planting of a refuge at a distance of no more than 500 m from the areas planted with GM maize seeds. If the refuge is planted in the same field as the GM maize, it must be in a contiguous block, not intermixed and must be clearly identified. The refuge scheme should be 90/10. In this scheme the farmer must plant 90% of the lot with GM maize seed and the remaining 10% with a conventional cultivar, if GM maize is resistant to insect attack and tolerant to herbicide application, the refuge may also be planted with a GM with herbicide tolerance (ICA, 2010d).

For GM cotton, refuge schemes should be 80/20 or 96/4. In the first, 80% of the lot will be planted with GM cotton seed and the remaining 20% with a conventional cultivar, and in the second, 96% with GM seed and 4% with a conventional cultivar. In both cases, if GM cotton is resistant to insect attack and tolerant to herbicide application, the refuge may also be planted with a GM cotton tolerant to herbicide application. In the refuge area the control of pests which are not targeted by these technologies can be carried out with any type of conventional insecticide except biological insecticides based on *Bacillus thuringiensis*. These schemes, which are widely used globally, were defined based on computer simulations of populations, analysis of the probability of crossing and aiming to delay the appearance and fixation of resistance alleles (ICA, unpublished data). In Colombia, the refuge scheme used for GM cotton is the farmer's choice and depends on the availability of area for planting, crop practices, pests and yields; management and monitoring by the agricultural authority (ICA) must be followed.

The management plan aims to: (i) control the use and commercialization of GM seeds, (ii) avoid the generation of resistance by the target pests through refuge strategies and monitoring of the baseline susceptibility to proteins that confer resistance to insect attack, (iii) ensure the effective use and management of herbicides in GM crops with herbicide tolerance to avoid the emergence of resistant weed species and (iv) prevent gene flow from GM crops to wild varieties by using forms of isolation.

The monitoring actions are carried out during all the stages of the crop from the presowing season, during the cultivation and at the end and closing of the harvest. During this time there is a verification of: (i) training, (ii) establishment of refuge schemes, (iii) monitoring the development of possible resistance by target pest species, (iv) weed management and (v) isolation.

Confined research with GMOs requires prior authorization by ICA, which decides if it authorizes it based on the information provided by the requesting Research Unit in relation to available infrastructure and equipment, waste management, description of the research group, parental and recipient organisms to be used, type of research to be carried out and containment and isolation measures to be used, in addition to the information set out in Annex I of the Cartagena Protocol.

2.3 Risk Assessment Process for Human Consumption

The first GM crop authorized for human consumption in Colombia was Bollgard™ cotton MON531, which was approved in 2003. Since that time and to date more than 100 events of genetic transformation have been approved for human consumption, as analysed in more detail forward. The country has specific regulations for GMO biosafety and their derived products for the production of food for human consumption since 1996 through regulations issued by the Ministry of Health and applied by INVIMA (National Institute for Drugs and Food Surveillance) (Castaño, 2015).

The Ministry of Health, through the evaluation carried out by the CTNSalud, is responsible for granting approval for human consumption in the country. GM crops evaluated are authorized as raw materials for the production of food for human consumption.

The possible effects on human health of foods derived from GM plants depend on the specific content of the food and have the same potential for risk or benefit as conventional foods. The safety assessment of GMOs is carried out case by case and step by step, which means that each evaluation is individual and specific for the criteria to be evaluated and the specific use, taking into account each stage of its development and the following criteria (FAO, 2007) also followed as principles by the Colombian CTNSalud:

- Concept of substantial equivalence (FAO/WHO, 1996; OECD, 1993a, 1993b; Shauzu, 2000).
- Use of a multidisciplinary approach, including evaluation of the characteristics of the new protein introduced.
- Consultation with national and international experts as well as regulatory authorities in other countries.
- Scientific basis.
- There is no zero risk.

The procedure adopted in Colombian regulation is similar to that followed in other countries, in accordance with the guidelines established in the Cartagena Protocol (Article 11, Annex II) and the *Codex Alimentarius*, mandatory in Colombia (CAC/GL44-2003, CAC/GL45-2003), and with the normative framework in force in Colombia that regulates activities carried out with food (Decree 3075/1997 updated by Resolution 2674/2003).

Based on the information provided by the applicant, CTNSalud does the respective risk assessment. All applications are submitted to a public consultation process in order to involve the community. This public consultation is carried out through the INVIMA website, as technical secretary of CTNSalud, and for no less than 60 days. The public consultation is carried out prior to the request being discussed in plenary session of the Committee and all comments or doubts that are received are discussed within the CTN in order to be taken into account (Hodson de Jaramillo & Castaño, 2013).

Studies to determine the safety of foods derived from GM crops include acute toxicity analyses according to the nature and function of the newly expressed protein, potential occurrence of unintentional effects, evaluations to establish the possible allergenicity of the new protein; and nutritional composition studies in which essential components, metabolites and food processing are analysed. All of these scientific studies are submitted by the applicant and evaluated by the competent national authorities. Subchronic toxicity analyses (90 day animal feeding studies) are not mandatory and their requirement depends on the assessment done by the Biosafety Committee. If the GM crop has been substantially modified or if there is an indication of unintended effects, the committee (CTNSalud) could ask for additional studies following OECD 408, 1995 guidelines and *Codex Alimentarius* (CAC/GL 44-2003 y CAC/GL 45-2003).

2.4 Labelling

Based on the fact that the country has a National Biosafety System for GMOs, and the risk assessments required to determine the safety of foods derived from GMOs are being developed, the regulations on the labelling of GMOs have been established under this principle.

Since 2011, by Resolution 4254, Colombia has a labelling regulation in force, which sets forth that labels or tags for foods derived from GMOs which are not substantially equivalent to their conventional counterparts in any of the next conditions must indicate the difference in the form of statements:

1. The values of nutritional components existing in food-containing GMOs, or having used raw materials that are GMOs, are not substantially equivalent to those of homologous conventional or food products already in the market.

2. Storage, preparation or cooking of a food-containing GMOs or having used raw materials that are GMOs differs thereby from equivalent homologous or food products already in the market.
3. Presence of allergens introduced as a result of genetic modification into food containing a GMO or having used raw materials being GMOs, and which are not expected by consumers.
4. Presence of differences in the organoleptic properties of a food as a consequence of genetic modification as compared to homologous conventional products.

This means that not all GMOs have to be labelled, only those in which a substantial difference is introduced because of the genetic modification. In these cases the required label is not just for the term GMO but for the specific change introduced. For example, if you have GM soybean with a high content of fatty acids which is used as an ingredient in the production of food, the label should state the difference with a statement such as: "Contains soybean with High omega-3 fatty acids obtained by genetic modification through modern biotechnology" (statement examples are under development by the Ministry of Health, Personal communication).

When statements are made relating to "GMO free" or "Does not contain GMOs" the aforementioned statements shall only be used if the manufacturer demonstrates and supports, through laboratory results validated by the GMO National Detection Laboratory from INVIMA, that the affirmation is true and not misleading, or by traceability systems that show that the claim is true and has technical support.

Importers of raw materials that are not packaged and that are or may contain GMOs, must provide product-related information, according to the parameters defined by the Cartagena Protocol, on the commercial invoice or by additional statements accompanying international shipments of GMOs for commercial production. Such statement uses the following terms: "This grain shipment contains GMOs intended for direct use as human food or for processing"; if the GMO is not authorized in Colombia for environmental release, it shall also state: "This shipment is not intended for intentional introduction into the environment".

Colombia has built a high institutional capacity, which is recognized in the international context, and is strengthened by two GMO detection laboratories, one lead by INVIMA (health sector) and another led between ICA and IAvH, representing agricultural and environmental authorities. These laboratories have the following aims: (1) develop and implement procedures and techniques for the detection and monitoring of GMOs, including raw materials and processed food; (2) design and adopt sampling plans and techniques for the different GM events approved in Colombia; (3) strengthen technical and human resources to respond to the country's commitments to the CPB; (4) formulate and develop interinstitutional, or in coordination with other laboratories, research projects for the detection, monitoring, and gene flow detection of GMOs and (5) support training activities for regulators in charge of GMOs in Latin America (Castaño, 2013).

The flowchart on Fig. 1 shows the steps to be taken to obtain the authorization of a GMO from research to the market or environmental release. Some steps are prerequisite to start any research with GMOs for agricultural purposes and their subsequent use.

3. ADOPTION OF GM CROPS IN COLOMBIA

Colombia had a leading role in the international agenda and policy formulation for the safe use of GMOs. It is therefore expected that the relatively solid institutions and clear implementation of international standards lead to the adoption of the technology in the country and its acceptance by producers, importers and the public.

There is, however, a marked asymmetry between the extent to which GM crops are grown in Colombia, both in area, crop species and events, and the number of approved events for human and animal consumption derived from imports. Several factors weigh into this asymmetry. Colombia is a net importer of cotton (CONALGODON, 2012), soybean and maize (FENALCE, 2016), which are the three GM crops of larger commercial importance worldwide. Available GM events of these crops have been developed for those countries with major planted areas, such as the United States, Brazil and Argentina. Other crops such as canola and sugar beet are more commonly grown in subtropical areas and are therefore not developed for Colombian agroenvironmental conditions.

The use of GM crops for use as food, feed and for commercial cultivation has been approved in Colombia for more than 15 years. New events have gone through the regulatory process to reach the market every year for this period. Nonetheless, the area cultivated has not kept a growing pace and the number of crops remains limited. This landscape may change with the development and eventual release of biotechnological crops developed by Colombian institutions, using local varieties and species of importance for national agriculture.

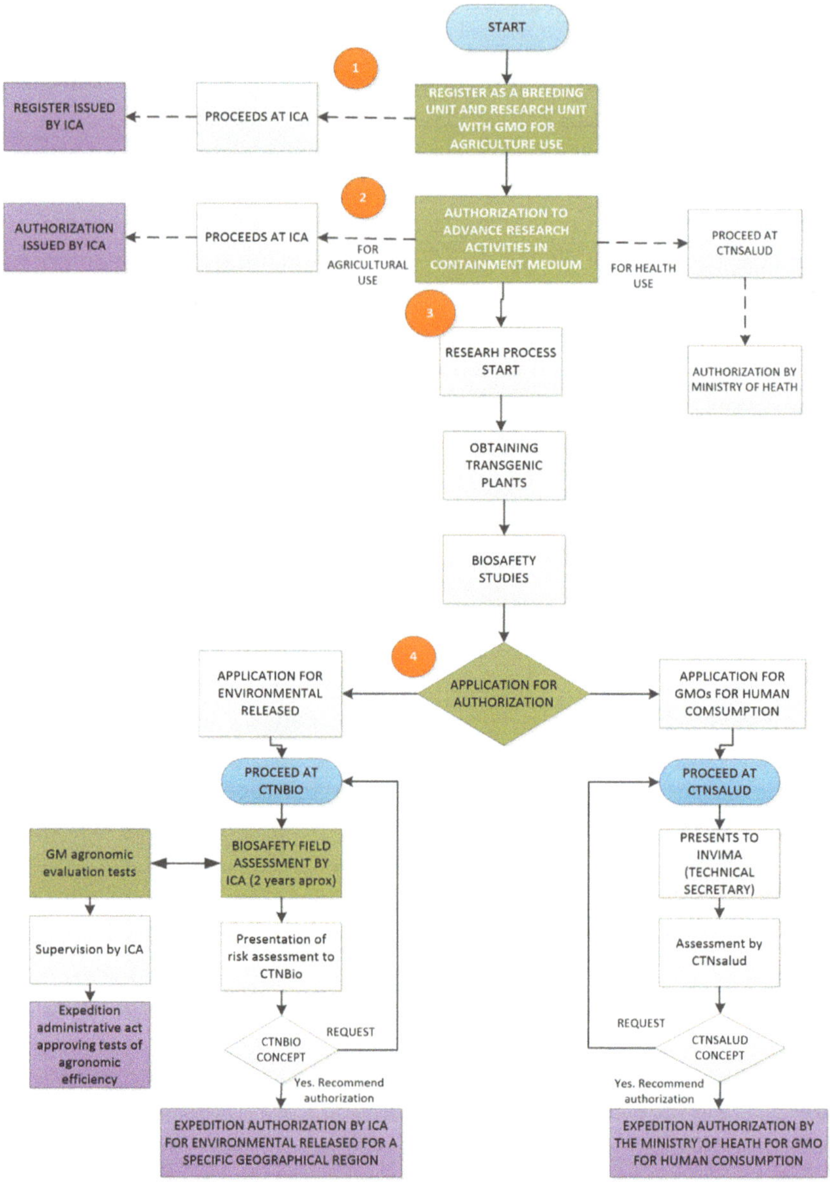

Fig. 1 Schematic description of the regulatory process for the approval of GMOs in Colombia for environmental release and human consumption. Own elaboration. (1) Any research group that is going to work with GMOs for agricultural purposes must first register as a Plant Breeding Unit with the ICA. (2) In order to advance research with GMOs in a confined environment, specific authorization must be granted by ICA (CTNBio) for research for agricultural purposes and by the Ministry of health (CTNSalud) if it is for health use purposes. (3) Once the respective authorizations have been obtained, the R&D process with GMOs can begin. (4) In order to commercialize or plant a GMO, authorizations must be obtained according to the use of the transformation event. If it is for human consumption the process will be advanced with the health sector to obtain authorization from the Ministry of Health, and if it is for planting or feed, from ICA.

3.1 Approved GM Crops

According to Colombian regulation, there are three general types of approval for GM crops depending on their intended use as food, feed or for their cultivation. These are issued by different authorities and are independent processes (see earlier).

The first approval for the commercial cultivation of a GM plant in Colombia was granted in 2000 for blue carnations which were first planted, in confined greenhouses, in 2002 (ICA, 2000). Environmental release of a GM crop was first approved for Bollgard™ cotton in 2003. When considering data available from national authorities (INVIMA, ICA) and international databases (*Sources*: cera-gmc.org/GMCropDatabase, bch.cbd.int, isaaa.org and FAO biosafety database), a total of 125 events are found to have received approval as of June 2017 (Table 1). It is important to note that not all of these are actually imported or cultivated, as is the case with the phosphinothricin-tolerant Liberty Link™ rice or Roundup Ready™ (RR) wheat which were approved but not commercialized. RR wheat is

Table 1 Number of GM Events Approved for Commercial Use in Colombia (Updated June 2017)

	Single/Hybrid		Use			
Crop	**Single**	**Hybrid**	**Food**	**Feed**	**Cultivation**	**Total**
Maize	20	41	59	36	13	61
Soybean	16	7	23	11	2	23
Cotton	12	7	17	9	7	19
Carnation	14				14	14
Rose	2				2	2
Rice	2		2	1		2
Canola	1		1			1
Flax	1			1		1
Sugar beet	1		1	1		1
Wheat	1		1			1
Total	70	55	104	59	38	125

Sources: ICA, Min Salud, INVIMA, bch.cbd.int, cera-gmc.org/GMCropDatabase, isaaa.org/gmapprovaldatabase/default.asp.

approved in Colombia only for human consumption as a result of the risk assessment carried out by the national authorities for this purpose in response to the request received. For commercial reasons GM wheat was never commercially developed or marketed, although the authorization has already been given by the competent authority and it is still in force to date.

GM events from 10 different species have gone through the regulatory process, of which 5 have been approved for cultivation, 7 for human food production and 6 for animal feed production (Table 1). Not surprisingly, maize, soybean and cotton have the largest number of approved events in all categories, given that they are also the most extensively cultivated GM crops worldwide (ISAAA, 2016). These are also the only crops for which some events have been approved for the three possible uses (11 maize events, 2 soybean events and 6 cotton events).

Most maize events (67%) are hybrids produced by crossing of between two and six single events (stacked) which in most cases have also been approved individually. The maize hybrids with the highest number of introduced genes are Syngenta's Agrisure® Duracade™ 5222 which has five insect (coleopteran and lepidopteran) resistance genes and two herbicide (glyphosate and phosphinothricin) tolerance genes; and maize 3272 × Bt11 × MIR604 × TC1507 × 5307 × GA21 which has four insect (coleopteran and lepidopteran) resistance genes, two herbicide (glyphosate and phosphinothricin) tolerance genes and a thermostable amylase gene for enhanced bioethanol production.

Since the approval for commercial cultivation, food and feed use of Bollgard™ cotton, cultivation of Roundup Ready™ cotton and food use of Yieldgard™ maize in 2003, every year at least on GM event has been approved in the country (*Sources*: INVIMA, ICA, cera-gmc.org/GMCropDatabase, bch.cbd.int, isaaa.org/gmapprovaldatabase/default.asp).

Approval of maize events has followed an increasing tendency (Fig. 2A), with a maximum number of approved events for a particular use in 2016 (14 events for human consumption). There was a lag of 4 years between the first authorization to import maize for food production and its approval for commercial cultivation due to the requirement of agronomic and biosafety tests in the potential growing areas. For Yieldgard™ maize, the import of seeds for these tests was also approved in 2003, but only until 2007 it was commercially grown (ICA, 2003, 2007). Since 2011, the number of approved maize events has been larger each year for human consumption than for feed use or cultivation (Fig. 2A).

Only two soybean GM events can be cultivated in Colombia: Roundup Ready™ soybean and MON87708 × MON89788. Despite this, these seeds

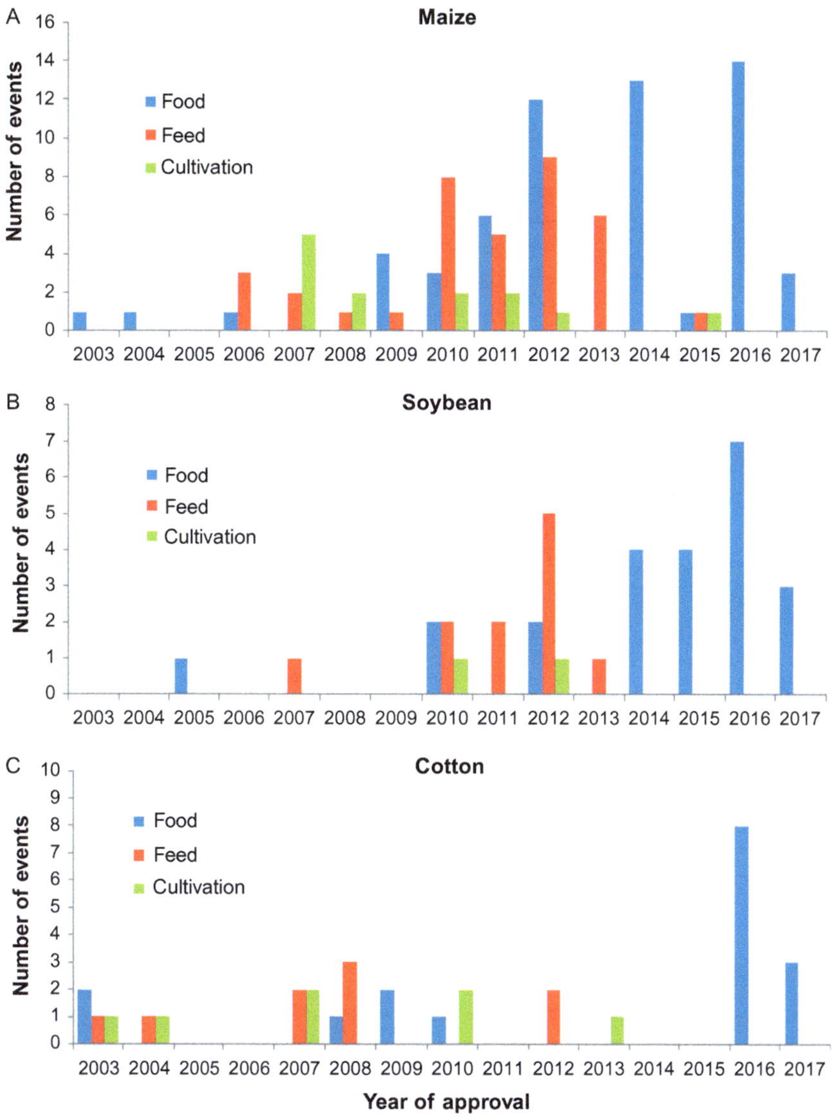

Fig. 2 Number of events approved for the first time each year for commercial use as food, feed or cultivation of maize (A), soybean (B) and cotton (C) in Colombia (Updated June 2017). *Own elaboration with data from ICA, INVIMA, bch.cbd.int, cera-gmc.org/GMCropDatabase, isaaa.org/gmapprovaldatabase/default.asp.*

have not yet been commercialized in the country for planting. Most GM soybean has been approved for use as feed and food since 2010, and in recent years there has been a relatively high number of events approved for human consumption (18 since 2014) (Fig. 2B).

Even though cotton was the first of the major GM crops introduced in Colombia, it only represents 15.2% of the total number of events available. Most of these events received their authorization between 2007 and 2010 (Fig. 2C), with only one event approved for cultivation since then, in 2013.

3.2 Introduced Traits

Out of the 125 total approved events from the 10 crops, 105 (84%) have at least one gene that confers herbicide tolerance (Table 2). Of these, 69 have another trait in addition to herbicide tolerance (not considering tolerance to more than one type of herbicide) such as insect resistance, modified contents or abiotic stress tolerance. Half (63) of the GM events approved in Colombia carry at least one gene which confers resistance to either coleopteran or lepidopteran insects, and 49 of these are present in maize, mostly in combination with tolerance to at least one type of herbicide.

Four maize events which have been approved for either human or animal consumption (Genuity® Droughtguard™ and three hybrids derived

Table 2 Number of Approved Events for Commercial Use in Colombia, by Crop and Introduced Characteristics (Updated June 2017)

	Introduced Characteristics								
Crop	**HT**	**IR**	**MC**	**AS**	**HT+IR**	**HT+AS**	**HT+MC**	**HT+IR+AS**	**HT+IR+MC**
Maize	8	6	2	1	39	1		2	2
Soybean	14	2	2		1		4		
Cotton	8	5			6				
Carnation							14		
Rose			2						
Rice	2								
Canola	1								
Flax	1								
Sugar beet	1								
Wheat	1								
Total	36	13	6	1	46	1	18	2	2

AS, Abiotic stress tolerance; HT, Herbicide tolerance; IR, Insect resistance; MC, Modified content.
Sources: ICA, INVIMA, bch.cbd.int, http://cera-gmc.org/GMCropDatabase, isaaa.org/gmapprovaldatabase/default.asp.

from it) are tolerant to drought stress. Two maize events with traits for nutritional improvement (higher fatty acid content, higher amino acid content), and two events with this trait in combination with insect resistance and tolerance to herbicides are approved. They have not, however, been approved for cultivation within the country.

Other than carnations and roses with modified colour, all GM crops approved for commercial cultivation are either herbicide tolerant or insect resistant or both. No other traits are present in the 13 maize, 2 soybean and 7 cotton events which can be grown.

3.3 Cultivation of GM Crops in Colombia

The worldwide trend in GM crops has been of increased adoption since their first commercial cultivation in 1996. With the exception of the period 2014–15, the total area dedicated to biotechnological crops has increased every year, up to a total of 185.1 MHa in 26 countries in 2016 (ISAAA, 2016). Of this area, 79% is planted in just three countries: the United States, Brazil and Argentina.

Colombia is considered one of the 18 biotech megacountries with 110,000 Ha of GM crops planted in 2016. It is ranked sixth in South America according to the area planted with GM crops, behind Brazil, Argentina, Paraguay, Uruguay and Bolivia, all of which plant soybeans as their main GM crop (ISAAA, 2016). Two soybean events have been approved for cultivation in Colombia but they are not currently grown even though Colombia has more than 30,000 Ha of this crop (FAOSTAT, 2017) and an important potential for growth (MinAgricultura, 2015) where GM soybean could be adopted.

Most of the area dedicated to biotechnological crops in Colombia corresponds to GM maize, with 100,000 Ha distributed over 23 of the 32 departments in the country in 2016 (ISAAA, 2016; agrobio.org/transgenicos-en-el-mundo-colombia-region-andina/). This is roughly 22% of the total area planted with maize. In the same year GM cotton was planted in 9800 Ha over six departments, but with 93% in just two of them: Tolima and Córdoba (agrobio.org/transgenicos-en-el-mundo-colombia-region-andina/). The adoption rate for cotton is higher than for maize, as the total GM area represents approximately 56% of the total area for this crop in Colombia.

When first introduced, 2000 Ha of GM cotton were planted. Until 2009, a gradual increase in the total area of GM crops was registered, initially for cotton and since 2007 for cotton and maize combined (Fig. 3). Between 2009 and 2011 the amount of GM crops in Colombia triplicated, mainly due to an increased adoption of GM maize. Since then, the area planted with GM cotton has shrunk back to levels comparable to 2003, while GM maize has continued to grow, but at a slower pace. The decrease in the area planted with GM cotton since 2012 responds to a general trend in the reduction of this crop in Colombia (Chaparro-Giraldo, 2015). The total area planted with cotton (conventional and GM) in 2004 was 74,106 Ha, and it fell to 44,500 Ha in 2012 (a reduction of 40%) (CONALGODON, 2012). According to farmers this decrease was due to high business and labour informality, high production costs, obsolete machinery, growth of both legal and illegal imports and changes in foreign markets (CONALGODON, 2012).

Blue carnations and roses have been grown in Colombia since 2002. The total area represented by these flowers is relatively small and has remained like this since they are only grown in containment in a few greenhouses in the central region of the country (Cundinamarca department) (*Source*: agrobio.org/transgenicos-en-el-mundo-colombia-region-andina/). The area planted responds to future demand and all the flowers produced are exported, mostly to Japan, Australia, New Zealand, Canada, the United States, Russia, the Netherlands and Spain (ICA, unpublished data). In 2016, 12 Ha of GM flowers were planted in Colombia (ISAAA, 2016).

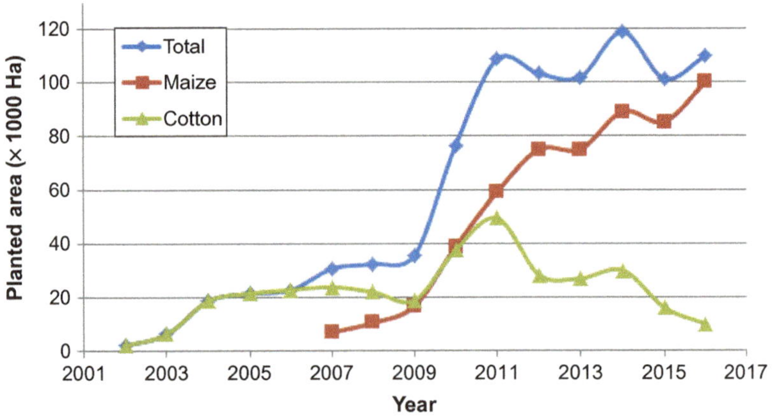

Fig. 3 Area planted with GM cotton and maize in Colombia. The total area planted with GM flowers has remained at 12 Ha since their introduction. This is not included in the figure because of the scale (Updated June 2017). *Own elaboration with data from ISAAA. (2016). Global status of commercialized biotech/gm crops: 2016.* ISAAA Brief No. 52. *Ithaca, NY: ISAAA and agrobio.org/transgenicos-en-el-mundo-colombia-region-andina/.*

It is worth noting that all the GM events commercialized in Colombia have been developed by agricultural biotechnology multinational companies, such as Monsanto, Syngenta, Bayer Crop Science, DuPont and Dow AgroSciences. Despite the efforts done in research in agricultural biotechnology in Colombia (see later), there are no commercially available GM crops developed by national seed companies or public institutions, and no GM varieties have been developed and commercialized specifically for local agroenvironmental conditions.

4. GENETIC ENGINEERING OF CROPS IN COLOMBIA

The first reports of research on genetic engineering of crops in Colombia were presented at the IV Congress of the Colombian Society for Plant Breeding and Crop Production, celebrated in 1995. Since then, there are several publications from the Biotechnology Institute of the Universidad Nacional de Colombia (UNC) (Chaparro-Giraldo, Acosta, Caro, & Peñaranda, 1995), CIAT (Lentini et al., 2003; Sarria et al., 1999; Sarria, Torres, Ángel, Chavarriaga, & Roca, 2000) and the Corporación Colombiana para Investigaciones Agrícolas (CORPOICA) (Reichel, Mariño, Kummert, Belalcázar, & Narváez, 1996). After 20 years, there is a limited use of GM technology in Colombia because of difficulties in funding and specific training, public policies and weakness of farmer associations. As an example, in 2016 the budget for basic and applied research of the National Department of Science, Technology and Innovation (COLCIENCIAS) was only 1.5 million US dollars, insufficient to meet the funding requirements of over 362 labs around the country.

Particular efforts have been made by groups working on GM crop development belonging to public and private organizations such as: CIAT, UNC, Sugarcane Research Center (Cenicaña), National Center for Coffee Research (Cenicafe), Corporation for Biological Research (CIB), CORPOICA, Universidad de Antioquia and Pontificia Universidad Javeriana.

4.1 Intellectual Property for GM Crops

GM crop development needs bioinformatics, tissue culture and molecular assays, and the commercial release requires interaction of experimental knowledge with legal issues such as IP and regulatory framework. Bioinformatic design of expression cassettes allows DNA sequence manipulation and their combination with other essential elements (promoters, terminators, restriction sites, tags, selection markers, etc.) for the synthesis of constructs and their introduction in transformation vectors (Chaparro-Giraldo & Flores, 2010). Two systems have been used for gene insertion into plant genomes: *Agrobacterium tumefaciens* (Otten, 2018; Zambryski, Tempe, & Schell, 1989) and biolistics (Sanford, 1988). Transgenic cells are obtained by means of tissue culture techniques including standardization of growth regulators, media, bacterium elimination by antibiotics, and selective agents or environmental conditions (temperature, relative humidity and photoperiod). Finally, molecular biology assays such as Southern Blot, Northern Blot, Western Blot, ELISA and PCR allow to determine transformation efficiency, transgene copies, tissue location (stems, leaves, flowers, etc.), mRNA levels or protein synthesis (Chaparro-Giraldo, 2005). In the IP context, patents contain claims on DNA, protein, promoter and regulation sequences, vectors (cloning, expression, transformation), bacterium strains, regeneration systems, GM organism or their parts, or cell lines (Hincapié & Chaparro-Giraldo, 2014). Five companies, which share licences between them, have an important control on GM plant patents: Monsanto, DuPont, Syngenta, Bayer and Dow AgroSciences (Wright & Pardey, 2006). This licensing limits biotechnology research (Pisano, 2006) and patent management can limit or even impede the delivery of a GM variety (Thomas, 2006).

A possible solution for patent problems (and other IP issues) is a Freedom to Operate (FTO) analysis which involves the study of the technologies and IP rights (breeders' title and commercial secret related to parental hybrids, among others). An FTO analysis requires the deconstruction or reverse engineering of the innovation, in order to establish the biological materials (genes, promoters or other DNA sequences, vectors, bacterial strains, genotypes, type of explants, etc.) and laboratory protocols (tissue culture, molecular techniques, transformation systems) to be used; also, geographical location for product distribution, biosafety regulation and legal registry of GM events (Kryder, Kowalski, & Krattiger, 2000). The first FTO analysis of a GM crop in Colombia was made on a GM potato developed by UNC and CIB. This study concluded that the most important problem was related to a Material Transfer Agreement (MTA) and legal agreements signed by universities, rather than to patents. The MTAs contain sections limiting the usage of biological materials to research in specific potato genotypes, without clear terms about IP (Hincapié & Chaparro-Giraldo, 2014).

The Public Intellectual Property Resource for Agriculture (PIPRA) based on University of California (Davis) is a public non-for-profit organization whose objective is agriculture innovation for humanitarian and commercial purposes. PIPRA is a consortium composed of 60 institutions (research organizations and public universities) from 17 countries. PIPRA has established a basis for the management of public IP (Delmer, Nottenburg, Graff, & Bennett, 2003) and the UNC, CIAT and CORPOICA are part of this consortium.

Laboratory and field studies are necessary for commercial delivery of GM crops, but this process is very expensive. According to McDougall (2011) and leading GM crop developers (BASF, Bayer Crop Science, Dow AgroSciences, DuPont/Pioneer Hi-Bred, Monsanto and Syngenta AG), development costs of a GM crop are 136 million US dollars distributed as follows: 31 million US dollars (23%) for gene identification, 69.9 million US dollars (51%) for product development, and 35.1 million US dollars (26%) for regulation and registration procedures. Administrative processes take at least 13 years. In Latin America these costs are smaller than in the United States, but in Colombia costs related to regulation are increased because independent field studies should be made for each ecological region with agricultural perspective. These are very high costs for public research institutions or farmers organizations.

4.2 GM Crop Research in Colombia

Research activities on GM crops must be authorized by official ICA resolutions in each case. CIAT has carried out research on the genetic transformation of *Stylosanthes guianensis* (forage legume) (Sarria et al., 1999). Several papers have been published on cassava transformation for phosphinothricin tolerance (Sarria et al., 2000), *bar* gene expression (Echeverry, Mancilla, Cortez, Chavarriaga, & Tohme, 2003), transgene analysis by Real-time PCR (Beltrán et al., 2009), provitamin A accumulation (Welsch et al., 2010), expression patterns (Beltrán et al., 2010), tissue culture studies (Ochoa, Chavarriaga, & López, 2012) and GM rice for hoja blanca virus (RHBV) resistance (Lentini et al., 2003). CIAT received authorizations for GM cassava small-scale trials: enhancement of starch quality by the *gbss1* gene from the same species (resolution 003854) (ICA, 2005a), production of GM events expressing the derived *ipt* gene of *A. tumefaciens* (resolution 003855) (ICA, 2005b), cyanide volatilization by events expressing *hnl* gene from cassava (resolution 003856) (ICA, 2005c). CIAT was also authorized to do research activities in confinement with GM varieties of rice and cassava (resolution 000858) (ICA, 2008b). In order to do this, several requirements are stated in the resolution which correspond to those established by Annex I of Law 740 (2002): (i) experimental plots should be located in surveilled areas, under controlled irrigation, with physical barriers of elephant grass, away from transited roads, crop rotation and weed control are to be implemented, and postharvest waste is to be incinerated (article 1); (ii) research activities must be developed in the experimental station and biosafety greenhouses, with equipment for insect control and wastewater treatment, the greenhouse soil must be sterilized, greenhouse access should be restricted, organic wastes must be incinerated and the wastewater treated with chlorinated disinfectants (article 2); (iii) CIAT must submit a report to ICA on a 6-monthly basis (article 3) and (iv) according to a precautionary principle, ICA can destroy all GM rice and cassava material without compensation or owner consent (articles 4 and 5 following the sanctions established by Decree 1840 of 1994). Finally, ICA authorized CIAT to carry out GM rice field trials for drought tolerance in Santa Rosa Experimental Station (Villavicencio-Meta) (resolution 004041) (ICA, 2010c).

Cenicaña received authorization for small-scale research using GM sugarcane for yellow leaf virus (SCYLV) tolerance (resolution 3995) (ICA, 2005d) including virus tolerance, agronomic traits and sucrose yield. Cenicaña must avoid, mitigate, correct and/or compensate for potential threats. If there is a negative environmental effect the GM material can be destroyed. Finally, Cenicaña must report to ICA on a 4-monthly basis. A paper reporting the transient expression of the *gus* gene in sugarcane using *A. tumefaciens* was published (Bonilla, Muñoz, & Ángel, 2008).

Cenicafe was authorized to carry out confined research trials using *Nicotiana benthamiana*, the fungus *Beauveria bassiana* and coffee (*Coffea arabica*), by resolution 2492 (ICA, 2010a). Cenicafe provided the information required in Annex I of Law 740 of 2002 establishing obligations related to lab management, planting areas, organic waste and sewage handling, and reports on a 6-monthly basis to ICA, and a declaration of GM material destruction if there are biosafety issues. Various GM-related projects were published by Cenicafe, including production of seedlings from protoplasts (Acuña & de Peña, 1991), optimization of coffee cell suspensions (Rosillo, Acuña, Gaitán, & de Peña, 2003) and transformation of *B. bassiana* using *Metarhizium anisopliae* genes (Gongora, 2004).

CIB was authorized to develop field trials in Rionegro (Antioquia) with Andean potato events expressing a *cry1Ac* gene, by resolution 004040 (ICA, 2010b). CIB must report on a 6-monthly basis following legal decree 1840 of 1994. Transformation for two potato cultivars was reported (Trujillo et al., 2001) and three Andean GM potato varieties with Guatemalan potato moth (*Tecia solanivora*) resistance expressing a *cry1Ac* gene were developed

(Valderrama et al., 2007). Expression studies were done on GM potato (var. Diacol Capiro) transformed with a *cry1Ac* gene (Vanegas-Araujo, Blanco-Martínez, & Chaparro-Giraldo, 2010).

The Ministry of Environment and Sustainable Development, IAvH and the Biodiversity Information System of Colombia organized a congress in 2008 about the state of research on GMOs in the country. CORPOICA presented studies on pea (var. Santa Isabel) using the reporter gene *gus* (Peñaranda, Ligarreto, & Núñez, 2013) and breeding of Colombian cotton varieties with GM varieties. University de Antioquia presented transformation assays of *Stevia rebaudiana* using gene *gus*. Pontificia Universidad Javeriana had studies on transformation of *Passiflora mollissima* using *A. tumefaciens* (Cancino et al., 2004).

The Plant Genetic Engineering Group (IGP) from UNC was registered as a research unit in plant breeding for rice and potato by resolution 003523 (ICA, 2008a). IGP has worked on tobacco, maize, soybean and cassava crops too. IGP has had support from the National Federation of Rice farmers (FEDEARROZ) and the National Federation of Cereal and Legumes farmers (FENALCE). IGP established a collaborative agreement with the Network of Agricultural Research Center (CENIRED) and the PIPRA consortium in 2012. The IGP group has published papers on different aspects of GM crop development including those establishing minimum inhibitory concentrations of kanamycin (Díaz-Granados & Chaparro-Giraldo, 2011) and hygromycin (Barbosa & Chaparro-Giraldo, 2011) in Colombian maize embryos, design of *cry* semisynthetic genes for rice (Díaz-Granados, Sandoval, & Chaparro-Giraldo, 2013) and potato (Erira-Mayorga, 2017), environmental effects of GM corn and cotton in Colombia (Ávila, Chaparro-Giraldo, & Reyes, 2011; Ávila-Méndez, Chaparro-Giraldo, Reyes-Moreno, & Silva-Castro, 2011; Reyes, Chaparro-Giraldo, & Ávila, 2010), gene flow from GM cotton (Rache-Cardenal, Mora-Oberlaender, & Chaparro- Giraldo, 2013) and maize (Blanco-Martínez, 2012; Chaparro-Giraldo, Blanco, & López-Pazos, 2015), and IP issues associated to GM soybean (Rojas, Palacio, Chaparro-Giraldo, & López-Pazos, 2017).

5. PERSPECTIVES IN GM CROP RESEARCH IN COLOMBIA

The development of commercial GM crops in Colombia by local institutions is a complex issue. This is due to problems typical of developing countries such as difficulties in the public financing of research and development activities and a relative weakness of the private agricultural sector. However, this objective may be achieved thanks to meaningful changes in the global status of IP rights as applied to GM crops brought by the expiry of patents conceded 20 or more years ago, recent developments in services for scientific research in life sciences, and the emergence of new concepts applied to bioeconomy.

Research in commercial GM crops may be blocked by a tangle of patents that cover most basic genetic engineering technologies (Kowalski, Ebora, Kryder, & Potter, 2002). These may include products based on genes or partial DNA sequences, regulatory elements, proteins and their functions, vectors used for gene transfer, GMOs and cellular lines derived from them, or processes for the development of GMOs (Lakshimikuraman & Phillips, 2005). Some reports document the negative effects of this situation, such as the early-stage problems in the commercial release of Golden Rice due to 40 patents involved in its development (Kowalski et al., 2002); although the issues related to Golden Rice patents were solved, other regulatory aspects have hindered the possibility of this product being available to growers and communities (Dubock, 2014). Due to the amount of patents restricting the use of basic technologies, the Strawberry Commission of the United States decided not to move forward with a project aimed at the development of a GM strawberry variety with resistance to a fungal disease (Thomas, 2006). Michigan State University had to destroy a line of GM foraging grass because of an unresolved dispute between two companies, one with ownership of the gene patent and another of the promoter sequence patent (Thomas, 2006). Studies carried out in Colombia show a similar tendency: at least 59 patents were found related to the development of a GM rice cultivar expressing gene *cry1Ac* (Diazgranados, Hincapié, & Chaparro-Giraldo, 2016); for a GM potato cultivar with the same introduced gene, 25 related patents were identified (Hincapié & Chaparro-Giraldo, 2014).

In this context, the development of GM crops in a country such as Colombia is not a viable process. However, a recent publication in Nature Biotechnology (Gruskhin, 2012) introduced the possible arrival of generic agricultural biotechnology and demonstrated that the panorama is subject to change. As an example, in 2014, the protection period of 29 patents from Monsanto was due to expire, including those protecting glyphosate-tolerance technology. This was later confirmed by another paper in the same journal (Jefferson, Graff, Chi-Ham, & Bennett, 2015) where the authors point to the emergence of agbiogenerics (generic GM crops) and reporting the commercial release in December 2014 of a generic glyphosate-tolerant soybean variety developed by the University of Arkansas, UA 5414RR. A key issue, according to the authors, is the way in which existing regulations apply to generic versions of a crop that incorporate previously patented components. They also point out that the dominant biotech companies, controlling most patents in this field, could still monopolize the market by means of their control of the data used in the regulatory process for commercial release.

As a result of the research experience in IGP, a strategy has been established for the development of agbiogenerics in Colombia, which can be transposed to countries of similar socioeconomic development: (1) FTO analysis; (2) design of expression cassettes and (3) generation of GM lines and regulatory process for commercial release.

5.1 Freedom to Operate Analysis

Professional attorney service fees are very high and national agencies do not cover them as part of R&D grants. A useful alternative to hiring an attorney in the decision-making process is to carry out FTO analyses by members of biotechnology research groups, bearing in mind that they have a limited legal validity (Zanga, Capell, Zhu, Christou, & Thangara, 2016). It is better to select mature technologies, which have been commercialized in various countries and that have gone through their regulatory processes and barriers. This should significantly decrease the costs associated with the regulatory dossier.

In order to follow this approach, the first step is to select the genotypes of interest within the crop and the desired characteristic to be introduced. For example, potato (*Solanum tuberosum* var. Pastusa Suprema) expressing resistance to the lepidopteran *T. solanivora* by the introduction of the *cry1Ac* gene. Next, an exhaustive list of biological materials and laboratory protocols to be used in the development of the GM crop is made. Biological materials include expression cassettes, vectors, bacterial strains and plant material (Diazgranados et al., 2016). In the expression cassettes, both the gene of interest and the selectable marker gene are considered, and, within each of these, their components: promoter region, coding sequence and terminator region. For plasmid vectors it is important to consider if they are to be used for expression assays, gene cloning or transformation. It is recommendable to minimize the use of these vectors in order to reduce the IP burden. In order to do this, expression cassettes can be designed in silico and their synthesis can be hired from a third party. When using *A. tumefaciens*, the specific strain must be considered too. Finally, the particular genotype or variety of the plant species to be transformed should also be taken into account. Protocols to be considered include: plant tissue culture, bacterial and plant genetic transformation, molecular characterization of transgenic bacteria and plants (Diazgranados et al., 2016). Growth, regeneration, selection and in vitro micropropagation media and conditions, as well as the plant (*Agrobacterium*, biolistic, etc.) and bacterial (heat shock, electroporation, etc.) transformation systems, and the molecular techniques for the characterization of the GMOs (PCR, Southern blot, Northern blot, Western blot, ELISA, etc.) should be known in advance.

The countries or regions where the GM crop is to be commercialized are another aspect to be determined. This is paramount for the FTO analysis. The process is relatively simple if it comprises one country, but this depends on the market for the crop. In some instances, regional blocks with common regulations can be considered, such as the CAN in South America (Colombia, Ecuador, Peru and Bolivia).

Once the list of materials and protocols, and the geographic range of commercial interest are established, an IP rights analysis is made for each item. The most common IP protection of GM crops comes from patents, which, depending on the jurisdiction, may cover nucleotide sequences and/or amino acid sequences up to whole organisms, including technical protocols. Breeders' rights as established by UPOV protect crop varieties in most countries and should also be considered. For patents, both global and local databases should be searched, some of which are open access (Jiménez-Barreto, Chaparro-Giraldo, Mora-Oberlaender, & Vargas-Sánchez, 2016). Recommended databases for global searches are: The Lens (https://www.lens.org/lens/), which specializes in plant genetic engineering; Patentscope (http://www.wipo.int/patentscope/en/) from the World Intellectual Property Organization; Espacenet (https://worldwide.espacenet.com/) from the European Patent Office; and the US Patent and Trademark Office (https://www.uspto.gov/patents-application-process/search-patents). Another interesting source of information and a good site to begin is Google's patent database (https://patents.google.com/). Queries can be made using key terms within the claims of the patents. For example, the name of the gene, the promoter region, the protein, the plasmid or the bacterial strain. It is also advisable to use the International Patent Classification (IPC) to get a context of the patents related to the GM crop in particular, including their time and geographic scope.

Local database searches are determined by the decision of where the GM crop is to be commercialized. In Colombia, the public database is managed by the Superintendence of Trade and Commerce (Superintendencia de Industria y Comercio—SIC) (http://sipi.sic.gov.co). Queries should be aimed at global results, by using search terms in the inventor, owner and title fields. Sometimes proprietary technology can be used in the development of generic versions of GM crops, given that patents are limited to a single country and for a certain amount of time, they may pose different exceptions or protect different objects according to local regulations, or they may have expired (Zanga et al., 2016). At this point in the process it is useful to consider strategies leading to a maximum FTO, such as redesign of expression cassettes, negotiation of licences for humanitarian purposes, negotiation of licences individually or as a consortium, or a combination of the above (Kryder et al., 2000).

A second type of IP right to be taken into consideration is the UPOV Plant Variety Protection. It is important to establish if the variety of the crop to be used for transformation is protected and the geographic and temporal limits of this protection. Furthermore, which UPOV Act it is based on, 1978 or 1991, as they have differences such as those of double protection (patents and breeders' title), farmers' privilege and breeders' exemption. Each country has a NCA in this matter, and a document which informs of the conceded breeders' rights. In Colombia the NCA is ICA and the decisions are published in the Gaceta de Variedades Vegetales.

5.2 Expression Cassette Design

FTO analysis is a criterion for the selection of the biological material and protocols to be used. Once the characteristics of the GM crop are defined, gene or protein sequences are captured from scientific literature and patents, based on the pertinence of the published studies, including positive results in field or laboratory trials (Carreño-Venegas & Chaparro-Giraldo, 2016). A useful source of information is Gene Bank, from the National Center for Biotechnology Information of the US (NCBI) (http://www.ncbi.nlm.nih.gov/gene). Each case is specific. As an example, for drought tolerance in model and crop plants, 105 scientific papers published between 1995 and 2013 were retrieved from the databases Academic Search Complete, Science Direct, Springer Journal, Journal Storage and Nature.com (Carreño-Venegas & Chaparro-Giraldo, 2016). Factors taken into consideration included the role of the gene in conferring drought tolerance and physiological parameters with which tolerance was evaluated: germinative capacity under drought conditions or simulated drought (in soil or in vitro), cell damage, photosynthetic activity, osmotic potential, phenology and yield. Drought-tolerance genes selected for the design of expression cassettes were *TaDREB3* AP2 (Lopato et al., 2006), *AtZAT10* (Meissner & Michael, 1997) and *CspB* (Willimsky, Bang, Fischer, & Marahiel, 1992), which come from wheat (*Triticum aestivum*), *Arabidopsis thaliana*, and *Bacillus subtilis*, respectively. Selection criteria included (i) strong empirical evidence of drought tolerance as evaluated by physiological parameters evaluated in vitro, in greenhouse or in field assays; (ii) yield in field trials for crops; (iii) absence of undesired pleiotropic effects or gene products acting on the early stages of stress signalling pathways and (iv) most recent studies.

Once the sequences to be used in the expression cassette are determined, open-access software can be used for in silico design. As an example, Gene Designer 2.0 allows free use of some of its functions (Villalobos, Ness, Gustafsson, Minshull, & Govindarajan, 2006). Design may include (Carreño-Venegas & Chaparro-Giraldo, 2016): (i) generation of the open reading frame (ORF), checked by using the "ORF finder" tool from NCBI; (ii) comparing the amino acid sequence generated by the designed ORF with that of the target protein; (iii) codon-use optimization of the coding sequence for the crop plant using the "DNA/Guide Random" option (Puigbo, Guzman, Romeu, & Garcia-Vallvé, 2007); (iv) including restriction enzyme recognition sites within the expression cassettes in order to be able to cut particular sequences or make restriction maps; (v) confirming the ORF after the restriction sites are included and (vi) confirming the identity using Blastx (BLAST®). It is also advisable to remove cryptic splicing sites and early termination sites, which can be done using Visual Gene Developer 1.3 (Jung & McDonald, 2011). As a result of this process, an in silico design of the expression cassette is generated, which can be used to hire its synthesis by a European, American or Chinese company. This may comprise only the expression cassette or its inclusion in a selected plasmid vector. It is recommended to request that the service contract does not include a "research only" clause.

In order to have an initial confirmation of the functionality of the expression cassettes, a model plant for which transformation protocols are available at the research lab can be transformed (Jiménez-Barreto & Chaparro-Giraldo, 2016). Many laboratories use *A. thaliana* or *Nicotiana tabacum*. After obtaining several independent transgenic lines, general transgene detection tests can be performed using PCR, and then an expression assay such as ELISA if possible, or RT-PCR as an alternative. The idea is to produce evidence of the correct function of the expression cassette.

5.3 Generation of GM Lines and Regulatory Process for Commercial Release

Establishing tissue-culture protocols is a key aspect of the research process because they are genotype-specific (Sairam et al., 2003). A regeneration system should be established. This is a particular combination and concentration of plant hormones and growth conditions that allow regenerating a whole plant from a single cell. Given this, the genotype to be used is first determined by its response to tissue culture and second by its response to the genetic transformation process. Based on these results, selection conditions are established according to the selection agent to be used (e.g. antibiotic, herbicide) in order to minimize false-positives. Then, transformation assays are carried out according to the method chosen (e.g. *Agrobacterium*, biolistic). If *A. tumefaciens* is used, factors to be considered include the concentration of the inoculum, the type of explant, coculture time and conditions, among others. Phenotypic expression of the desired characteristic depends on the insertion site in the host genome. Therefore, several independent transgenic lines should be obtained in order to select those with better performance in the field. Primary regenerated transformants obtained in regeneration and selection media should be molecularly characterized in order to establish the number of inserts and the expression level of the recombinant protein. Plant populations are preferably obtained from single-copy events and appropriate phenotypic expression, according to the biological context of interest. Finally, these populations are taken to field conditions where homozygous genotypes can be selected using molecular or mendelian tools. All along the process, information relevant for the regulatory process is generated.

Several issues should be considered in a parallel and simultaneous manner in order to approach the regulatory process. The first concerns the environmental and health safety requirements of national regulations for the commercial release of a GM event (detailed earlier). It is at this stage that the initial precaution of working with mature technologies that have been used in different countries becomes particularly relevant. Well-known technologies with a history of safe use facilitate the process and reduce costs because both scientific and legal documents may be available and can be cited to resolve key aspects of the regulatory process (e.g. health safety).

A second relevant process is the request of IP protection for the GM variety by UPOV breeders' rights. Here, Distinction, Homogeneity and Stability (DHS) criteria and whether it is an essentially derived variety are issues to be addressed. The same data can be used to solve regulatory aspects concerning the commercial registry of crop varieties, which require trials in different locations for at least two crop cycles. During this period, an FTO analysis can be formally addressed together with an IP attorney to prevent legal issues forthcoming.

Recent studies by the International Potato Center on the commercial release of GM late blight-resistant potato varieties demystify the costs associated with the commercial use of GM crops, estimating them at 1.6 million US dollars and a time frame of 8 years (Schiek et al., 2016). This is in contrast with values cited by biotechnology companies of 136 million US dollars and 13 years for the same purpose (McDougall, 2011).

REFERENCES

Acuña, J. R., & de Peña, M. (1991). Plant regeneration from protoplasts of embryogenic cell suspensions of *Coffea arabica* L. cv. caturra. *Plant Cell Reports, 10,* 345–348.

Artunduaga, S. R., Silva, C., Guerrero, J. L., Vargas, S., Patarroyo, M. F., & Huertas, D. (1997). *Protocolo sobre bioseguridad: énfasis en plantas transgénicas.* Instituto Colombiano Agropecuario (ICA). 67 p.

Ávila, K., Chaparro-Giraldo, A., & Reyes, G. (2011). Environmental effect of conventional and GM crops of cotton and corn. *Agronomía Colombiana, 29,* 341–348.

Ávila-Méndez, A., Chaparro-Giraldo, A., Reyes-Moreno, G., & Silva-Castro, C. (2011). Production cost analysis and use of pesticides in the transgenic and conventional corn crop [*Zea mays* (L.)] in the valley of San Juan (Tolima). *GM Crops, 2,* 163–168.

Barbosa, I. D., & Chaparro-Giraldo, A. (2011). Concentración mínima inhibitoria de higromicina B en callos embriogénicos de arroz (*Oryza sativa* L.). *Revista Colombiana de Biotecnología, 13,* 193–198.

Beltrán, J., Jaimes, H., Echeverry, M., Ladino, Y., López, D., Duque, C., et al. (2009). Quantitative analysis of transgenes in cassava plants using real-time PCR technology. *In Vitro Cellular & Developmental Biology—Plant, 45,* 48–56.

Beltrán, J., Prías, M., Al-Babili, S., Ladino, Y., López, D., Beyer, P., et al. (2010). Expression pattern conferred by a glutamic acid-rich protein gene promoter in field-grown transgenic cassava (*Manihot esculenta* Crantz). *Planta, 231,* 1413–1424.

Blanco-Martínez, J. T. (2012). *Monitoreo del flujo de genes de cultivos transgénicos de maíz a razas locales y variedades comerciales de maíz en el Valle de San Juan, Tolima.* [Master thesis] (pp. 1–72). Bogotá: Universidad Nacional de Colombia.

Bonilla, M. L., Muñoz, J. E., & Ángel, S. F. (2008). Expresión transitoria del GUS en caña de azúcar usando *Agrobacterium tumefaciens*. *Acta Agronómica, 57,* 161–166.

Cancino, G. O., Gill, M. I. S., Anthony, P., Davey, M. R., Lowe, K. C., & Power, J. B. (2004). Naringenin enhanced efficiency of GUS activity in *Passiflora mollissima* (H.B.K.) Bailey. *Vol. 9* (pp. 47–57). Universitas Scientarium.

Carreño-Venegas, A., & Chaparro-Giraldo, A. (2016). Design of expression cassettes that confer drought and glufosinate tolerance to maize (Zea mays). *Acta Bilógica Colombiana, 21*(3), 555–570.

Castaño, A. (2013). Regulación para liberación comercial de cultivos genéticamente modificados con uso en alimentación humana. In A. Chaparro (Ed.), *Propiedad intelectual y regulación en biotecnología vegetal.* Bogotá: Universidad Nacional de Colombia.

Castaño, A. (2015). Alimentos derivados de los cultivos genéticamente modificados -¿Nuevos, Seguros para la salud humana, se consumen? *Revista Colombiana de Pediatría, 48*, 68–74.

Céleres. (2015). *Los beneficios agronómicos, económicos y socio ambientales de la biotecnología agrícola en Colombia: El caso del algodón, del maíz y de la soya genéticamente modificados.* Uberlândia: Céleres®; Agro-Bio.

Chaparro-Giraldo, A. (2005). Elementos básicos para entender la tecnología transgénica. In A. Chaparro-Giraldo (Ed.), *Introducción a la ingeniería genética de plantas* (pp. 11–22). Bogotá: Unibiblos.

Chaparro-Giraldo, A. (2015). La ingeniería genética de plantas en Colombia: Un camino en construcción. *Acta Biológica Colombiana, 20*(2), 13–22.

Chaparro-Giraldo, A., & Flores, E. (2010). FEDEARROZ y la Universidad Nacional construyen las bases de la investigación en líneas transgénicas de arroz. *Arroz, 58*, 36–39.

Chaparro-Giraldo, A., Acosta, O., Caro, M., & Peñaranda, J. (1995). Establecimiento de una metodología para el mejoramiento no convencional de tabaco (*Nicotiana tabacum*) mediante transformación con Agrobacterium tumefaciens. In *Memorias del IV Congreso Nacional de la Sociedad Colombiana de Fitomejoramiento y Producción de Cultivos* (p. 134). Chinchina: Cenicafé.

Chaparro-Giraldo, A., Blanco, J., & López-Pazos, S. A. (2015). Evidence of gene flow between transgenic and non-transgenic maize in Colombia. *Agronomía Colombiana, 33*, 294–304.

Colombia. (2002). *Ley 740 de 2002. Por medio de la cual se aprueba el "Protocolo de Cartagena sobre Seguridad de la Biotecnología del Convenio sobre la Diversidad Biológica"* (p. 44816). Diario Oficial.

Colombia. (2005). *Decreto 4525 de 2005: Marco regulatorio de los organismos vivos modificados - OVM- de acuerdo con lo establecido por la Ley 740 de 2002.* Bogotá: Ministerio de Agricultura y Desarrollo Rural.

CONALGODON. (2012). *Avances de la estrategia exportadora para el algodón de Colombia.* Confederación Colombiana del Algodón.

Delmer, D. P., Nottenburg, C., Graff, D. G., & Bennett, A. (2003). Intellectual property resources for international development in agriculture. *Plant Physiology, 133*, 1666–1670.

Díaz-Granados, C., & Chaparro-Giraldo, A. (2011). Concentración mínima inhibitoria de Kanamicina para callos de cuatro variedades colombianas de arroz. *Revista Colombiana de Biotecnología, 13*, 210–220.

Díaz-Granados, C., Sandoval, A. M., & Chaparro-Giraldo, A. (2013). Diseño de un gen semisintético *cry1Ac* y análisis de la estructura de la proteína traducida. *Biotecnología en el Sector Agropecuario y Agroindustrial, 2*, 155–164.

Diazgranados, C., Hincapié, V., & Chaparro-Giraldo, A. (2016). Aproach study of freedom to operate for transgenic line rice in Colombia. *Revista Colombiana de Biotecnología, 18*(1), 165–172.

Dubock, A. (2014). The politics of golden rice. *GM Crops & Food, 5*(3), 210–222.

Echeverry, M., Mancilla, L. I., Cortez, D. F., Chavarriaga, P., & Tohme, J. (2003). Evaluación preliminar de la expresión del gen *bar* en plantas transgénicas de yuca (*Manihot esculenta*) mantenidas en reproducción vegetativa por cerca de diez años. *Revista de la Asociación Colombiana de Ciencias Biológicas, 15*(1), 43–52.

Erira-Mayorga, N. (2017). *Desarrollo de casetes de expresión para variedades de papa (*Solanum tuberosum *L.), basados en el gen cry1Ba1 de* Bacillus thuringiensis [Master thesis] (pp. 1–93). Bogotá: Universidad Nacional de Colombia.

FAO (Food and Agricultural Organization of the United Nations). (2007). *Instrumentos de la FAO sobre la Bioseguridad.* Roma: Organización de las Naciones Unidas para la Agricultura y la Alimentación. Rome: FAO.

FAO/WHO. (1996). *Joint FAO/WHO expert consultation on biotechnology and food safety.* Rome www.fao.org/ag/agn/food/pdf/biotechnology.pdf; Last accessed December 17th 2017.

FAOSTAT. (2017). *Crop production statistics.* Food and Agriculture Organization of the United Nations. Retrieved from http://www.fao.org/faostat/en/#data.

FENALCE. (2016). *Indicadores cerealistas, 2016A.* Retrieved from http://www.fenalce.org/archivos/IndicadorCerealista2016A.pdf; Last accessed December 17th 2017.

Gongora, C. E. (2004). Transformación de *Beauveria bassiana* cepa (Bb9112) con los genes de la proteína verde fluorescente y la proteasa pr1A de *Metarhizium anisopliae. Revista Colombiana de Biotecnología, 30,* 15–21.

Gruskhin, D. (2012). Agbiotech 2.0. *Nature Biotechnology, 30*(3), 211–214.

Hincapié, V. P., & Chaparro-Giraldo, A. (2014). Estudio de la libertad de operación para una línea genéticamente modificada de papa (*Solanum tuberosum* L.). *Revista Colombiana de Biotecnología, 16,* 119–128.

Hodson de Jaramillo, E., & Castaño, A. (2013). *Proceso de adopción de OGM en Colombia, sistema nacional de bioseguridad.* Toronto: The International Union of Food Science and Technology. *Retrieved from* http://worldfoodscience.com/article/proceso-de-adopci%C3%B3n-de-ogm-en-colombia-sistema-nacional-de-bioseguridad-abstract-english#sthash.SaWwxAQy.dpuf.

Hodson de Jaramillo, E., Castaño, A., & Uscátegui, M. A. (2012). Biotecnología Agrícola Moderna, Organismos Genéticamente Modificados y Bioseguridad. Módulo de Aprendizaje Auto dirigido - Plan de formación de la Rama Judicial. *Consejo Superior de la Judicatura, Sala Administrativa.* Bogotá: Escuela Judicial "Rodrigo Lara Bonilla".

ICA. (1998a). *Acuerdo 0013 del 22 de diciembre de 1998 "Por el cual se crea el Consejo Técnico Nacional (CTN) para introducción, producción, liberación y comercialización de Organismos Modificados Genéticamente (OMG) de uso agrícola"* Bogotá: Instituto Colombiano Agropecuario.

ICA. (1998b). *Resolución 3492 del 22 de diciembre de 1998 "Por la cual se reglamenta y se establece el procedimiento para la introducción, producción, liberación y comercialización de Organismos Modificados Genéticamente (OMG) y se dictan otras disposiciones"* Bogotá: Instituto Colombiano Agropecuario.

ICA. (2000). *Resolución 1219 del 18 de mayo de 2000 "Por la cual se autoriza la introducción de plantas de clavel modificado genéticamente"* Bogotá: Instituto Colombiano Agropecuario.

ICA. (2001). *Resolución 2935 del 23 de octubre de 2001 "Por la cual se se reglamenta y establece el procedimiento de bioseguridad para la introducción, producción, liberación, comercialización, investigación, desarrollo biológico y control de calidad de Organismos Modificados Genéticamente, OMG, de interés en salud y producción pecuaria, sus derivados y productos que los contengan"* Bogotá: Instituto Colombiano Agropecuario.

ICA. (2002). *Acuerdo 004 del 5 de Abril de 2002 "Por el cual se crea el Consejo Técnico Nacional de Bioseguridad Pecuario (CTN Pecuario)"* Bogotá: Instituto Colombiano Agropecuario.

ICA. (2003). *Resolución 1697 del 7 de julio de 2007 "Por la cual se autorizan la importación de semilla de maíz blanco y amarillo para la realización de proyectos de bioseguridad para determinar los posibles riesgos de la tecnología YieldGard (B.t.) en maíz en Colombia"* Bogotá: Instituto Colombiano Agropecuario.

ICA. (2005a). *Resolución 003854 del 16 de diciembre de 2005 "Por el cual se autoriza al Centro Internacional de Agricultura Tropical, CIAT, a realizar actividades de investigación con plantas de yuca genéticamente modificada en pequeña escala en campo"* Bogotá: Instituto Colombiano Agropecuario.

ICA. (2005b). *Resolución 003855 del 16 de diciembre de 2005 "Por el cual se autoriza al Centro Internacional de Agricultura Tropical, CIAT, a realizar actividades de investigación con plantas de yuca genéticamente modificada en pequeña escala en campo"* Bogotá: Instituto Colombiano Agropecuario.

ICA. (2005c). *Resolución 003856 del 16 de diciembre de 2005 "Por el cual se autoriza al Centro Internacional de Agricultura Tropical, CIAT, a realizar actividades de investigación con plantas de yuca genéticamente modificada en pequeña escala en campo"* Bogotá: Instituto Colombiano Agropecuario.

ICA. (2005d). *Resolución 3995 del 23 de diciembre de 2005 "Por la cual se autoriza al Centro de Investigación de la Caña de Azúcar de Colombia, Cenicaña, a continuar con actividades de investigación en pequeña escala en campo con plantas de caña de azúcar modificadas genéticamente"* Bogotá: Instituto Colombiano Agropecuario.

ICA. (2006). *Resolución 000946 del 17 de abril de 2006 "Por la cual se establece el procedimiento ante el ICA de solicitudes de Organismos Vivos Modificados, OVM; se aprueba el Reglamento Interno del Comité Técnico Nacional de Bioseguridad, CTNBio para OVN con fines exclusivamente agrícolas, pecuarios, pesqueros, plantaciones forestales comerciales y agroindustriales y se dictan otras disposiciones"* Bogotá: Instituto Colombiano Agropecuario.

ICA. (2007). *Resolución 465 del 26 de febrero de 2007 "Por la cual se autorizan siembras de maíz con la tecnología Yieldgard® (MON 810)"* Bogotá: Instituto Colombiano Agropecuario.

ICA. (2008a). *Resolución 003523 del 14 de octubre de 2008 "Por la cual se ordenó el registro como Unidad de Investigación en Fitomejoramiento de materiales convencionales y modificados genéticamente de las especies arroz y papa de la UNC con sede en el Grupo de Ingeniería Genética de Plantas (IGP)"* Bogotá: Instituto Colombiano Agropecuario.

ICA. (2008b). *Resolución 00858 del 18 de marzo de 2008 "Por el cual se autoriza al Centro Internacional de Agricultura Tropical, CIAT, a realizar actividades de investigación en medio confinado con Organismos Genéticamente Modificados de Arroz y Yuca"* Bogotá: Instituto Colombiano Agropecuario.

ICA. (2010a). *Resolución 002492 del 26 de julio de 2010 "Por la cual se autoriza a la Federación Nacional de Cafeteros de Colombia—Centro Nacional de Investigación del Café (Cenicafe), a realizar actividades de investigación en medio confinado con Organismos Genéticamente Modificados de Tabaco Enano* "Nicotiana benthamiana", *Hongo* "Beauveria bassiana" *y Plantas de Café* "Coffea arabica"". Bogotá: Instituto Colombiano Agropecuario.

ICA. (2010b). *Resolución 004040 del 6 de diciembre de 2010 "Por la cual se autoriza a la Corporación para Investigaciones Biológicas, CIB, a realizar ensayos de campo bajo condiciones controladas de papa-Bt OVM (resistencia a* Tecia solanivora*), en la estación experimental "La Selva" de Corpoica—Rionegro, Antioquia".* Bogotá: Instituto Colombiano Agropecuario.

ICA. (2010c). *Resolución 004041 del 6 de diciembre de 2010 "Por la cual se autoriza al Centro Internacional de Agricultura Tropical, CIAT, a realizar ensayos de campo bajo condiciones controladas, con plantas de arroz genéticamente modificado para tolerancia a sequía, en la estación experimental de Santa Rosa Ubicada en Villavicencio-Meta"* Bogotá: Instituto Colombiano Agropecuario.

ICA. (2010d). *Resolución 2894 del 6 de septiembre de 2010 "Por medio de la cual se implementa el Plan de manejo, bioseguridad y seguimiento para siembras controladas de maíz genéticamente modificado"* Bogotá.

ISAAA. (2016). *Global status of commercialized biotech/GM crops: 2016.* ISAAA Brief No. 52 Ithaca, NY: ISAAA.

Jefferson, D. J., Graff, D. G., Chi-Ham, C. L., & Bennett, A. B. (2015). The emergence of agbiogenerics. *Nature Biotechnology*, *33*(8), 819–823.

Jiménez-Barreto, J. P., & Chaparro-Giraldo, A. (2016). In silico design and functional assessment of semisynthetic genes that confer tolerance to phosphinothricin. *Revista Colombiana de Biotecnología, 18*(2), 90–96.

Jiménez-Barreto, J., Chaparro-Giraldo, A., Mora-Oberlaender, J., & Vargas-Sánchez, J. E. (2016). Molecular characterization and freedom to operate analysis of maize hybrids from genetically modified and Colombian varieties. *Agronomía Colombiana, 34*(3), 309–316.

Jung, S. K., & McDonald, K. (2011). Visual gene developer: A fully programmable bioinformatics software for synthetic gene optimization. *BMC Bioinformatics, 12*(1), 1–13.

Kowalski, S., Ebora, B., Kryder, D., & Potter, R. (2002). Transgenic crops, biotechnology and ownership rights: What scientists need to know? *The Plant Journal, 31*(4), 407–421.

Kryder, R., Kowalski, S., & Krattiger, A. (2000). *The intellectual and technical property components of pro-vitamin a Rice (GoldenRiceTM): A preliminary freedom-to-operate review*. ISAAA Briefs No. 20 (p. 56). Ithaca, NY: ISAAA.

Lakshimikuraman, M., & Phillips, P. (2005). Patenting of biotechnological innovations. *Asian Biotechnology and Development Review, 17*(2), 25–44.

Lentini, Z., Lozano, I., Tabares, E., Fory, L., Domínguez, J., Cuervo, M., et al. (2003). Expression and inheritance of hypersensitive resistance to rice hoja blanca virus mediated by the viral nucleocapsid protein gene in transgenic rice. *Theoretical and Applied Genetics, 106*, 1018–1026.

Lopato, S., Bazanova, N., Morran, S., Milligan, A. S., Shirley, N., & Langridge, P. (2006). Isolation of plant transcription factors using a modified yeast one-hybrid system. *Plant Methods, 2*(1), 3.

McDougall, P. (2011). *The cost and time involved in the discovery, development and authorisation of a new plant biotechnology derived trait*. Midlothian, UK: Crop Life International.

Meissner, R., & Michael, A. J. (1997). Isolation and characterisation of a diverse family of *Arabidopsis* two and three-fingered C2H2 zinc finger protein genes and cDNAs. *Plant Molecular Biology, 33*(4), 615–624.

MinAgricultura. (2015). *Colombia Siembra. Ministerio de Agricultura y Desarrollo Rural*. Retrieved from https://www.minagricultura.gov.co/ColombiaSiembra.

Ochoa, J. C., Chavarriaga, P., & López, C. (2012). Embriogénesis somática y producción de callo embriogénico friable de dos cultivares de yuca (*Manihot esculenta* Crantz). *Revista Colombiana de Biotecnología, 14*, 20–27.

OECD. (1993a). *Safety evaluation of foods derived by modern biotechnology: Concepts and principles*. Paris.

OECD. (1993b). *Safety considerations for biotechnology: Scale-up of crop plants*. http://www.oecd.org/science/biotrack/1958527.pdf.

Otten, L. (2018). How agrobacterium, a natural genetic engineer, became a tool for modern agriculture. *Advances in Botanical Research, 86*, 17–44.

Peñaranda, M. I., Ligarreto, G. A., & Núñez, V. M. (2013). Estudios de transformación genética de arveja voluble cultivar Santa Isabel. *Revista Colombiana de Biotecnología, 15*, 29–37.

Pisano, G. P. (2006). Can science be a business? Lessons from biotech. *Harvard Business Review, 84*, 1–12.

Puigbo, P., Guzman, E., Romeu, A., & Garcia-Vallvé, S. (2007). OPTIMIZER: A web server for optimizing the codon usage of DNA sequences. *Nucleic Acids Research, 35*(Suppl. 2), W126–W131.

Rache-Cardenal, L. Y., Mora-Oberlaender, J., & Chaparro- Giraldo, A. (2013). Study of gene flow from GM cotton varieties in "El Espinal" (Tolima—Colombia). *Acta Biológica Colombiana, 18*, 489–498.

Reichel, H., Mariño, L., Kummert, J., Belalcázar, S., & Narváez, J. (1996). Caracterización molecular del gen de la proteína de la cápside de dos aislamientos del virus del mosaico del pepino (CMV), obtenidos de plátano y banano (*Musa* spp.). *Revista Corpoica, 1*, 1–5.

Reyes, G., Chaparro-Giraldo, A., & Ávila, K. (2010). Efecto ambiental de agroquímicos y maquinaria agrícola en cultivos transgénicos y convencionales de algodón. *Revista Colombiana de Biotecnología, 12*, 151–162.

Rojas, A., Palacio, J., Chaparro-Giraldo, A., & López-Pazos, S. A. (2017). Patents and genetically modified soybean for glyphosate resistance. *World Patent Information, 48*, 47–51.

Rosillo, A. G., Hodson de Jaramillo, E., & Carrizosa-Pardo, M. S. (2007). Analysis of the biosafety regulatory framework for genetically modified agricultural crops in Colombia. Comp In *Disertaciones del Programa de Postgrado en Bioseguridad Binas UNIDO-UdeC. Desarrollo de Capacidades para evaluación y gestión del riesgo y monitoreo de organismos genéticamente modificados (OGM): Tomo II*, Bogotá: Instituto de Investigaciones de Recursos Biológicos Alexander von Humboldt.

Rosillo, A. G., Acuña, J. R., Gaitán, A. L., & de Peña, M. (2003). Optimised DNA delivery into *Coffea arabica* suspension culture cells by particle bombardment. *Plant Cell, Tissue and Organ Culture, 74*, 45–49.

Sairam, G., Franklin, R., Hassel, B., Smith, K., Meeker, N., Kashikar, M., et al. (2003). A study on the effect of genotypes, plant growth regulators and sugars in promoting plant regeneration via organogenesis from soybean cotyledonary nodal callus. *Plant Cell, Tissue and Organ Culture, 75*(1), 79–85.

Sanford, J. C. (1988). The biolistic process—A new concept in gene transfer and biological delivery. *Trends in Biotechnology, 6*, 229–302.

Sarria, R., Calderon, A., Thro, A. M., Torres, E., Mayer, J. E., & Roca, W. M. (1999). *Agrobacterium*-mediated transformation of *Stylosanthes guianensis* and production of transgenic plants. *Plant Science, 96*, 119–127.

Sarria, R., Torres, E., Ángel, F., Chavarriaga, P., & Roca, W. M. (2000). Transgenic plants of cassava (*Manihot esculenta*) with resistant to Basta obtained by *Agrobacterium*-mediated transformation. *Plant Cell Reports, 19*, 339–344.

Schiek, B., Hareau, G., Baguma, Y., Medakker, A., Douches, D., Shotkoski, F., et al. (2016). Demystification of GM crop costs: Releasing late blight resistant potato varieties as public goods in developing countries. *International Jornal of Biotechnology, 14*(2), 112–131.

Secretariat of the Convention on Biological Diversity. (2000). *Cartagena protocol on biosafety to the convention on biological diversity: Text and annexes*. Montreal: Secretariat of the Convention on Biological Diversity.

Shauzu, M. (2000). The concept of substantial equivalence in safety assessment of foods derived from genetically modified organisms. *AgBiotechNet, 20*(2), 1–4.

Silva, C. (2001). Experiencia colombiana en la aplicación de la reglamentación sobre bioseguridad agrícola. *Revista Colombiana de Biotecnología, 3*(2), 17–22.

Thomas, Z. (2006). Agricultural biotechnology and proprietary rights, challenges and policy options. *The Journal of World Intellectual Property, 8*, 711–734.

Trujillo, C., Rodríguez-Arango, E., Jaramillo, S., Hoyos, R., Orduz, S., & Arango, R. (2001). One-step transformation of two Andean potato cultivars (*Solanum tuberosum* L.). *Plant Cell Reports, 20*, 637–641.

USDA-United States Department of Agriculture. (2014). Available in: https://www.ers.usda.gov/data-products/international-macroeconomic-data-set/international-macroeconomic-data-set/#Historical Data Files. last consulted January 2018.

Valderrama, A. M., Velásquez, N., Rodríguez, E., Zapata, A., Zaidi, M. A., Altosaar, I., et al. (2007). Resistance to *Tecia solanivora* (Lepidoptera: Gelechiidae) in three transgenic andean varieties of potato expressing *Bacillus thuringiensis* Cry1Ac protein. *Journal of Economic Entomology, 100*, 172–179.

Vanegas-Araujo, P. A., Blanco-Martínez, J. T., & Chaparro-Giraldo, A. (2010). Expresión de la proteína Cry1ac en tejidos de líneas transgénicas de papa (*Solanum tuberosum* spp. *andígena*) var. Diacol capiro. *Acta Biológica Colombiana, 15*, 101–114.

Villalobos, A., Ness, J., Gustafsson, C., Minshull, J., & Govindarajan, S. (2006). Gene designer: A synthetic biology tool for constructing artificial DNA segments. *BMC Bioinformatics*, 7, 285–292.

Welsch, R., Arango, J., Bär, C., Salazar, B., Al-Babili, S., Beltrán, J., et al. (2010). Provitamin A accumulation in cassava (*Manihot esculenta*) roots driven by a single nucleotide polymorphism in a phytoene synthase gene. *Plant Cell*, *22*, 3348–3356.

WHO/FAO. (2000). *Safety aspects of genetically modified foods of plant origin*. In *Report of a joint FAO/WHO expert consultation on foods derived from biotechnology*. Available in: http://www.fao.org/fileadmin/templates/agns/pdf/topics/ec_june2000_en.pdf. Consulted January 2018.

Willimsky, G., Bang, H., Fischer, G., & Marahiel, M. A. (1992). Characterization of cspB, a *Bacillus subtilis* inducible cold shock gene affecting cell viability at low temperatures. *Journal of Bacteriology*, *174*(20), 6326–6335.

Wright, B. D., & Pardey, P. (2006). The evolving rights to intellectual property protection in the agricultural biosciences. *International Journal of Technology and Globalisation*, *2*, 12–29.

Zambryski, P., Tempe, J., & Schell, J. (1989). Transfer and function of T-DNA genes from *Agrobacterium*, Ti and Ri plasmids in plants. *Cell*, *56*, 193–201.

Zanga, D., Capell, T., Zhu, C., Christou, P., & Thangara, H. (2016). Freedom-to-operate analysis of a transgenic multivitamin corn variety. *Plant Biotechnology Journal*, *14*, 1225–1240.

FURTHER READING

United Nations. (2014). Available in: https://esa.un.org/unpd/wup/DataQuery/. Consulted January 2018.

CHAPTER SEVEN

Recombinant Therapeutic Molecules Produced in Plants

Qiang Chen[1]
School of Life Sciences, Arizona State University, Tempe, AZ, United States
Center for Immunotherapy, Vaccines and Cancer Therapy, Arizona State University, Tempe, AZ, United States
[1]Corresponding author: e-mail address: qiang.chen.4@asu.edu

Contents

Abstract

Research on the use of plants for production of protein-based therapeutics has increased tremendously since the initial experiments in the early 1990s. Plant-based expression systems offer several production advantages of low cost, rapidity, scalability, and a significantly lower chance of contamination with prion or mammalian viruses. In addition, the capability of plants in producing homogeneous N-glycans allows the development of novel therapeutics with superior efficacy and safety to their mammalian cell-produced counterparts. Various plant species have been used to develop and produce vaccines, antibodies, and pharmaceutical enzymes against a myriad of diseases by multiple expression technologies. While most of these plant-made therapeutics are in preclinical development, many have progressed into human clinical study phases and several have been approved by regulatory agencies. The current status and recent advancement of plant-based expression systems and key clinical products will be presented in this chapter. The remaining challenges and future directions for the field of plant-made therapeutics will be discussed.

Advances in Botanical Research, Volume 86
ISSN 0065-2296
https://doi.org/10.1016/bs.abr.2017.11.006

1. INTRODUCTION

The potential of plants as a system to develop protein-based therapeutics is highlighted by the success story of ZMapp as a life-saving drug during the Ebola outbreak of 2014–16 (see Section 6). The majority of human therapeutics including vaccine and therapeutic proteins are currently produced in mammalian and microbial cell cultures. However, these technology platforms require prohibitive upfront capital investment and have limited scalability, thus, inadequate to meet the increasing consumer demand. Growing evidence demonstrates that plant-based expression technologies are viable alternatives to resolve these problems. It has been demonstrated that plants can produce large volumes of proteins efficiently and sustainably with lower manufacturing costs (Chen & Davis, 2016; Tuse, Tu, & McDonald, 2014). It is also safer to use plants to produce human therapeutics because they are less likely to introduce human or animal pathogens to the manufacture stream compared to systems based on mammalian cells (Chen, 2008). Unlike bacteria-based systems, plants share a similar endomembrane system and a secretory pathway with human cells, enabling the efficient assembly of multiple subunit proteins and posttranslational modifications required for most therapeutics (Chen, 2011). Innovations in plant molecular biology have armed current plant expression platforms with features beyond the traditional advantages of low cost, high scalability, increased safety, and eukaryotic protein modification. The development of novel expression vectors allows the production of therapeutics at unprecedented speed to control potential pandemics or bioterrorism attacks (Chen, Zhang, & Santi, 2014). Plant-derived proteins are now able to exhibit unique and uniform mammalian posttranslational modifications through genetic engineering of plant hosts, providing opportunity to develop therapeutics with far superior potency than those produced in mammalian cell cultures (Chen, 2016). Furthermore, the unique plant cellular structure allows plant cells to be natural capsules in preserving the medicinal properties of pharmaceuticals they produce (Chen & Davis, 2016). As a result, encapsulated ingredients in dried plants can maintain their pharmacological efficacy several years after storage at room temperature (Lakshmi, Verma, Yang, Lloyd, & Daniell, 2013). This feature eliminates the need for cold storage and reduces the production cost, allowing these pharmaceuticals to be widely used around world.

2. PRODUCTION STRATEGIES FOR THERAPEUTICS IN PLANTS

Three strategies have been developed to produce therapeutics in plants: creating stable transgenic plants with transgene integrated into the nuclear genome, creating stable plant lines with a transformed chloroplast genome, and transiently expressing the transgene in nontransgenic plants (Fig. 1) (Chen, 2011). For the first two strategies, the gene(s) coding for the target biologic is (are) cloned into an expression vector and delivered into plant cells. Cells with a stably integrated transgene in their nuclear or chloroplast genome are selected to generating transgenic plants. Transgenic plant lines will be screened for their traits in producing the target product including transgene expression level, biologic accumulation and identity, and transgene stability. In the third strategy of transient expression, the transgene for the biologic is also introduced into plant cells. However, instead of selecting cells with copies of the stably integrated transgene, the transgene is allowed to be actively transcribed and translated to produce the biologic while being in the cell transiently (Komarova et al., 2010).

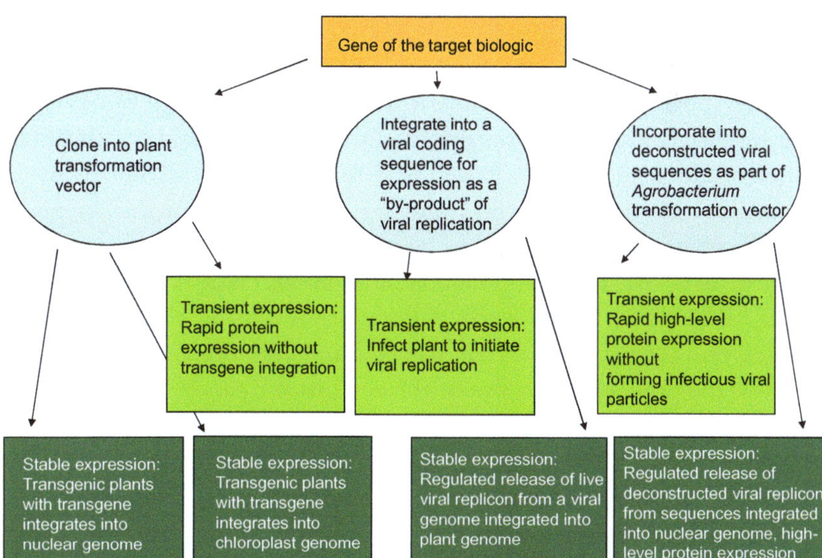

Fig. 1 Strategies for the production of protein therapeutics in plants.

All three plant expression strategies have their advantages and limitations for the development and production of therapeutics. Transgenic plants with a stably integrated transgene provide a permanent genetic source for the cost-effective propagation of plants and commercial scale production of therapeutics. However, the process of selecting and generating stable transgenic plants can take several months to a year, and thus, severely limiting its applications where the biologic candidate needs to be quickly produced. Because transgene can be integrated into regions of the genome with various transcriptional activities, wide fluctuation of gene expression levels among individual plants is common, lowering the chance of identifying high producers of therapeutics (Chen & Lai, 2014b). In addition, the occurrence of transgene silencing can affect the quality of the seed bank of the selected high producers (Chen, 2011). Public concern over genetically modified (GM) plants further complicates its current application. In spite of these challenges, however, once accomplished, the Mendelian inheritance of the transgene may allow the stable expression of the biologic over many generations and can lead to the propagation and stock-up of master seed banks. Therefore, stable transgenic plants provide the most scalable technology for very large-scale commercial production of therapeutics. Compared with nuclear-transformed plants, chloroplast-transformed (transplastomic) plants offer several advantages in both the technical and regulatory areas. For example, the prokaryotic genome of chloroplast allows the targeting of the transgene to a specific genomic position through homologous recombination. As a result, the "position effect" of nuclear transformation, which causes inconsistent level of transgene expression, is eliminated. This and the shear large number of chloroplasts in each cell lead to the significant higher level of accumulation of therapeutics in transplastomic plants than that in nuclear transgenic plants (Daniell, 2006). The transplastomic strategy is a preferred alternative for ensuring transgene containment and addressing public concerns for GM plants because the maternal transmission of chloroplasts in most flowering plants reduces the risk of transgene escape through pollen. However, the prokaryotic nature of chloroplasts also renders them incapable of performing the extensive posttranslational modifications required for many therapeutics. This limits the application of this strategy to niche products where such modifications are not essential for their function.

Because transient expression does not require the integration of transgene into a plant genome, this strategy eliminates the time-consuming process of transgenic plant generation and selection and, thereby, accelerates the production speed. For the same reason, transient expression removes the position effects of transgene integration and offers higher levels of biologic protein accumulation (Chen & Lai, 2014a). The rapid and high-yield nature of this strategy makes it an optimal platform for developing new therapeutics and/or obtaining the initial material (mg – g level) for preclinical or Phase I trial testing. An additional benefit of this strategy is the versatility and flexibility for producing personalized therapeutics and vaccines against cancers or against pathogens that have rapid antigenic drift and/or multiple strains with unpredictable epidemics. This advantage also arms transient systems with the "surge" capability to rapidly produce therapeutics to address a bioterror event. This strategy also offers advantages from a public acceptance and regulatory perspective. It does not produce stable transgenic plants or an intact replication-competent plant virus (see Section 4). As a result, the risk of transgene spreading through pollen, seeds, or virus is eliminated. However, the scale-up production of therapeutics using this strategy poses a bigger challenge. Because no genetically stable seed bank is produced, scale-up is no longer just a matter of increasing acreage, but requiring transgene delivery to be repeated for each batch of biologic production. Thus, transient expression is currently limited to the applications of research and development or small-scale production. Nevertheless, the vastly improved speed and yield, low risk of environmental and biosafety impact, along with the new development of expression vectors that will resolve its scalability issue, will position it to be the most promising plant-based strategy for the development and production of therapeutics.

3. TRANSGENE DELIVERY INTO PLANTS AND AGROINFILTRATION

Introducing transgenes into plant cells can be accomplished by either direct delivery by gene gun or indirect delivery through using *Agrobacterium tumefaciens* or plant viruses depending on the expression vector and the host plants (Rivera, Gomez-Lim, Fernandez, & Loske, 2012). The advantage of direct delivery with gene gun resides in its versatility because it can be used to deliver transgene to both nuclear and chloroplast genomes and, at least in theory, can be applied to any plant host species independent of the vector (Rivera et al., 2012). *Agrobacteria* is a group of soil bacteria that transfer a portion of their tumour-inducing plasmid (Ti plasmid) DNA (T-DNA) into plant cells as a part of their pathogenesis. The Ti plasmid can be modified into delivery vectors by replacing pathogenic genes in T-DNA with gene of interest for therapeutics. When incubated with susceptible plant hosts, transgene is transferred from Agrobacterial to plant cells along with the rest of T-DNA through the natural interaction between *A. tumefaciens* and its plant hosts (see chapter "How *Agrobacterium*, a natural genetic engineer, became a tool for modern agriculture" by Otten in this volume). Compared to the direct transgene delivery methods, the indirect delivery via *A. tumefaciens* has a narrower plant host range restricted to dicotyledonous and a few monocotyledonous plants (Klee, Horsch, & Rogers, 1987). It also requires the transgene to be cloned into a specific vector derived from a modified Ti plasmid. However, *Agrobacterium*-based delivery is generally advantageous over gene gun in controlling transgene copy number and integration into the genome, transformation efficiency, as well as in transgene expression and stability (Rivera et al., 2012). It is speculated that the coevolution of *Agrobacterium* and its plant hosts may favour the insertion of transgenes into more transcriptionally active genomic loci, leading to its high level of expression of the biologic protein (Chen & Lai, 2014a; Klee et al., 1987).

Agroinfiltration is the most efficient way to introduce transgene into plants for recombinant protein production. It can be simply accomplished by injecting transgene-carrying *Agrobacterium* in the infiltration medium into the leaf with a needleless syringe (Fig. 2A). Syringe infiltration offers the flexibility of introducing multiple transgene constructs into different leaf areas (Fig. 2B), allowing multiple assays to be performed on a single leaf (Leuzinger et al., 2013; Vaghchhipawala, Rojas, Senthil-Kumar, & Mysore, 2011). Vacuum infiltration is a more scalable agroinfiltration

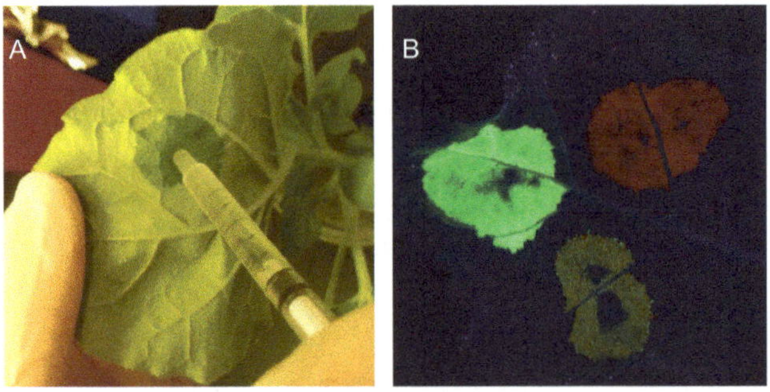

Fig. 2 Syringe infiltration of *N. benthamiana* leaves with *Agrobacterium tumefaciens. A. tumefaciens* carrying the gene of green fluorescent protein (GFP) or red fluorescent protein (DsRed) were syringe-infiltrated into a *N. benthamiana* leaf (A) and GFP or DsRed expression was observed 4 days postinfiltration under UV light (B). The *yellow spot* in (B) indicates the leaf area that was coinfiltrated with Agrobacteria carrying the expression vectors of GFP and DsRed.

method (Leuzinger et al., 2013). Aerial parts of plants are submerged into an *Agrobacterium* culture and the air in the interstitial space of the leaves is pulled out by a vacuum pump. The subsequent release of the vacuum allows agrobacteria in the medium to enter the intercellular space once occupied by the air and, therefore, achieve agroinfiltration (Fig. 3). Vacuum infiltration has a broader host range than syringe infiltration, allowing efficient agroinfiltration in plant species that are unamenable to syringe infiltration (Chen et al., 2014, 2010; Takata & Eriksson, 2012).

The scalability of vacuum infiltration beyond the desiccator prototype was demonstrated by a vacuum chamber that is able to infiltrate up to 1.2 tons of plant biomass per day, allowing for the production of up to 75 g of therapeutics per greenhouse lot (Fig. 4) (Gleba, Tusé, & Giritch, 2014; Leuzinger et al., 2013; Pogue et al., 2010). It was shown that the

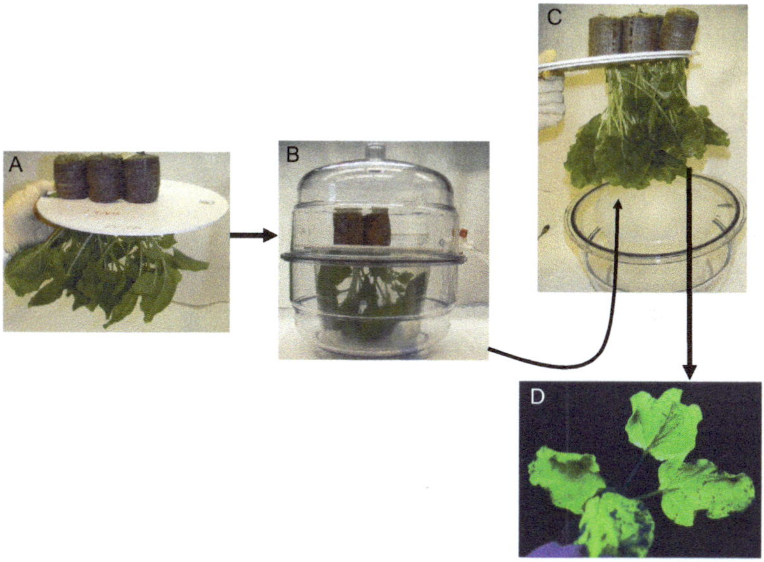

Fig. 3 Vacuum infiltration of *N. benthamiana* leaves with *Agrobacterium tumefaciens*. 6-week-old plants are placed upside down on the desiccator plate (A). The entire leaf and stem system is then submerged into a tub that is filled with suspensions of *A. tumefaciens* harbouring the target protein-expressing vectors; and the tub is placed in a vacuum desiccator that was connected to a vacuum pump (B). Agroinfiltration was achieved by applying and releasing a vacuum (C). Target protein (GFP) expression can be observed in all leaves 4–10 days after agroinfiltration (D). *From Chen, Q. (2008). Expression and purification of pharmaceutical proteins in plants. Biological Engineering Transactions, 1(4), 291–321. Copyright 2008 American Society of Agricultural and Biological Engineers.*

Fig. 4 *N. benthamiana* plant growth (A) and agroinfiltration (B) at commercial production scale.

production yield and the temporal expression patterns of biologic proteins were similar, regardless if they were produced in this scale-up vacuum chamber or with the desiccator (He, Lai, Brock, & Chen, 2012; Huang, Chen, Hjelm, Arntzen, & Mason, 2009; Lai & Chen, 2012; Lai et al., 2010). The only limitation of vacuum infiltration is its inapplicability to field-grown plants because the process requires the inversion of plants to submerge into the agrobacteria mixture. To overcome this challenge, spray-based agroinfiltration has been developed to fill this need. It has been shown that when *Agrobacterium* with enhanced infectivity is sprayed onto leaves with surfactant or abrasives, transgene can be delivered to >90% cells in leaves (Gleba et al., 2014). Therefore, this method provides a simple and scalable process of transgene delivery into field-grown plants, allowing transient expression on a large agricultural scale. Furthermore, genetic engineering of *Agrobacterium* has produced strains with low survivability in field conditions, reducing the risk of spreading of the transgene into the environment (Gleba et al., 2014). Collectively, these studies demonstrate the enormous scalability potential of vacuum and spray agroinfiltration and support their application for commercial manufacturing of human therapeutics.

4. DECONSTRUCTED PLANT VIRAL VECTORS ARE OPTIMAL FOR TRANSIENT PRODUCTION OF THERAPEUTICS

Plant viruses have been explored to develop vectors for gene delivery into plants (Fig. 1). High level of therapeutic protein production is expected with whole plant virus-based vectors due to the robust replication and/or transcription abilities of plant viruses (Lico, Chen, & Santi, 2008). However, whole plant viruses have several limitations as vectors for recombinant protein production. These include the inability to produce large proteins due to sensitivity to viral genome deletion or substitution or the requirement to produce viral genome by a difficult in vitro process (Lico et al., 2008).

"Deconstructed" viral vectors represents the new generation of vectors for therapeutic production in plants (Fig. 1). Deconstruction of viral genomes has effectively eliminated unnecessary genomic segments but preserved components for robust replication and transcription. The reduction in genome size allows the insertion of large transgenes, making the vectors more versatile for producing therapeutics of various sizes. As a result, deconstructed viral vectors combine the robustness of full viral vectors and the versatility of nonviral vectors. Deconstructed viral vectors are especially advantageous for RNA virus because they can be delivered in the form of DNA which will be transcribed and spliced into autonomous replicons in plant cells (Lico et al., 2008). This eliminates the tedious process of generating RNA vectors in vitro. More importantly, agroinfiltration can now be used to deliver deconstructed genome of RNA virus. Furthermore, the viral systemic spreading function is no longer needed as agroinfiltration can deliver vectors to almost all cells on the entire plant (Leuzinger et al., 2013). This not only allows the deletion of genes for systemic spreading to accommodate even larger transgene insertion but also alleviates the concern for transgene loss during systemic spreading. Agroinfiltration also broadens the range of plant hosts susceptible to viral vector delivery and allows for the delivery of vectors that are not mechanically transmissible in nature. In terms of biosafety regulations, deconstructed vectors do not produce infectious plant viruses, eliminating the public concerns for the spreading of recombinant plant viruses. Therefore, the development and application of deconstructed viral vectors not only overcome the limitations of full viral vectors but further enhance the robustness and versatility of the transient expression system.

MagnICON system offers the most popular deconstructed vectors used today. They are derived from tobacco mosaic virus (TMV) and potato virus X (PVX) and can be used either in the modular or in the fully assembled form depending on the application (Giritch et al., 2006). The MagnICON system has been tested at various production scales, from a few plants in a laboratory to 1.2 tons plant material/day in industry. Collectively, they demonstrated that very high-level accumulation of protein therapeutics (up to 5 mg/g leaf fresh weight (LFW)) can be achieved within 7–10 days postinfiltration (DPI). They include therapeutics of all sizes and complexity, ranging from small subunit vaccines to large tetravalent antibodies (Bendandi et al., 2010; Chen, 2015; Giritch et al., 2006; Lai & Chen, 2012; Lai et al., 2010, 2014; Phoolcharoen, Bhoo, et al., 2011).

One of the major limitations of the current MagnICON vectors is their inability to produce therapeutics with more than two different polypeptides. "Competing replicons" is the root cause of this issue because codelivery of viral vectors built on the same viral backbone often results in early segregation and subsequent preferential amplification of only one of the vectors in a single cell (Lico et al., 2008). For therapeutics containing two different polypeptides, such as monoclonal antibodies (MAbs) made up by both the heavy (HC) and light chains (LC), only one of the subunits (i.e. HC or LC) will be produced in a single cell and the assembled full molecule will not be produced if genes for both subunits are expressed by the TMV vector. However, if the two genes are built on vectors derived from two viruses that are not competing with each other, then the issue is resolved. This is how MagnICON vectors can produce fully assembled MAbs because the HC and LC genes are built on the TMV and PVX backbone, respectively (Giritch et al., 2006). However, for expressing proteins with three or more distinct subunits with the MagnICON system, it requires the identification of additional viruses that are noncompeting with neither TMV nor PVX, a very difficult if not impossible task (Chen & Lai, 2013; Thuenemann, Meyers, Verwey, Rybicki, & Lomonossoff, 2013).

To circumvent this problem, a noncompeting vector system based on bean yellow dwarf virus (BeYDV), a monopartite virus in the *Geminiviridae* family, was developed (Chen, He, Phoolcharoen, & Mason, 2011). BeYDV requires only two *cis*-acting elements (the long intergenic region (LIR) and the short intergenic region (SIR)), and one single viral protein (replication-associated protein (Rep)) to produce very high copy numbers of its genome by rolling circle replication (Fig. 5) (Chen et al., 2011; Huang et al., 2009). This leads to the simplicity of the BeYDV-based deconstructed vectors (geminiviral vectors) with the transgene cassette inserted between the LIR and the SIR. In the first generation of geminiviral vectors, the transgene and the Rep protein are supplied in two separate modules (Chen et al., 2011; Huang et al., 2009). It was demonstrated that high-level production of therapeutics, such as hepatitis B core antigen (HBcAg) (a vaccine candidate), can be achieved by coagroinfiltration of the two modules (Huang et al., 2009). In the second generation of geminiviral vectors, the modules of Rep and transgene were combined into a single vector system with the transgene, which obviates the need to generate several vector modules and manufacture multiple inocula of *A. tumefaciens* strains. This is especially of importance in reducing capital and operational cost for large-scale manufacturing of therapeutics. Most importantly, the geminiviral system

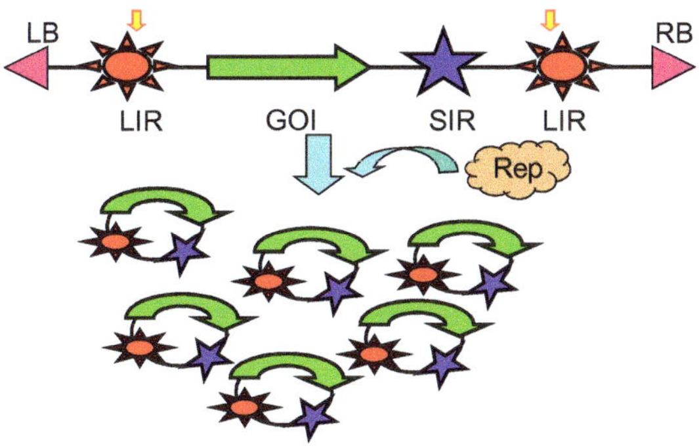

Fig. 5 Geminiviral BeYDV vector for the production of human therapeutics in plants. The *left* (LB, *pink triangle*) and *right border* (RB, *pink triangle*) delineate the T-DNA construct that will be transferred into plant cells by *Agrobacterium*. Upon delivery into plant cells, expression of Rep gene produces the Rep protein (*brown cloud*) that nicks the LIRs (*red stars*) in the T-DNA to release a single-stranded DNA molecule (between the two *yellow arrows*). This DNA molecule recircularizes and is copied to make double-stranded DNAs that can replicate by the rolling circle mechanism to produce very high copy numbers of DNA templates (*circles*) and, in turn, abundant mRNAs of gene of interest (GOI) for the translation of the recombinant protein. *Blue star*, SIR; *brown cloud*, Rep protein; *green arrow*, gene of interest; *pink triangle*, LB and RB of the T-DNA; *red stars*, LIRs.

is noncompeting and permits the efficient expression and assembly of MAbs when the LC and HC genes are cloned into the same BeYDV backbone (Huang et al., 2010). This leads to the third generation of geminiviral vectors, a single vector system that contains multiple replicon cassettes with each encoding for a distinct polypeptide. It was demonstrated that an independent replicon was generated from each expression cassette upon agroinfiltration into plant cells. These independent replicons can reach very high copy numbers without competing with the replication of other replicons or the production of other proteins. Thus, high-level accumulation of each protein subunit and the overall production of the fully assembled multisubunit protein can be achieved (Huang et al., 2010). Studies have also shown that geminiviral vectors are speedier than the MagnICON system, and can rapidly express protein therapeutics in plants within 4 DPI (Huang et al., 2009, 2010; Lai, He, Engle, Diamond, & Chen, 2012). In addition, they may also have broader plant host ranges than those of the MagnICON system (He et al., 2012; Huang et al., 2009, 2010; Lai et al., 2012). Overall, the geminiviral replicon system enables the production of multiple heterosubunit proteins in plants, while retaining the speed and high production capacity of other deconstructed viral systems.

Utilizing the enhancing ability of untranslated regions (UTRs) of plant viral genes is another approach to develop robust deconstructed vectors. For example, a vector based on the 5′ and 3′ UTRs of cowpea mosaic virus (CPMV) RNA-2 has allowed the expression and assembly of Bluetongue virus-like particles (VLPs) in *Nicotiana benthamiana* that requires coexpression of four different protein components (Thuenemann et al., 2013). Furthermore, because certain versions of the CPMV-based vectors are replication independent, they also may have a broader host range, allowing production of therapeutics in plants that are not compatible with replication-dependent vector systems. Another interesting UTR-based vector system is derived from tobacco yellow dwarf mastrevirus (Dugdale et al., 2013). The most important feature of this vector system is its ability in bridging the two platforms of therapeutic production in plants: transient expression and stable transgenic plants. Thus, this vector system represents "bridge" vectors that allow the stable inheritance of the transgene and a robust yet controlled transient expression of a biologic protein upon the induction with a specific chemical signal (Dugdale et al., 2013). Specifically, the transgene is stably integrated into the plant genome but remains silent to allow the accumulation of biomass. Once sufficient plant material is amassed, a transition from the growth phase to the phase of biologic production will be triggered by a chemical signal to induce the replication of replicons. As a result, the bridge vector system effectively combines the strengths of both the stable and transient expression systems and potentially offers a complete platform for the rapid assessment of biologic candidates and their transition to a large-scale commercial production. Collectively, the development of deconstructed viral vectors marks a significant advancement in transient expression technology and positions it as a competitive platform for commercial production of therapeutics.

5. A VARIETY OF PLANTS CAN BE USED FOR THERAPEUTIC PROTEIN PRODUCTION

A diversity of plant species have been explored for the production of therapeutics. The most common plants used for therapeutic protein production are from the genus of *Nicotiana* such as tobacco (*N. tabacum*) and the related *N. benthamiana*. *Nicotiana* plants are prolific seed producers and can produce large amount of biomass rapidly, making them the optimal plants for scale-up production of therapeutics (Chen & Lai, 2013). They are also permissive to the replication of a variety of plant virus-based vectors and amenable to agroinfiltration, making them favourable for therapeutic production through transient expression (Chen & Lai, 2014a). The readily available large-scale processing infrastructures of traditional tobacco industry are another advantage and can be quickly adapted for processing biologic-containing biomass. Since plants in the *Nicotiana* genus are nonfood crops, the risk of contamination of human food chain by transgenic plants is relatively low. Because of these advantages, *Nicotiana* plants have been used for the production of therapeutics of various nature, sizes, and applications. Some of these therapeutics have progressed to the stage of human clinical testing (Table 1). As a result, the US Food and Drug Administration (FDA) and other regulatory agencies are acquainted with these plant hosts, facilitating their acceptance as organisms for therapeutic protein production (Bendandi et al., 2010; McCormick et al., 2008). Several examples are given below to demonstrate the advantages and versatility of the *Nicotiana* plant hosts.

Nicotiana plant-produced MAbs were used to treat non-Hodgkin's lymphoma (NHL) in the first of its kind demonstration clinical trial. In NHL, a group of blood cancers, each clone of malignant B cells is unique because it expresses a unique cell surface immunoglobulin (Ig). As a result, standard treatments are ineffective, calling for patient-specific cancer treatments. However, the production of personalized therapeutics requires an expression system with speed and flexibility. In contrast to cell-based production systems, plant production platforms based on agroinfiltration are optimal to meet this demand. MAbs and single-chain variable fragment (scFv) of MAbs derived from each patient's own tumour were used as the vaccine candidate. When expressed in *N. benthamiana* plants, patient-specific scFvs and MAbs were produced at high levels in leaves within 2 weeks of agroinfiltration (Bendandi et al., 2010). Furthermore, the manufacturing process is robust, requiring less than 12 weeks from biopsy to vaccination (Bendandi et al., 2010). This study showcased the capacity of *N. benthamiana* in rapidly generating multiple patient-specific cancer vaccines.

N. benthamiana plant also offers opportunities to produce therapeutics with superior functionalities (biosuperiors or biobetters). An example is given below to demonstrate the flexibility of *N. benthamiana* plant in turning a major drawback of the plant-based system into a major advantage in the case of plant N-glycosylation. N-glycosylation has an important impact on pharmacokinetics, antigen/receptor binding, stability, effector functions, and efficacy of the target therapeutics. As such, therapeutics produced in recombinant organisms may have differences in N-glycosylation from the

Table 1 Plant-Derived Therapeutics That Have Been Approved or Reached Clinical Trial Stages

Product/Application	Pathogen/Disease Target	Plant Hosts	Development Status
ELELYSO/therapeutic	Gaucher's disease	Carrot and tobacco cells	FDA approved
Oral GCD/therapeutic	Gaucher's disease	Carrot cells	Phase II
α-Galactosidase/therapeutic	Fabry disease	Carrot cells	Phase I/II
Acetylcholinesterase/therapeutic	Nerve gas poisoning	Carrot cells	Phase I
ZMapp (MAb-based)/therapeutic	Ebola virus infection	*N. benthamiana*	Phase I/compassion use
MB66 (MAb-based)/therapeutic	HIV/HSV	*N. benthamiana*	Phase I
HA VLP/vaccine	Pandemic influenza	*N. benthamiana*	Phase II/III/emergency use
HA VLP/vaccine	Seasonal influenza	*N. benthamiana*	Phase I
E. coli heat-liable toxin/vaccine	Traveller's diseases	Potato	Phase I
HepB surface antigen/vaccine	Hepatitis B	Tobacco	Phase I
NVCP VLP/vaccine	Norovirus	*N. benthamiana*	Phase I
NHL/vaccine	NHL cancer	*N. benthamiana*	Phase II
Colicins/food additive	*E. coli* food poisoning	*N. benthamiana*, spinach	GRAS
HIV antibody/microbicide	HIV	Tobacco	Phase I
HA/vaccine	Influenza	*N. benthamiana*	Phase I
Lactoferrin/therapeutic	Diarrhoea	Rice	Phase II
α-Interferon (Locteron)/therapeutic	Hepatitis C	Duckweeds	Phase II/IIb
CarolRX (MAb-based)/prevention	Tooth decay	Tobacco	Approved as medical device
Gastric lipase/therapeutic	Cystic fibrosis	Corn	Phase II
Lactoferrin/therapeutic	Gastrointestinal diseases	Corn	Phase I
Insulin/therapeutic	Diabetes	Safflower	Phase I
HepB surface antigen/therapeutic	Hepatitis B	Lettuce	Phase I
Rabies glycoprotein/vaccine	Rabies virus	Spinach	Phase I

native molecules produced in humans and could be immunogenic. For example, N-glycans produced in plants may carry unique plant-specific β-1,2-xylose and core α-1,3-fucose residues and lack terminal β1,4-Gal and *N*-acetylneuraminic acid (Neu5Ac) residues, despite the overall similarity of N-glycosylation between plant and mammalian cells (Chen, 2016). These differences may alter the stability and efficacy of plant-produced therapeutics, as well as cause adverse effects. Sceptics of plant-based expression systems often cite the glycan differences as the main reason for their lack of confidence in plant biotechnology. However, glycoengineering has largely silenced such criticism. Due to the long history of using *Nicotiana* plants as models for genetic manipulations, the readily available genetic information and tools have facilitated the success of glycoengineering in creating "humanized" plant lines. These humanized *Nicotiana* plant lines not only have eliminated the concerns associated with plant-specific glycans but are also able to produce therapeutics with tailor-designed glycans, namely, biobetters that have superior safety and efficacy over their mammalian cell-produced counterparts (Chen & Davis, 2016; Chen & Lai, 2014b).

For example, a double knockdown (ΔXF) *N. benthamiana* line was created by using an RNA interference strategy to suppress the production of the two plant-specific glycans (Strasser et al., 2008). Results indicated that anti-Ebola MAbs produced in the ΔXF plant line had no detectable plant-specific N-glycans. More significantly, the plant-produced MAb contained only a single true mammalian glycoform GnGn with a high degree (90%) of glycan uniformity that cannot be matched by mammalian cells (Zeitlin et al., 2011). The lack of fucose and the high homogeneity of plant-derived MAbs have led to their higher affinity to the Fc gamma receptor (FcγR) III, improved antibody-dependent cell-mediated cytotoxicity, and enhanced potency against Ebola virus over the mammalian cell-produced MAbs (Zeitlin et al., 2011). The superior potency of plant-produced MAbs was further demonstrated in a challenge study with nonhuman primates, in which plant-produced MAbs were far more protective against a lethal Ebola challenge than those produced in mammalian cells (Olinger et al., 2012). In a remarkable and exciting development, these plant-made MAbs were recently used to treat human Ebola patients and showed promising results (Check Hayden & Reardon, 2014; Chen & Davis, 2016). Similarly, ΔXF plant-derived anti-West Nile virus (WNV) and anti-Dengue virus (DENV) MAbs also displayed enhanced efficacy or safety in comparison to their mammalian counterparts (Dent et al., 2016; Lai et al., 2014).

In addition to eliminated plant-specific glycans, the amenability of *Nicotiana* plants allowed the introduction of mammalian genes for glycoenzymes to further humanize their N-glycosylation pathway. As a result, a portfolio of *N. benthamiana* lines has been generated to produce proteins with various defined mammalian N-glycoforms (Chen, 2016; Loos et al., 2014; Strasser et al., 2008). The availability of these *Nicotiana* plant lines allows the production of therapeutics with a tailor-made N-glycosylation pattern based on functional needs. Therefore, *Nicotiana* plants provide a superior system for producing therapeutics with defined carbohydrate moieties and a platform for evaluating their functional contributions to the overall properties of the target protein.

In addition to *Nicotiana* plants, a diversity of plants and their leaves, seeds, fruits, and tubers have been explored to express therapeutics. Leafy plants such as lettuce, alfalfa, spinach, petunia, cotton, grapevine, switchgrass, radish, and *Arabidopsis thaliana* share certain advantages with *Nicotiana* plants in producing therapeutics. Quick biomass generation and processing are examples of these advantages. However, the amenability to all three plant expression platforms is the most advantageous characteristic of the leafy plant hosts. This flexibility increases the chance of success for many therapeutics, since the options of using stable nuclear or chloroplast transformation and viral vector-based transient expression are available to identify an optimal expression platform based on the property of the biologic candidate. Drawbacks of leafy plant hosts include the potential interference of therapeutics on plant growth and biologic stability. Since therapeutics are accumulated in leaves, some of them may affect the photosynthetic activity of plants, thereby interfering with plant growth and development. This may lead to poor accumulation of plant biomass and reduce the yield of the target biologic. This issue can be easily circumvented by using transient expression systems as biomass production has already accomplished before agroinfiltration. For applications where stable transgenic plants have to be used to express "toxic" therapeutics, inducible promoters such as those in the bridge vectors can be used to drive the expression of the biologic. As such, the biologic is not accumulated during plant growth, but will only be produced upon application of specific inducer agent postbiomass harvest (Chen, 2011). Leafy hosts are not the optimal hosts in situations that the therapeutics needs to be stored for a long period of time in the tissue before they can be processed. Therapeutics accumulated in leaves are generally in an aqueous environment and thus tend to be not as stable as those accumulated in seeds (see below). As a result, therapeutics produced in leaves have to be quickly transported to a processing facility at a controlled temperature and extracted before proteolysis occurs. They can also be processed by freezing or drying for storage. Regulatory concerns for food or feed leafy plants are the contamination of human food chain, and the potential exposure of leaf-eating herbivores to plant-made drugs for nonfood, nonfeed leafy hosts. These risks, however, can be managed by using the transient expression platform or using bridge vectors with inducible promoters to drive the expression of the transgene so that therapeutics can only start to accumulate in a controlled environment post biomass harvest (Chen, 2011).

The special utility of non-*Nicotiana* leafy hosts in producing therapeutics can be demonstrated by lettuce. Despite the aforementioned advantages of *Nicotiana* hosts, higher levels of phenolics and alkaloids such as nicotine are accumulated in their leaves. These compounds foul purification resins and are difficult to remove from the target therapeutics in downstream processing (Hey & Zhang, 2012; Roque, Lowe, & Taipa, 2004). Thus, there is a need for plant hosts that produce lower levels of phenolics and alkaloids yet retain the robustness of *Nicotiana* hosts for biologic production. Lettuce (*Lactuca sativa*) is already produced in large scales commercially with a speed of biomass generation and yield comparable to those of *Nicotiana* plants. Importantly, lettuce produces negligible quantities of phenolics and alkaloids and, therefore, foregoes the challenge of their removal during downstream processing. Results showed that vaccines and therapeutics can be expressed in lettuce leaves with levels and functional activities similar to those produced in *N. benthamiana* plants (Chen, 2013; Lai et al., 2012). The superiority of lettuce over *Nicotiana* plants was demonstrated by the case of producing VLPs. Although VLPs are promising vaccine candidates against a broad range of viral diseases (Chen & Lai, 2013), their porous nature renders them prone to trap contaminants inside. Consequently, it is very difficult to remove plant secondary metabolites from preparations of *Nicotiana* plant-produced VLPs (Lai & Chen, 2012). If not resolved, this problem will diminish the vast potential of plant-produced VLPs. Lettuce-derived VLPs can meet the purity specifications of FDA without the extra need for elimination of phenolics and alkaloids (Lai et al., 2012).

Seeds are natural desiccators and can preserve functional proteins for a long period of time through a natural drying process during maturation. As a result, seeds of maize (corn), rice, sunflower, *Arabidopsis thaliana*, wheat, barley, tobacco, oilseed rape, and a variety of beans have been explored to express therapeutics that require long-term storage in biomass before processing (Maruyama et al., 2014; Morandini et al., 2011; Takaiwa &

Yang, 2014). It was shown that MAbs produced in corn seeds remain stable after 3 years of storage at room temperature and retain their biological activities (Stoger et al., 2000). Compared with leafy plants, only nuclear stable transformation is currently feasible for scalable seed production of therapeutics. In respect of biosafety concerns, the necessity of pollination for seed production increases the risk for transgene escape into the environment through pollen- and seed-eating animals. The risk via pollen, however, can be remediated by taking both biological and mechanical measures such as using male sterile plants or removing pollen-producing structures (e.g. maize detasseling) (Daniell, 1999). The utility of seeds for biologic production has been demonstrated by maize. Maize has been a model for genetic studies. As such, the availability of a vast genetic and molecular knowledge base allows the successful production of a variety of recombinant proteins. For example, maize seed-produced MAb (2G12), a promising anti-HIV microbicide candidate, showed equivalent potency in neutralizing HIV as its counterpart produced in mammalian cells (Ramessar et al., 2008). Rice has also been successfully used in producing important human biologics such as lactoferrin and lysozyme in the United States and human serum albumin in China (Bethell, 2007; He et al., 2011).

Plants that produced fruits and tubers have also been explored for the production of therapeutics (Chen & Lai, 2014a). Along with seeds, these plant hosts provide the opportunity for a novel approach of biologic delivery through oral route by ingesting palatable plant tissues/organs. Production and delivery of conventional therapeutics require a costly downstream extraction and purification process, as well as continuous refrigeration (the "cold chain") for transport and storage, and sterile needles for injection. Oral delivery of therapeutics in its unpurified form will overcome the aforementioned challenges, but remains elusive due to the denaturation and degradation of therapeutics in the digestive system, and their inability to cross the gut epithelium and subsequent delivery to target cells. Plant cells, however, can help to resolve these issues by bioencapsulation. Specifically, the glycosidic bonds in the carbohydrates of the plant cell wall are resistant to human digestive enzymes, thereby protecting expressed biologic proteins from acids and enzymes in the stomach and allowing them to enter the gut lumen where they are enzymatically released by gut commensal bacteria (Kwon & Daniell, 2015). Furthermore, it has been recently demonstrated that plant cell-encapsulated therapeutics can cross the gut epithelium to reach circulation via oral delivery (Kwon & Daniell, 2015;

Su, Sherman, et al., 2015; Su, Zhu, et al., 2015). In addition, therapeutics encapsulated in plant cells has been found to maintain their efficacy several years after storage at room temperature (Lakshmi et al., 2013). These results suggest that therapeutics produced in seeds, fruits, and tubers can be delivered to patients orally as fresh products, or stored as an ambient temperature-stable dry product that can be later delivered to patients by simply ingesting. In both scenarios, the logistic challenges of "cold chain" and medical supply constrains can be circumvented to allow practical implementation of healthcare programs in source-poor regions of the world (Hiwasa-Tanase & Ezura, 2016). Excitingly, recent research demonstrated further advantages of the orally delivered biologic drugs: they offer the flexibility of targeting the therapeutics to different cell types after crossing the gut epithelium. For example, different tags can be fused to the therapeutics, so they can be delivered to different cell types to either induce tolerance against inhibitory antibody production associated with their injection, or reach the circulation to treat diabetes, hypertension, or other metabolic diseases (Kwon & Daniell, 2015; Su, Sherman, et al., 2015; Su, Zhu, et al., 2015). The last hurtles against orally delivery of therapeutics produced in seeds, fruits, and tubers to become licenced drugs are the public concerns for GMO and regulatory requirement of strictly controlled dosage content per unit of drug containing material. Nevertheless, as new expression vectors have allowed more consistent biologic accumulation per unit of plant tissue, this strategy may eventually offer an attractive future option for biologic delivery in both the developed and developing world.

6. PLANT-PRODUCED THERAPEUTICS ARE PROMISING DRUGS

6.1 Monoclonal Antibodies

MAbs produced in mammalian cell culture have achieved remarkable pharmaceutical and financial success with several blockbuster therapeutic products. However, mammalian cell-produced MAbs are prohibitively expensive, making them unaffordable for >75% population of countries around the globe (Walsh, 2014). Despite the complexity of the antibody molecule that requires the assembly of four polypeptides and glycosylation, MAbs and their derivatives including secretory IgAs, tetravalent MAbs, recombinant immune complex (RIC), single-domain fragments, scFvs, and diabodies have been produced in many plant species, and several of them have now reached human clinical trials with ZMapp being used compassionately in human patients (compassionate use, also called "expanded access," is the use of an investigational medical product that has not been approved by FDA) (Table 1) (Chen & Davis, 2016; Chen & Lai, 2014b, 2015; He et al., 2014; Lai et al., 2014; Phoolcharoen, Bhoo, et al., 2011; Phoolcharoen, Dye, et al., 2011).

The growing promise of plant-made MAbs is highlighted by the success story of ZMapp as a potentially life-saving drug in treating two American health aid workers during the Ebola outbreak of 2014–16. Dr. Kent Brantly and Nancy Writebol's Ebola infection escalated to the point where they thought they were dying. Remarkably, their condition quickly improved after receiving an experimental drug called ZMapp, a cocktail of three MAbs produced in *N. benthamiana* plants (Fig. 6) (Lyon et al., 2014). The early work of plant-made MAbs against Ebola that contributed to the eventual development of ZMapp started with expressing three MAbs (6D8, 13C6, and 12F6) in plants (Arntzen, 2015). It was shown that fully assembled MAbs can be produced with geminiviral vectors within 4 DPI up to 0.5 mg/g LFW in both *N. benthamiana* and lettuce, and purified to greater than 95% homogeneity by simple chromatographic methods (Fulton, Lai, Chen, & Zhang, 2015;Huang et al., 2010; Lai et al., 2012). Both tobacco- and lettuce-produced MAbs bind specifically to Ebola GP antigen, suggesting that they are functionally active (Huang et al., 2010; Lai et al., 2012). The in vivo efficacy of these MAbs was later established in mice (Zeitlin et al., 2011). Subsequently, a cocktail of the three plant-made MAbs (6D8, 13C6, and 12F6), called MB-003, was shown to be protective in nonhuman primate models (Olinger et al., 2012). Excitingly, these animal studies revealed that plant-derived individual anti-Ebola MAb and the MB-003 cocktail both have a superior potency to their mammalian-produced counterparts, most likely due to their homogeneous GnGn N-glycosylation pattern (Olinger et al., 2012; Zeitlin et al., 2011).

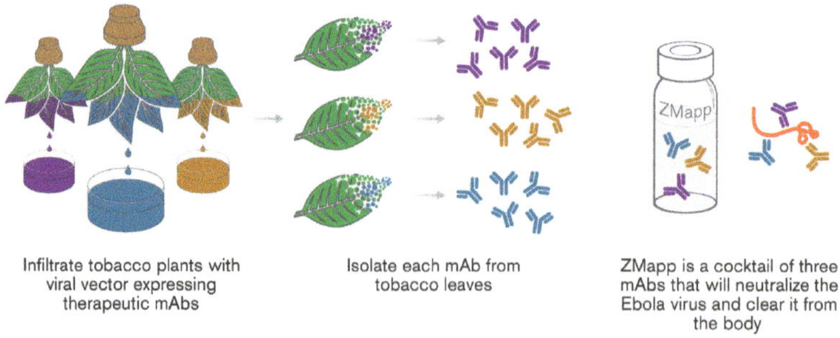

Fig. 6 Overview of the production of ZMapp in *N. benthamiana* plants.

This paved the way for formulating plant-produced MAbs into ZMapp and its compassionate use in Brantly, Writebol, and five other human patients during the Ebola outbreak. A subsequent report confirmed that ZMapp is able to rescue 100% of rhesus macaques even when given 5 days postlethal Ebola challenge (Qiu et al., 2014). A clinical trial has recently been concluded in the United States, Liberia, Sierra Leone, and Guinea for ZMapp by the US National Institutes of Health. The trial compared an optimized standard of care to an identical regimen plus ZMapp. The results indicated that mortality in the ZMapp-treated participants was lower, and participants had a more rapid elimination of virus from the bloodstream (http://mappbio.com/leafbio-announces-conclusion-of-zmapp-clinical-trial/). While the trial did not ultimately enrol enough patients to produce definitive results, researchers said the drug was well tolerated and showed promise. The application for licensure of ZMapp is being vigorously pursed as a treatment for Ebola.

The advantages of plant-made MAbs can be also showcased by developing safer and more efficacious therapeutics against flaviviruses. Flavivirus belongs to a genus of viruses in the *Flaviviridae* family that includes WNV, DENV, Zika virus (ZIKV), Japanese and St. Louis encephalitis virus, tick-borne encephalitis virus, and yellow fever virus. Even though they shared a similar positive-sense single-stranded RNA genome and a general enveloped structure, each of them has a unique mechanism of pathogenies. This presents difficult challenges for the development of MAb-based therapeutics. For example, WNV is a neurotropic virus and can spread to the brain and spinal cord, causing encephalitis, meningitis, and paralysis. The neuroinvasive nature of WNV requires an MAb to overcome its inability to cross the blood–brain barrier (BBB) to reach the central nervous system where the infection occurs. To address this major impediment, a plant-derived MAb (E16) against WNV was developed with various unique N-glycoforms (He et al., 2012, 2014; Lai et al., 2010, 2014). E16 can be robustly produced in both *N. benthamiana* and lettuce plants, easily purified to homogeneity with a scalable and cGMP-compliant method, and that a single dose of plant-made E16 protected mice from lethal infection of WNV, even given 4 days postinfection (Lai et al., 2010). However, peripheral delivery of E16 likely will have a limited window of efficacy due to its inability to efficiently cross the BBB and accumulate in the brain at concentrations sufficient for neutralization. In response to this need, bifunctional MAbs have been developed in plants that have shown promise in gaining the ability to cross the BBB but retaining a potent ability in neutralizing WNV (Chen, Lai, & He, 2016).

The unique aspect of DENV pathogenesis is antibody-dependent enhancement of infection (ADE). ADE occurs because subneutralizing concentrations of antibodies (including therapeutic MAbs) form complexes with the infecting DENV that bind to FcγR-bearing myeloid cells (monocytes, macrophages, and dendritic cells), resulting in increased virus uptake and infection (Morens, 1994). DENV has four serotypes and MAbs against one serotype are often nonneutralizing or subneutralizing against other serotypes, leading to ADE with high levels of proinflammatory cytokines in blood and tissues, vascular leakage, and shock syndrome. Therefore, ADE renders anti-DENV MAb-treated subjects more susceptible to serious infection, presenting a major hurtle for the development of anti-DENV MAb therapeutics. Recent ZIKV outbreaks have been linked to the development of severe foetal abnormalities that include microcephaly and Guillain–Barré syndrome (Attar, 2016; Cao-Lormeau et al., 2016). ZIKV and DENV are closely related and antibodies against one ZIKV often cross-react with different serotypes of DENV, strongly indicating that ADE may occur between these two cocirculating viruses. This has serious implications for the safety of therapeutics against both viruses. Since the binding of human IgG antibodies to FcγR is highly sensitive to the nature of the N-linked glycosylation they carry, the ADE challenge was tackled by utilizing a portfolio of *N. benthamiana* plant lines that allow the production of MAbs with tailor-made mammalian N-glycans. Results demonstrated that in contrast to mammalian cell-made anti-DENV antibodies, all plant-made glycovariants lost ADE activity (Fig. 7) (Dent et al., 2016; He et al., 2014). These results

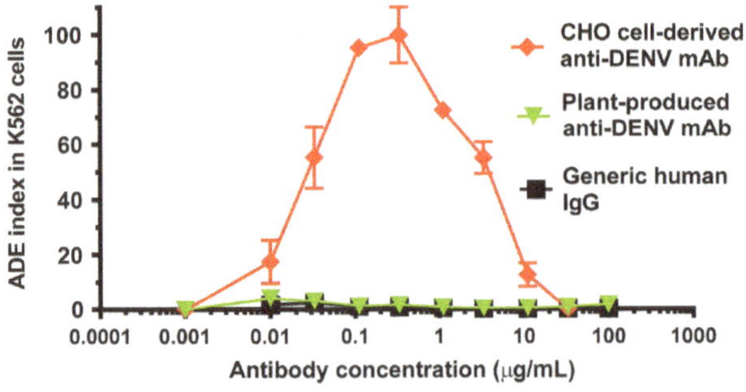

Fig. 7 A plant-derived anti-Dengue virus MAb forgoes the side effect of enhancing the secondary infection of heterologous strain of Dengue virus (ADE). Serial dilutions of an anti-DENV MAb variants or a generic human IgG (negative control) were mixed with DENV and added to FcγR-expressing K562 cells. Forty-eight hours later, cells were fixed, permeabilized, and stained with anti-DENV E antibody 4G2 and analysed by flow cytometry for DENV infection of cells.

demonstrate the potential of plant-made MAbs and their variants as effective yet safe therapeutics against DENV, ZIKV, and other ADE-prone viruses.

6.2 Vaccines

Inadequate immunization contributes to the majority of death of people in the world. There is an urgent need for affordable vaccines especially in developing countries. Plant-based production systems are in the unique position to produce low-cost vaccines. As discussed earlier, fresh or dried plant parts containing vaccine therapeutics may present ambient temperature-stable edible vaccines to circumvent the needs for the costly downstream process, the cold chain, and medical supplies associated with injection. Since the expression of hepatitis B surface antigen in transgenic tobacco in 1992, an increasing number of vaccine candidates have been produced in a variety of plant hosts and vector systems (Chen, 2008; Chen et al., 2009). Collectively, these studies demonstrated that: (1) plants are capable to express a broad portfolio of vaccines; (2) plant-derived antigens were able to induce protective humoral and cell-mediated immune responses; and (3) plant-based systems offer numerous advantages over the traditional platforms for vaccine production. The following examples demonstrate some of these advantages.

The development of plant-made Norwalk virus (NV) capsid protein (NVCP) VLP as a vaccine candidate illustrates the progress and remaining challenges of the plant-made therapeutic field. NVCP was first expressed in transgenic tobacco, potato, and tomato plants (Herbst-Kralovetz, Mason, & Chen, 2010). It required up to a year to obtain plants even with low NVCP-expressing levels (Santi, Huang, & Mason, 2006). Nevertheless, assembled VLPs were observed (Santi et al., 2006). To eliminate the long lead time for generating stable transgenic plants, the transient expression systems were used for NVCP VLP production, which allowed the production of fully assembled VLPs at high levels within 4–12 days of infiltration in *N. benthamiana* or lettuce plants, at least an 80-fold greater accumulation than in transgenic tobacco and tomato (Chen et al., 2011; Lai et al., 2012; Santi et al., 2008). Studies also demonstrated the feasibility of using commercially produced lettuce for high-level and rapid VLP production, allowing the access to unlimited quantities of inexpensive plant material for industry-scale production (Lai et al., 2012). At this point, remaining challenges to the commercialization of plant-made VLPs as a vaccine included the lack of scalable downstream processing procedures, and the uncertainty of regulatory compliance for production processes. These were common obstacles shared by other plant-made therapeutics at that time (Chen, 2008, 2011; Chen & Davis, 2016; Chen & Lai, 2015). To overcome these challenges, a novel and scalable extraction and purification scheme was later developed for efficiently recovering NVCP VLPs from the plant tissue (Lai & Chen, 2012). Moreover, the upstream and downstream production processes in this scheme are cGMP compliant and produced high-quality NVCP VLPs that meet all the preset release specifications in identity, purity, potency, and safety (Lai & Chen, 2012). This provided the first precedent of producing a plant-derived vaccine at scale under cGMP regulations in an academic setting and should streamline its future approval as a licenced human vaccine.

Due to the enteropathogenic nature of NV and other viruses in the *Norovirus* genus, an ideal vaccine should induce norovirus-specific gut mucosal immunity. Thus, one of the research focuses was to examine the potential of plant-derived NVCP VLP as an oral mucosal vaccine in the form of minimally processed plant material. Indeed, NVCP VLPs were shown to be acid and protease resistant and stable in the oral-gastrointestinal environment (Ausar, Foubert, Hudson, Vedvick, & Middaugh, 2006; Jiang, Wang, Graham, & Estes, 1992). In addition, NVCP VLP-expressing potato tubers elicited specific intestinal IgA as well as serum IgG and responses when they were used to feed mice (Santi et al., 2006). Excitingly, similar mucosal and systemic immune responses were also observed in human volunteers in a Phase I clinical trial, indicating the immunogenicity and safety of NVCP VLP as vaccines in humans (Chen & Lai, 2013; Chen et al., 2009). Tomato fruits were also used to express NVCP VLPs as they are more palatable and their production can be readily adapted from the well-established greenhouse culture practice and fruit-processing technologies of the food industry (Chen, 2011, 2013). It was shown that oral delivery of freeze-dried tomatoes stimulated strong serum anti-NVCP faecal mucosal IgA and IgG and responses in 100% of mice (Zhang, Buehner, Hutson, Estes, & Mason, 2006). As discussed earlier, transient expression in lettuce has overcome the low expression levels of NVCP VLP in transgenic potato and tomato (Chen et al., 2016; Lai et al., 2012). In addition, uncooked lettuce leaves are already consumed as salad and present another attractive plant tissue for the delivery of oral NVCP VLP vaccines. Oral immunization of mice with transiently expressed NVCP VLPs evoked strong and balanced systemic IgG1/IgG2a response, as well as significant NVCP-specific vaginal and faecal mucosal

IgA responses in all immunized mice in the absence of any adjuvant (Santi et al., 2008). The effect of various adjuvants including toll-like receptors (TLRs) on systemic and mucosal (nasal, bronchoalveolar, salivary, gastrointestinal, faecal, and vaginal) immunity when codelivered orally or nasally with the cGMP-purified VLPs has also been evaluated. It was shown that cholera toxin enhanced NVCP-specific immune response significantly when codelivered orally (Herbst-Kralovetz et al., 2010). In addition, the inclusion of TLR9 and TLR7 agonists in oral or nasal immunization stimulated stronger systemic and mucosal IgG and IgA responses than with VLP alone, with nasal delivery producing the most potent response and being able to induce IgA response in distal mucosa (Hjelm, Kilbourne, & Herbst-Kralovetz, 2014). There is still no licenced norovirus vaccine on the market for human use. However, plant-based systems have provided a platform for the cost-effective and rapid manufacture of VLP vaccines for noroviruses that can induce strong humoral, mucosal, and cellular immune response in animal models, presenting an important step forward for VLP vaccines to become a commercial reality (Chen, 2013; Chen & Lai, 2013; Lai & Chen, 2012; Lai et al., 2012). The robustness and scalability of the VLP expression in lettuce, coupled with the unlimited nature of plant material generation, provide a production platform for orally deliverable NVCP VLPs that is low cost, safe, and amenable to large-scale manufacturing. A new Phase I human clinical trial with NVCP VLPs purified from plants is being planned.

Another successful case that has demonstrated the benefit of plant-based systems is the development of influenza VLP-based vaccines. Specifically, it has demonstrated the superiority of plant systems in simplicity and speed over other manufacturing platforms for controlling potential pandemics of infectious diseases. To be an effective pandemic flu vaccine, it must be produced in the shortest timeframe to halt the spread of the new strain. The simplest vaccine candidate for influenza can be VLPs that are comprised of only haemagglutinin (HA). This vaccine candidate requires the minimal knowledge (i.e. HA-coding sequence) and minimal adaption of manufacturing processes for the new viral strain, thereby imposing fewer constraints on process and product characterization and reducing the risk of failure of process adaptation (D'Aoust et al., 2010). However, HA-VLPs cannot be produced in animal cells because HA binds to the sialylated glycoproteins on the cell surface (Chen, Leser, Morita, & Lamb, 2007). Since plant glycoproteins are not sialylated, HA-VLP production becomes feasible. Other problems plagued the traditional production systems for influenza vaccines, including the short supply of eggs due to massive culling of chickens and the lethality of some vaccines to embryonated eggs, can be also avoided by using plants (Chen & Davis, 2016). Importantly, plant production foregoes the need for strain adaptation, greatly shortening the time required for vaccine production of new viral strains. These advantages were demonstrated in a real-life scenario when an unexpected outbreak of a novel A/H1N1 influenza virus occurred. In stark contrast with current manufacturing technologies which rely on strain adaptation and consequently require >6 additional months before vaccine production can be initiated, it took only 10 days to obtain infiltrated *N. benthamiana* that expressed high levels of HA and another 5 days to obtain the first purified lot of the vaccine from the date that the HA sequence of this new strain was determined (D'Aoust et al., 2010). Plant-derived HA-VLPs have been found to have equivalent if not superior potency compared with vaccines produced in eggs (D'Aoust et al., 2010). The plant-derived vaccine candidates against this and other various influenza strains (e.g. H5N, H1N1) have been tested in Phase I and Phase II human clinical trials (Table 1) with results of being safe, well tolerated, and the potency being among the most effective of the industry (http://www.medicago.com). These studies demonstrated that plant-based platforms are an ideal system to produce therapeutics in response to emerging or reemerging pathogens with unpredictable and frequent genetic drift or stockpile for bioterrorism threats.

The advantage of plant-made vaccines was also demonstrated by the development of safer and efficacious vaccines against flaviviruses, especially the recently emerged ZIKV. Several ZIKV vaccine candidates are being developed using inactivated whole virus, DNA, or RNA that express the virus' Envelope (E) glycoprotein as the antigen. These candidates successfully induce the production of virus-targeted antibodies that protected animals against ZIKV challenges (Abbink et al., 2016; Larocca et al., 2016; Pardi & Weissman, 2017). However, their use potentially may predispose vaccinated subjects to infection by DENV. In response, our research group has devised several protein subunit vaccines based on the E protein of ZIKV (He et al., 2014; Yang, Dent, Lai, Sun, & Chen, 2017; Yang, Lai, Sun, & Chen, 2017; Yang, Sun, Lai, Hurtado, & Chen, 2017). We demonstrated that a VLP that displays the ZIKV E protein domain III (zEDIII-VLP) can be produced quickly and enriched to homogeneity in large quantities from plants. zEDIII-VLPs are shown to be highly immunogenic in eliciting potent humoral and cellular responses that exceed the threshold correlated with protective immunity against multiple strains of ZIKV

(Yang, Lai, et al., 2017). Notably, zEDIII-VLP-induced antibodies did not enhance the infection of DENV in Fc gamma receptor-expressing cells. This result highlights the potential of plant-made vaccines in offsetting the concern of ZIKV vaccines in sensitizing people to subsequent DENV infection.

The success of developing various MAbs in plants also facilitates innovation in vaccine development. For example, plant-derived anti-Ebola MAbs allowed the development of novel vaccines against Ebola. Currently, there is no licenced vaccine available to prevent this detrimental disease in humans. Vaccines in development include candidates based on Venezuelan equine encephalitis virus replicon particles (VRPs) (Pushko et al., 2000) and adenovirus (Richardson et al., 2009). Recently, a vesicular stomatitis virus-vectored vaccine candidate has shown potent efficacy in preventing Ebola in a clinical trial in Guinea, west Africa (Henao-Restrepo et al., 2017). Since these vaccine candidates are based on genetically modified live viruses, they tend to lose potency over time and demand stringent storage conditions. These tendencies are inconsistent with the requirement for biodefense stockpiles for which long-term stability and minimal requirement for cold chain are essential. The availability of anti-Ebola surface glycoprotein 1 (GP1) MAbs encouraged the development of novel Ebola vaccine candidates based on RIC. Specifically, the Ebola GP1 is genetically fused to the C-terminus of 6D8 MAb which specifically recognizes a linear epitope on GP1. When coexpressed with 6D8 LC, this HC-GP1 polypeptide should assemble into an IgG-like structure with two antigen (GP1)-binding sites at its antigen-binding (Fab) domains, and their cognate antigen (GP1) on the opposite end of the molecule. When the Fab domains of the molecule bind to GP1 on other molecules, they should form a repeated array called RIC. A RIC should behave similarly to a native immune complex formed between an antigen and an antibody, which has been shown to have enhanced binding to C1q and FcγRs than antigen alone, to have enhanced their pick up by antigen-presenting cells and to promote the efficient presentation of antigen peptides on major histocompatibility complex class I and II (Regnault et al., 1999). The enhanced endocytosis, the improved antigen presentation, and thereby enhanced T cell activation suggest the potential of RIC as effective vaccine candidate against Ebola. Studies revealed that despite its large size, the HC-GP1 fusion polypeptide was successfully expressed in plants and correctly assembled into a tetrameric IgG-like structure with the LC (Phoolcharoen, Bhoo, et al., 2011). Analysis with an MAb which recognizes a conformational epitope on GP1 demonstrated that GP1 folded properly to form an authentic antigen in the fusion protein. Furthermore, results from dynamic light scattering and size-exclusion chromatography revealed that RIC was indeed formed as predicted and each RIC contains four fusion proteins (Phoolcharoen, Bhoo, et al., 2011). Functional analysis demonstrated that plant-derived RIC had enhanced binding to C1q than the parent 6D8 MAb and evoked Ebola-specific antibody response with titres comparable to those of the VRP-GP1 control (Phoolcharoen, Bhoo, et al., 2011; Phoolcharoen, Dye, et al., 2011). Ultimately, the efficacy of the plant-derived RIC was demonstrated in a mouse model with 80% of mice protected against a lethal challenge of Ebola (Fig. 8) (Phoolcharoen, Dye, et al., 2011). These studies, therefore, demonstrated the potential of RIC as an effective vaccine strategy for filoviruses. Furthermore, the plant production of RIC also overcame the long-standing technical hurdle of producing GP1 due to its toxicity to the production host cell. Our fusion strategy not only resolved the issue of GP1 production but also provides an IgG scaffold that allows its easy purification. As a result, plant-derived RIC also enhances filovirus vaccine manufacturing, making it highly relevant to a national biothreat reduction stockpile.

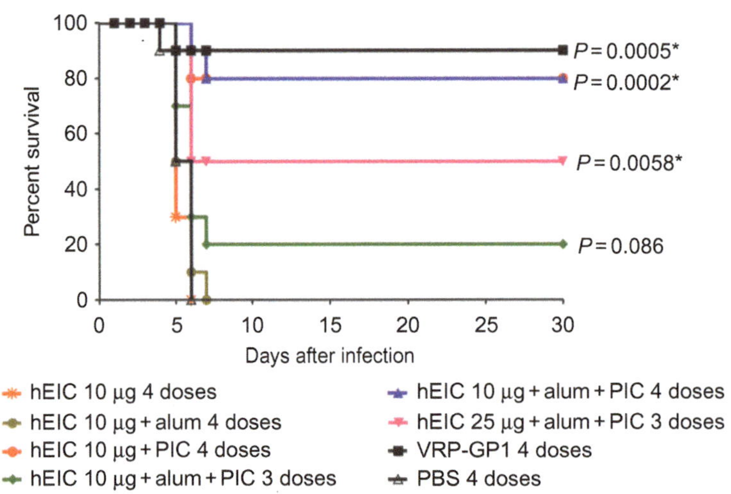

Fig. 8 Plant-produced RIC significantly protected mice from Ebola lethal challenge. Vaccination with three or four doses of RIC with or without candidate adjuvants was performed in mice. After vaccination, mice were challenged i.p. with 1000 PFU (30,000 LD50) of mouse-adapted Ebola virus on day 84. Survival was recorded for 30 days. Statistically significant differences ($P<0.05$) in survival were determined by Log rank test and comparisons were made between vaccination groups relative to PBS control as indicated (*). *hEIC*, humanized antibody Ebola recombinant immune complex; *PIC*, polyinosinic–polycytidylic acid; *VRP-GP1*, recombinant RNA replicons that express GP1 (positive control).

6.3 Plant-Produced Therapeutic Enzymes and Antimicrobials

The application of plant production platforms has reached far beyond the traditional realm of vaccines and therapeutic MAbs. A vast number of non-MAb nonvaccine proteins with human health-related applications have been produced in plants. We will use two examples below to demonstrate the broad application of the plant production technology.

Enterohaemorrhagic *Escherichia coli*-contaminated foods are among the leading causes of bacterial enteric infections in the world. Currently, treatments with heat, organic acid, and antibiotics are the predominant methods in controlling pathogenic bacteria in food. These treatments are not preferred because they either permanently alter the taste and texture of food or promote the emergence of antibiotic-resistant bacterial strains. Colicins are a group of antimicrobial proteins produced by bacteria to inhibit the growth of *E. coli*. Recently, it was shown that most colicins can be expressed up to 30% of total soluble protein or 3 mg/g LFW in both tobacco and edible leafy plants (Schulz et al., 2015). Furthermore, when applied to food products, the plant-derived colicins were shown to be highly and broadly active against all major pathogenic *E. coli* strains that cause food poisoning even at low concentrations. The production cost was estimated to be $1.00 per g of purified colicins, indicating its commercial viability (Schulz et al., 2015). Importantly, plant-produced colicins match those of the colicinogenic strains of bacteria that are members of human healthy microflora, and any impurities from the plant production process would be normal components of food vegetables. This will eliminate potential issues in biosafety regulation. Indeed, FDA's "no questions" response letter to the commercialization request of plant-made colicins signals its potential regulatory approval. The commercialization of this product will significantly reduce bacterial enteric infections worldwide. More broadly, plant-made colicins may present a new class of nonantibiotic antibacterials that could be used to combat the worldwide problem of emerging super-resistant bacterial strains.

The advantages of plant-based systems for producing non-MAb therapeutics have been further demonstrated in the case of glucocerebrosidase (GCD), a therapeutic enzyme for treating Gaucher's disease. GCD was originally extracted from human placentas and later produced in mammalian cell cultures. However, placenta-derived GCD has a limited supply and has been banned due to biosafety issues. Mammalian cell culture can produce a large quantity of GCD, but it carries undesirable N-glycans, thus requiring a complex and costly in vitro N-glycan processing to achieve the optimal efficacy. In contrast, the carrot cell-produced GCD already contains the required glycoform, eliminating the costly N-glycan processing and resulting in better and more consistent efficacy (Zimran et al., 2011). As a result, FDA and other regulatory agencies have approved the use of plant-produced GCD (commercial name: ELELYSO) to treat Gaucher's patients (Fox, 2012). This is the very first plant-made therapeutics ever approved by the FDA and it has been marketed in the United States, Canada, and many Latin American countries. A new version of GCD has been developed for oral delivery. Results from Phase I and Phase II human clinical trials showed that when patients were given a daily drink containing lyophilized GCD-expressing carrot cells, the levels of GCD in their blood circulation matched those of healthy individuals (Shaaltiel et al., 2015). The ability of intact GCD to cross the intestinal epithelial barrier was demonstrated by both in vitro and in vivo models in this study. Even though the exact mechanism of how GCD passes the intestinal barrier awaits further investigation, it was speculated that the relatively lipophilic nature and the high-mannose glycans on its surface may have enhanced its permeability across the intestinal barrier (Shaaltiel et al., 2015). The needle-free delivery of GCD should improve patients' quality of life, encourage treatment compliance, and allow the implementation of therapy in areas where medical supplies are limited.

The FDA approval of ELELYSO has far-reaching significance for the field of plant-made therapeutics. Since the launch of this field, there has been significant scepticism that a biologic product made in plants was ever going to be licenced. This leads to a lingering criticism of the lack of approved human products in major biologic categories after more than 25 years of active research and development. The approval of ELELYSO as a therapeutic drug, along with the compassionate use of ZMapp in human patients and the approval of a plant-made MAb (CaroRx) as a medical device to prevent tooth decay, has silenced this criticism. The approval of ELELYSO by the FDA has also paved a clear regulatory pathway specific for plant-made therapeutics and should streamline the approval of plant-made therapeutics and vaccines that have shown safety and efficacy in human clinical trials (Ma et al., 2015).

7. CONCLUSION

In the past decade, significant progress has been made in the field of plant-made therapeutics especially in expression vector optimization, downstream processing, and plant host engineering. As a result, plant production systems not only offer the traditional advantages of proper eukaryotic protein modification, low cost, high scalability, and increased safety, but they also allow the production of therapeutics at unprecedented speed to control potential pandemics or with specific posttranslational modifications for superior potency or safety. These advantages make plants a superior alternative production system for therapeutic production. For example, plants would be a system of choice to produce biosimilars to meet the increasing global demand of therapeutics because of their large capacity to rapidly generate therapeutics at low cost, easily incorporate posttranslational modifications, and low contamination risks compared with animal or human pathogens. For similar reasons, plants also offer an optimal system to produce therapeutics with an extraordinarily large-quantity demand but with relative low profit margins. Plants will be also essential for producing safer and more effective biobetters due to the flexibility of producing therapeutics with specific and homogeneous mammalian glycoforms that cannot currently be easily produced by other cell culture systems. For the ultimate adoption and success of plant-made therapeutics, it is crucial to address the public concern for GMOs and vastly expand the approval pipeline in several key classes of therapeutics such as MAbs and vaccines within the next decade. The FDA approval of ELELYSO and the success of ZMapp marked a new era for the field of plant-made therapeutics and is a promising sign for the potential of plants in providing efficacious, safe, and affordable therapeutics.

ACKNOWLEDGEMENTS

The author wish to thank Huafang Lai, Ming Yang, Haiyan Sun, Johnathan Hurtado, Adrian Esqueda, Collin Jugler, and other current and former members of the author's laboratory for their contribution to the research cited in this chapter. Research in the author's laboratory is supported in part by NIH-NIAID Grants U01 AI075549, 1R21AI101329, and R33AI101329 to Q.C.

REFERENCES

Abbink, P., Larocca, R. A., De La Barrera, R. A., Bricault, C. A., Moseley, E. T., Boyd, M., et al. (2016). Protective efficacy of multiple vaccine platforms against Zika virus challenge in rhesus monkeys. *Science*, *353*(6304), 1129. Retrieved from http://science.sciencemag.org/content/353/6304/1129.abstract.

Arntzen, C. (2015). Plant-made pharmaceuticals: From 'edible vaccines' to Ebola therapeutics. *Plant Biotechnology Journal*, *13*(8), 1013–1016. https://doi.org/10.1111/pbi.12460.

Attar, N. (2016). ZIKA virus circulates in new regions. *Nature Reviews Microbiology*, *14*(2), 62. https://doi.org/10.1038/nrmicro.2015.28.

Ausar, S. F., Foubert, T. R., Hudson, M. H., Vedvick, T. S., & Middaugh, C. R. (2006). Conformational stability and disassembly of Norwalk virus like particles: Effect of pH and temperature. *The Journal of Biological Chemistry*, *281*(28), 19478–19488. https://doi.org/10.1074/jbc.M603313200.

Bendandi, M., Marillonnet, S., Kandzia, R., Thieme, F., Nickstadt, A., Herz, S., et al. (2010). Rapid, high-yield production in plants of individualized idiotype vaccines for non-Hodgkin's lymphoma. *Annals of Oncology, 21*(12), 2420–2427. https://doi.org/10.1093/annonc/mdq256.

Bethell, D. R. (2007). *Generally recognized as safe (GRAS) notification for lactoferrin (human) from rice for use as an ingredient in pediatric medical foods and oral rehydration solutions.*

Cao-Lormeau, V.-M., Blake, A., Mons, S., Lastère, S., Roche, C., Vanhomwegen, J., et al. (2016). Guillain-Barré syndrome outbreak associated with Zika virus infection in French Polynesia: A case-control study. *The Lancet, 387*(10027), 1531–1539. https://doi.org/10.1016/S0140-6736(16)00562-6.

Check Hayden, E., & Reardon, S. (2014). Should experimental drugs be used in the Ebola outbreak? *Nature (London)*. https://doi.org/10.1038/nature.2014.15698.

Chen, Q. (2008). Expression and purification of pharmaceutical proteins in plants. *Biological Engineering, 1*(4), 291–321.

Chen, Q. (2011). Expression and manufacture of pharmaceutical proteins in genetically engineered horticultural plants. In B. Mou & R. Scorza (Eds.), *Transgenic horticultural crops: Challenges and opportunities—Essays by experts* (pp. 83–124). Boca Raton: Taylor & Francis.

Chen, Q. (2013). Virus-like particle vaccines for Norovirus gastroenteritis. In M. Giese (Ed.), *Molecular vaccine: Vol. 1* (pp. 153–181). Vienna: Springer. https://doi.org/10.1007/978-3-7091-1419-3_8.

Chen, Q. (2015). Plant-made vaccines against West Nile virus are potent, safe, and economically feasible. *Biotechnology Journal, 10*(5), 671–680. https://doi.org/10.1002/biot.201400428.

Chen, Q. (2016). Glycoengineering of plants yields glycoproteins with polysialylation and other defined N-glycoforms. *Proceedings of the National Academy of Sciences of the United States of America, 113*(34), 9404–9406. https://doi.org/10.1073/pnas.1610803113.

Chen, Q., & Davis, K. (2016). The potential of plants as a system for the development and production of human biologics F1000Research, 5(912), doi: 10.12688/f11000research.18010.12681. Retrieved from http://f1000r.es/6nf.

Chen, Q., Dent, M., Hurtado, J., Stahnke, J., McNulty, A., Leuzinger, K., et al. (2014). Transient protein expression by agroinfiltration in lettuce. In R. Menassa, J. Macdonald, & I. Kolotilin (Eds.), *Recombinant proteins from plants, 2.*

Chen, Q., Dent, M., Hurtado, J., Stahnke, J., McNulty, A., Leuzinger, K., et al. (2016). Transient protein expression by agroinfiltration in lettuce. *Methods in Molecular Biology, 1385*, 55–67. https://doi.org/10.1007/978-1-4939-3289-4_4.

Chen, X., Equi, R., Baxter, H., Berk, K., Han, J., Agarwal, S., et al. (2010). A high-throughput transient gene expression system for switchgrass (Panicum virgatum L.) seedlings. *Biotechnology for Biofuels, 3*, 9.

Chen, Q., He, J., Phoolcharoen, W., & Mason, H. S. (2011). Geminiviral vectors based on bean yellow dwarf virus for production of vaccine antigens and monoclonal antibodies in plants. *Human Vaccines*, 7(3), 331–338. Retrieved from http://www.ncbi.nlm.nih.gov/entrez/query.fcgi?cmd=Retrieve&db=PubMed&dopt=Citation&list_uids=21358270.

Chen, Q., & Lai, H. (2013). Plant-derived virus-like particles as vaccines. *Human Vaccines & Immunotherapeutics, 9*(1), 26–49. Retrieved from http://www.landesbioscience.com/journals/vaccines/article/22218/.

Chen, Q., & Lai, H. (2014a). Gene delivery into plant cells for recombinant protein production. *BioMed Research International, 2014*, 10. https://doi.org/10.1155/2014/932161.

Chen, Q., & Lai, H. (2014b). Plant-derived monoclonal antibodies as human biologics for infectious disease and cancer. In K. L. Hefferon (Ed.), *Plant-derived pharmaceuticals: Principles and applications for developing countries* (pp. 42–75). Croydon, UK: CABI.

Chen, Q., & Lai, H. (2015). The growing potential of plant-made monoclonal antibodies. *Drug Target Review, 2*(4), 41–44. Retrieved from http://www.drugtargetreview.com/8059/drug-target-review-journal/issue-4-2015/protein-expression-the-growing-potential-of-plant-made-monoclonal-antibodies/.

Chen, Q., Lai, H., & He, J. (2016). Bispecific monoclonal antibody therapeutics against West Nile virus with improved CNS penetration, US patent # US 9499608 B2. Retrieved from http://patft.uspto.gov/netacgi/nph-Parser?Sect2=PTO1&Sect2=HITOFF&p=1&u=/netahtml/PTO/search-bool.html&r=1&f=G&l=50&d=PALL&RefSrch=yes&Query=PN/9499608.

Chen, B. J., Leser, G. P., Morita, E., & Lamb, R. A. (2007). Influenza virus hemagglutinin and neuraminidase, but not the matrix protein, are required for assembly and budding of plasmid-derived virus-like particles. *Journal of Virology, 81*(13), 7111–7123. Retrieved from http://www.ncbi.nlm.nih.gov/entrez/query.fcgi?cmd=Retrieve&db=PubMed&dopt=Citation&list_uids=17475660.

Chen, Q., Mason, H., Mor, T., Cardineau, G. A., Arntzen, C., & Tacket, C. O. (2009). Subunit vaccines produced using plant biotechnology. In M. M. Levine (Ed.), *New generation vaccines* (4th ed., pp. 306–315). New York: Informa Healthcare USA, Inc.

Chen, Q., Zhang, C., & Santi, L. (2014). Plant-made biologics. *BioMed Research International, 2014*, 3. https://doi.org/10.1155/2014/418064.

Daniell, H. (1999). GM crops: Public perception and scientific solutions. *Trends in Plant Science, 4*(12), 467–469. https://doi.org/10.1016/S1360-1385(99)01503-4.

Daniell, H. (2006). Production of biopharmaceuticals and vaccines in plants via the chloroplast genome. *Biotechnology Journal, 1*(10), 1071–1079. Retrieved from http://www.ncbi.nlm.nih.gov/entrez/query.fcgi?cmd=Retrieve&db=PubMed&dopt=Citation&list_uids=17004305.

D'Aoust, M. A., Couture, M. M., Charland, N., Trepanier, S., Landry, N., Ors, F., et al. (2010). The production of hemagglutinin-based virus-like particles in plants: A rapid, efficient and safe response to pandemic influenza. *Plant Biotechnology Journal, 8*(5), 607–619. Retrieved from http://www.ncbi.nlm.nih.gov/entrez/query.fcgi?cmd=Retrieve&db=PubMed&dopt=Citation&list_uids=20199612.

Dent, M., Hurtado, J., Paul, A. M., Sun, H., Lai, H., Yang, M., et al. (2016). Plant-produced anti-dengue virus monoclonal antibodies exhibit reduced antibody-dependent enhancement of infection activity. *Journal of General Virology, 97*(12), 3280–3290. https://doi.org/10.1099/jgv.0.000635.

Dugdale, B., Mortimer, C. L., Kato, M., James, T. A., Harding, R. M., & Dale, J. L. (2013). Plant activation: An inducible, hyperexpression platform for recombinant protein production in plants. *The Plant Cell Online, 25*(7), 2429–2443. https://doi.org/10.1105/tpc.113.113944.

Fox, J. L. (2012). First plant-made biologic approved. *Nature Biotechnology, 30*(6), 472. https://doi.org/10.1038/nbt0612-472.

Fulton, A., Lai, H., Chen, Q., & Zhang, C. (2015). Purification of monoclonal antibody against Ebola GP1 protein expressed in Nicotiana benthamiana. *Journal of Chromatography A, 1389*(0), 128–132. https://doi.org/10.1016/j.chroma.2015.02.013.

Giritch, A., Marillonnet, S., Engler, C., van Eldik, G., Botterman, J., Klimyuk, V., et al. (2006). Rapid high-yield expression of full-size IgG antibodies in plants coinfected with noncompeting viral vectors. *Proceedings of the National Academy of Sciences of the United States of America, 103*(40), 14701–14706. Retrieved from http://www.ncbi.nlm.nih.gov/entrez/query.fcgi?cmd=Retrieve&db=PubMed&dopt=Citation&list_uids=16973752.

Gleba, Y., Tusé, D., & Giritch, A. (2014). Plant viral vectors for delivery by agrobacterium. In K. Palmer & Y. Gleba (Eds.), *Plant viral vectors: Vol. 375* (pp. 155–192). Springer Berlin Heidelberg.

He, J., Lai, H., Brock, C., & Chen, Q. (2012). A novel system for rapid and cost-effective production of detection and diagnostic reagents of West Nile virus in plants. *Journal of Biomedicine and Biotechnology*, *2012*, 1–10. https://doi.org/10.1155/2012/106783.

He, J., Lai, H., Engle, M., Gorlatov, S., Gruber, C., Steinkellner, H., et al. (2014). Generation and analysis of novel plant-derived antibody-based therapeutic molecules against West Nile virus. *PLoS One*, *9*(3). e93541. https://doi.org/10.1371/journal.pone.0093541.

He, Y., Ning, T., Xie, T., Qiu, Q., Zhang, L., Sun, Y., et al. (2011). Large-scale production of functional human serum albumin from transgenic rice seeds. *Proceedings of the National Academy of Sciences of the United States of America*, *108*(47), 19078–19083. https://doi.org/10.1073/pnas.1109736108.

He, J., Peng, L., Lai, H., Hurtado, J., Stahnke, J., & Chen, Q. (2014). A plant-produced antigen elicits potent immune responses against West Nile virus in mice. *BioMed Research International*, *2014*, 10. https://doi.org/10.1155/2014/952865.

Henao-Restrepo, A. M., Camacho, A., Longini, I. M., Watson, C. H., Edmunds, W. J., Egger, M., et al. (2017). Efficacy and effectiveness of an rVSV-vectored vaccine in preventing Ebola virus disease: Final results from the Guinea ring vaccination, open-label, cluster-randomised trial. *The* Lancet, *389*(10068), 505–518. https://doi.org/10.1016/S0140-6736(16)32621-6.

Herbst-Kralovetz, M., Mason, H. S., & Chen, Q. (2010). Norwalk virus-like particles as vaccines. *Expert Review of Vaccines*, *9*(3), 299–307. https://doi.org/10.1586/erv.09.163.

Hey, C., & Zhang, C. (2012). Process development for antibody purification from tobacco by protein A affinity chromatography. *Chemical Engineering & Technology*, *35*(1), 142–148. https://doi.org/10.1002/ceat.201100259.

Hiwasa-Tanase, K., & Ezura, H. (2016). Molecular breeding to create optimized crops: From genetic manipulation to potential applications in plant factories. *Frontiers in Plant Science*, 7, 539. https://doi.org/10.3389/fpls.2016.00539.

Hjelm, B. E., Kilbourne, J., & Herbst-Kralovetz, M. M. (2014). TLR7 and 9 agonists are highly effective mucosal adjuvants for norovirus virus-like particle vaccines. *Human Vaccines & Immunotherapeutics*, *10*(2), 410–416. https://doi.org/10.4161/hv.27147.

Huang, Z., Chen, Q., Hjelm, B., Arntzen, C., & Mason, H. (2009). A DNA replicon system for rapid high-level production of virus-like particles in plants. *Biotechnology and Bioengineering*, *103*(4), 706–714. Retrieved from https://doi.org/10.1002/bit.22299.

Huang, Z., Phoolcharoen, W., Lai, H., Piensook, K., Cardineau, G., Zeitlin, L., et al. (2010). High-level rapid production of full-size monoclonal antibodies in plants by a single-vector DNA replicon system. *Biotechnology and Bioengineering*, *106*(1), 9–17. Retrieved from https://doi.org/10.1002/bit.22652.

Jiang, X., Wang, M., Graham, D. Y., & Estes, M. K. (1992). Expression, self-assembly, and antigenicity of the Norwalk virus capsid protein. *Journal of Virology*, *66*(11), 6527–6532.

Klee, H., Horsch, R., & Rogers, S. (1987). Agrobacterium-mediated plant transformation and its further applications to plant biology. *Annual Review of Plant Physiology*, *38*, 467–486.

Komarova, T. V., Baschieri, S., Donini, M., Marusic, C., Benvenuto, E., & Dorokhov, Y. L. (2010). Transient expression systems for plant-derived biopharmaceuticals. *Expert Review of Vaccines*, *9*(8), 859–876. https://doi.org/10.1586/erv.10.85.

Kwon, K.-C., & Daniell, H. (2015). Low-cost oral delivery of protein drugs bioencapsulated in plant cells. *Plant Biotechnology Journal*, *13*(8), 1017–1022. https://doi.org/10.1111/pbi.12462.

Lai, H., & Chen, Q. (2012). Bioprocessing of plant-derived virus-like particles of Norwalk virus capsid protein under current Good Manufacture Practice regulations. *Plant Cell Reports*, *31*(3), 573–584. Retrieved from https://doi.org/10.1007/s00299-011-1196-6.

Lai, H., Engle, M., Fuchs, A., Keller, T., Johnson, S., Gorlatov, S., et al. (2010). Monoclonal antibody produced in plants efficiently treats West Nile virus infection in mice. *Proceedings of the National Academy of Sciences of the United States of America, 107*(6), 2419–2424. https://doi.org/10.1073/pnas.0914503107.

Lai, H., He, J., Engle, M., Diamond, M. S., & Chen, Q. (2012). Robust production of virus-like particles and monoclonal antibodies with geminiviral replicon vectors in lettuce. *Plant Biotechnology Journal, 10*(1), 95–104. Retrieved from https://doi.org/10.1111/j.1467-7652.2011.00649.x.

Lai, H., He, J., Hurtado, J., Stahnke, J., Fuchs, A., Mehlhop, E., et al. (2014). Structural and functional characterization of an anti-West Nile virus monoclonal antibody and its single-chain variant produced in glycoengineered plants. *Plant Biotechnology Journal, 12*(8), 1098–1107. https://doi.org/10.1111/pbi.12217.

Lakshmi, P. S., Verma, D., Yang, X., Lloyd, B., & Daniell, H. (2013). Low cost tuberculosis vaccine antigens in capsules: Expression in chloroplasts, bio-encapsulation, stability and functional evaluation in vitro. *PLoS One, 8*(1), e54708. https://doi.org/10.1371/journal.pone.0054708.

Larocca, R. A., Abbink, P., Peron, J. P. S., Zanotto, P. M., Iampietro, M. J., Badamchi-Zadeh, A., et al. (2016). Vaccine protection against Zika virus from Brazil. *Nature, 536*(7617), 474–478. https://doi.org/10.1038/nature18952.

Leuzinger, K., Dent, M., Hurtado, J., Stahnke, J., Lai, H., Zhou, X., et al. (2013). Efficient agroinfiltration of plants for high-level transient expression of recombinant proteins. *Journal of Visualized Experiments* 77. https://doi.org/10.3791/50521. Retrieved from http://www.jove.com/video/50521.

Lico, C., Chen, Q., & Santi, L. (2008). Viral vectors for production of recombinant proteins in plants. *Journal of Cellular Physiology, 216*(2), 366–377. Retrieved from http://www.ncbi.nlm.nih.gov/entrez/query.fcgi?cmd=Retrieve&db=PubMed&dopt=Citation&list_uids=18330886.

Loos, A., Gruber, C., Altmann, F., Mehofer, U., Hensel, F., Grandits, M., et al. (2014). Expression and glycoengineering of functionally active heteromultimeric IgM in plants. *Proceedings of the National Academy of Sciences of the United States of America, 111*(17), 6263–6268. https://doi.org/10.1073/pnas.1320544111.

Lyon, G. M., Mehta, A. K., Varkey, J. B., Brantly, K., Plyler, L., McElroy, A. K., et al. (2014). Clinical care of two patients with Ebola virus disease in the United States. *New England Journal of Medicine, 371*(25), 2402–2409. https://doi.org/10.1056/NEJMoa1409838.

Ma, J. K. C., Drossard, J., Lewis, D., Altmann, F., Boyle, J., Christou, P., et al. (2015). Regulatory approval and a first-in-human phase I clinical trial of a monoclonal antibody produced in transgenic tobacco plants. *Plant Biotechnology Journal, 13*(8), 1106–1120. https://doi.org/10.1111/pbi.12416.

Maruyama, N., Fujiwara, K., Yokoyama, K., Cabanos, C., Hasegawa, H., Takagi, K., et al. (2014). Stable accumulation of seed storage proteins containing vaccine peptides in transgenic soybean seeds. *Journal of Bioscience and Bioengineering, 118*(4), 441–447. https://doi.org/10.1016/j.jbiosc.2014.04.004.

McCormick, A. A., Reddy, S., Reinl, S. J., Cameron, T. I., Czerwinkski, D. K., Vojdani, F., et al. (2008). Plant-produced idiotype vaccines for the treatment of non-Hodgkin's lymphoma: Safety and immunogenicity in a phase I clinical study. *Proceedings of the National Academy of Sciences of the United States of America, 105*(29), 10131–10136. https://doi.org/10.1073/pnas.0803636105.

Morandini, F., Avesani, L., Bortesi, L., Van Droogenbroeck, B., De Wilde, K., Arcalis, E., et al. (2011). Non-food/feed seeds as biofactories for the high-yield production of recombinant pharmaceuticals. *Plant Biotechnology Journal, 9*(8), 911–921. https://doi.org/10.1111/j.1467-7652.2011.00605.x.

Morens, D. M. (1994). Antibody-dependent of enhancement of infection and the pathogenesis of viral disease. *Clinical Infectious Diseases*, *19*, 500–512.

Olinger, G. G., Pettitt, J., Kim, D., Working, C., Bohorov, O., Bratcher, B., et al. (2012). Delayed treatment of Ebola virus infection with plant-derived monoclonal antibodies provides protection in rhesus macaques. *Proceedings of the National Academy of Sciences of the United States of America*, *109*(44), 18030–18035. https://doi.org/10.1073/pnas.1213709109.

Pardi, N., & Weissman, D. (2017). Nucleoside modified mRNA vaccines for infectious diseases. In T. Kramps & K. Elbers (Eds.), *RNA vaccines: Methods and protocols* (pp. 109–121). New York, NY: Springer New York.

Phoolcharoen, W., Bhoo, S. H., Lai, H., Ma, J., Arntzen, C. J., Chen, Q., et al. (2011). Expression of an immunogenic Ebola immune complex in Nicotiana Benthamiana. *Plant Biotechnology Journal*, *9*(7), 807–816. Retrieved from http://www.ncbi.nlm.nih.gov/entrez/query.fcgi?cmd=Retrieve&db=PubMed&dopt=Citation&list_uids=21281425.

Phoolcharoen, W., Dye, J. M., Kilbourne, J., Piensook, K., Pratt, W. D., Arntzen, C. J., et al. (2011). A nonreplicating subunit vaccine protects mice against lethal Ebola virus challenge. *Proceedings of the National Academy of Sciences of the United States of America*, *108*(51), 20695–20700. https://doi.org/10.1073/pnas.1117715108.

Pogue, G., Vojdani, F., Palmer, K. C., Hiatt, E., Hume, S., Phelps, J., et al. (2010). Production of pharmaceutical-grade recombinant aprotinin and a monoclonal antibody product using plant-based transient expression systems. *Plant Biotechnology Journal*, *8*(5), 638–654. Retrieved from https://doi.org/10.1111/j.1467-7652.2009.00495.x.

Pushko, P., Bray, M., Ludwig, G. V., Parker, M., Schmaljohn, A., Sanchez, A., et al. (2000). Recombinant RNA replicons derived from attenuated Venezuelan equine encephalitis virus protect guinea pigs and mice from Ebola hemorrhagic fever virus. *Vaccine*, *19*(1), 142–153. Retrieved from http://www.ncbi.nlm.nih.gov/entrez/query.fcgi?cmd=Retrieve&db=PubMed&dopt=Citation&list_uids=10924796.

Qiu, X., Wong, G., Audet, J., Bello, A., Fernando, L., Alimonti, J. B., et al. (2014). Reversion of advanced Ebola virus disease in nonhuman primates with ZMapp. *Nature*, *514*(7520), 47–53. https://doi.org/10.1038/nature13777.

Ramessar, K., Rademacher, T., Sack, M., Stadlmann, J., Platis, D., Stiegler, G., et al. (2008). Cost-effective production of a vaginal protein microbicide to prevent HIV transmission. *Proceedings of the National Academy of Sciences of the United States of America*, *105*(10), 3727–3732. https://doi.org/10.1073/pnas.0708841104.

Regnault, A., Lankar, D., Lacabanne, V., Rodriguez, A., Thery, C., Rescigno, M., et al. (1999). Fcgamma receptor-mediated induction of dendritic cell maturation and major histocompatibility complex class I-restricted antigen presentation after immune complex internalization. *Journal of Experimental Medicine*, *189*(2), 371–380. Retrieved from http://www.ncbi.nlm.nih.gov/entrez/query.fcgi?cmd=Retrieve&db=PubMed&dopt=Citation&list_uids=9892619.

Richardson, J. S., Yao, M. K., Tran, K. N., Croyle, M. A., Strong, J. E., Feldmann, H., et al. (2009). Enhanced protection against Ebola virus mediated by an improved adenovirus-based vaccine. *PLoS One*, *4*(4), e5308. Retrieved from http://www.ncbi.nlm.nih.gov/entrez/query.fcgi?cmd=Retrieve&db=PubMed&dopt=Citation&list_uids=19390586.

Rivera, A., Gomez-Lim, M., Fernandez, F., & Loske, A. (2012). Physical methods for genetic plant transformation. *Physics of Life Reviews*, *9*(3), 308–345. Retrieved from http://www.sciencedirect.com/science/article/pii/S1571064512000498.

Roque, A. C. A., Lowe, C. R., & Taipa, M. A. (2004). Antibodies and genetically engineered related molecules: Production and purification. *Biotechnology Progress*, *20*(3), 639–654. Retrieved from http://pubs3.acs.org/acs/journals/doilookup?in_doi=10.1021/bp030070k.

Santi, L., Batchelor, L., Huang, Z., Hjelm, B., Kilbourne, J., Arntzen, C. J., et al. (2008). An efficient plant viral expression system generating orally immunogenic Norwalk virus-like particles. *Vaccine, 26*(15), 1846–1854. https://doi.org/10.1016/j.vaccine.2008.01.053.

Santi, L., Huang, Z., & Mason, H. (2006). Virus-like particles production in green plants. *Methods, 40*(1), 66–76. Retrieved from http://www.ncbi.nlm.nih.gov/entrez/query.fcgi?cmd=Retrieve&db=PubMed&dopt=Citation&list_uids=16997715.

Schulz, S., Stephan, A., Hahn, S., Bortesi, L., Jarczowski, F., Bettmann, U., et al. (2015). Broad and efficient control of major foodborne pathogenic strains of Escherichia coli by mixtures of plant-produced colicins. *Proceedings of the National Academy of Sciences of the United States of America, 112*(40), E5454–E5460. https://doi.org/10.1073/pnas.1513311112.

Shaaltiel, Y., Gingis-Velitski, S., Tzaban, S., Fiks, N., Tekoah, Y., & Aviezer, D. (2015). Plant-based oral delivery of β-glucocerebrosidase as an enzyme replacement therapy for Gaucher's disease. *Plant Biotechnology Journal, 13*(8), 1033–1040. https://doi.org/10.1111/pbi.12366.

Stoger, E., Vaquero, C., Torres, E., Sack, M., Nicholson, L., Drossard, J., et al. (2000). Cereal crops as viable production and storage systems for pharmaceutical scFv antibodies. *Plant Molecular Biology, 42*(4), 583–590. Retrieved from http://www.ncbi.nlm.nih.gov/entrez/query.fcgi?cmd=Retrieve&db=PubMed&dopt=Citation&list_uids=10809004.

Strasser, R., Stadlmann, J., Schahs, M., Stiegler, G., Quendler, H., Mach, L., et al. (2008). Generation of glyco-engineered Nicotiana benthamiana for the production of monoclonal antibodies with a homogeneous human-like N-glycan structure. *Plant Biotechnology Journal, 6*(4), 392–402. https://doi.org/10.1111/j.1467-7652.2008.00330.x.

Su, J., Sherman, A., Doerfler, P. A., Byrne, B. J., Herzog, R. W., & Daniell, H. (2015). Oral delivery of acid alpha glucosidase epitopes expressed in plant chloroplasts suppresses antibody formation in treatment of Pompe mice. *Plant Biotechnology Journal, 13*(8), 1023–1032. https://doi.org/10.1111/pbi.12413.

Su, J., Zhu, L., Sherman, A., Wang, X., Lin, S., Kamesh, A., et al. (2015). Low cost industrial production of coagulation factor IX bioencapsulated in lettuce cells for oral tolerance induction in hemophilia B. *Biomaterials, 70*, 84–93. https://doi.org/10.1016/j.biomaterials.2015.08.004.

Takaiwa, F., & Yang, L. (2014). Development of a rice-based peptide vaccine for Japanese cedar and cypress pollen allergies. *Transgenic Research, 23*(4), 573–584. https://doi.org/10.1007/s11248-014-9790-3.

Takata, N., & Eriksson, M. E. (2012). A simple and efficient transient transformation for hybrid aspen (Populus tremula x P. tremuloides). *Plant Methods, 8*(1), 30. Retrieved from http://www.plantmethods.com/content/8/1/30.

Thuenemann, E. C., Meyers, A. E., Verwey, J., Rybicki, E. P., & Lomonossoff, G. P. (2013). A method for rapid production of heteromultimeric protein complexes in plants: Assembly of protective bluetongue virus-like particles. *Plant Biotechnology Journal, 11*(7), 839–846. Retrieved from https://doi.org/10.1111/pbi.12076.

Tuse, D., Tu, T., & McDonald, K. (2014). Manufacturing economics of plant-made biologics: Case studies in therapeutic and industrial enzymes. *BioMed Research International, 2014*(256135), 1–16. https://doi.org/10.1155/2014/256135.

Vaghchhipawala, Z., Rojas, C. M., Senthil-Kumar, M., & Mysore, K. S. (2011). Agroinoculation and agroinfiltration: Simple tools for complex gene function analyses. *Methods in Molecular Biology, 678*, 65–76. Retrieved from http://www.ncbi.nlm.nih.gov/entrez/query.fcgi?cmd=Retrieve&db=PubMed&dopt=Citation&list_uids=20931373.

Walsh, G. (2014). Biopharmaceutical benchmarks 2014. *Nature Biotechnology, 32*(10), 992–1000. https://doi.org/10.1038/nbt.3040.

Yang, M., Dent, M., Lai, H., Sun, H., & Chen, Q. (2017). Immunization of Zika virus envelope protein domain III induces specific and neutralizing immune responses against Zika virus. *Vacccine*, *35*(33), 4287–4294. https://.doi.org/10.1016/j.vaccine.2017.04.052.

Yang, M., Lai, H., Sun, H., & Chen, Q. (2017). Virus-like particles that display Zika virus envelope protein domain III induce potent neutralizing immune responses in mice. *Scientific Reports*, 7(1), 7679. https://doi.org/10.1038/s41598-017-08247-9.

Yang, M., Sun, H., Lai, H., Hurtado, J., & Chen, Q. (2017). Plant-produced Zika virus envelope protein elicits neutralizing immune responses that correlate with protective immunity against Zika virus in mice. *Plant Biotechnology Journal*. https://doi.org/10.1111/pbi.12796.

Zeitlin, L., Pettitt, J., Scully, C., Bohorova, N., Kim, D., Pauly, M., et al. (2011). Enhanced potency of a fucose-free monoclonal antibody being developed as an Ebola virus immunoprotectant. *Proceedings of the National Academy of Sciences of the United States of America*, *108*(51), 20690–20694. https://doi.org/10.1073/pnas.1108360108.

Zhang, X., Buehner, N. A., Hutson, A. M., Estes, M. K., & Mason, H. S. (2006). Tomato is a highly effective vehicle for expression and oral immunization with Norwalk virus capsid protein. *Plant Biotechnology Journal*, *4*(4), 419–432. Retrieved from http://www.ncbi.nlm.nih.gov/entrez/query.fcgi?cmd=Retrieve&db=PubMed&dopt=Citation&list_uids=17177807.

Zimran, A., Brill-Almon, E., Chertkoff, R., Petakov, M., Blanco-Favela, F., Munoz, E. T., et al. (2011). Pivotal trial with plant cell-expressed recombinant glucocerebrosidase, taliglucerase alfa, a novel enzyme replacement therapy for Gaucher disease. *Blood*, *118*(22), 5767–5773. https://doi.org/10.1182/blood-2011-07-366955.

CHAPTER EIGHT

Genome Editing in Agricultural Biotechnology

Maxence Pfeiffer*, Francis Quétier[†,‡], Agnès Ricroch[*,§,1]

*AgroParisTech, Paris, France
[†]University of Evry Val d'Essonne, Evry, France
[‡]Génopole-Evry, Evry, France
[§]University of Paris-Sud, Sceaux, France
[1]Corresponding author: e-mail address: agnes.ricroch@u-psud.fr

Contents

Abstract

Genome editing with engineered nucleases represents a specific and efficient tool to generate useful novel phenotypes in crops with an economic interest by base additions, deletions, gene replacement or transgene insertion. These techniques generate

Advances in Botanical Research, Volume 86
ISSN 0065-2296
https://doi.org/10.1016/bs.abr.2017.11.020

phenotypic variation in plants that can be indistinguishable from those obtained through natural means or conventional mutagenesis. The rapid development of these new techniques of plant breeding leads to several issues concerning the regulatory status of plants edited by engineered nucleases. This chapter aims at providing some keys to answer these issues. The intellectual property and legislation of genetically modified organisms (GMOs) in several countries including European Union and major countries such as the USA, China, Brazil, Argentina and India are discussed. A scientific description of these new editing techniques and of recently edited plants is included. From a technical point of view, edited plants should only be considered as GMOs in the current EU legislation of GMO in the case of transgene insertion, while the best regulatory issue might be a product-based approach.

1. INTRODUCTION

Molecular plant breeding techniques used in the past displayed two main limits. The first concern resulted from the very low transformation yield, which required the use of marker genes to select the transformed plants. In addition to the gene of interest to be inserted, modified or knocked down, a number of scientists used in the past a second gene (marker gene) to identify the transformed plants among the numerous primary plants (T_0) regenerated after the transformation step. In those days, the efficiency of transformation indeed ranged to a few percent only. As most of the selection markers were herbicide or antibiotic tolerance genes, regulation policies made a strong pressure to avoid using this strategy, and two main technologies were designed to comply with this safety recommendation.

The first one is the Cre-Lox deletion system in which the herbicide tolerance gene is flanked on both sides by a sequence named *Lox* which will be specifically recognized by the *Cre* recombinase: when activated by supplied oestradiol or tetracycline (compounds which are not present in plants), *Cre* deletes the stretch of DNA delimited by the two *Lox* sequences (Quétier, 2016). The second strategy is based on simple genetics: it takes advantage of a transformation system where the gene of interest and the marker gene to identify transformed plants are each carried by a distinct plasmid vector. T_0 plants are selfed and natural crossing-over induces the independent segregation of the two loci, provided that they are not too close from each other. This procedure is known as "purification by meiosis" and is widely used in plant breeding. The screening is based on Southern DNA hybridization and later by PCR.

The second and huge concern was due to the transformation process, which has been for a long time a random insertion along any chromosome; this may disrupt a resident gene and, consequently, may bring on unwanted phenotypic side effects. In addition, for breeding, the breeder has to examine the phenotype because of the unpredictable expression of selected genes. Contrary to mouse transgenesis where multicopies of a transgene result in increased amounts of the corresponding protein, the insertion of multicopies in a plant genome may lower the gene expression by gene silencing.

For all these reasons, it was necessary to design new transformation techniques allowing targeting specific sequences, with high efficiency. The group of Bernard Dujon in France discovered in 1986 (Colleaux et al., 1986) the endonuclease I-Sce I, which displayed a 18-bp recognition site (Colleaux, d'Auriol, Galibert, & Dujon, 1988). The same team showed in 1993 that I-Sce I created targeted double-strand breaks (DSBs) in *Nicotiana plumbaginifolia* protoplasts (Puchta, Dujon, & Hohn, 1993) and highlighted in 1994 that DSBs can be repaired, in the presence of a template sequence, via double homologous recombination (Puchta, Dujon, & Hohn, 1996). These works brought the concept of genome editing, which means in situ genomic modifications and corrections, by using a nuclease attached to a molecule enabling to target specific area in the genome. Kim and Chandrasegaran provided the proof of concept of such a chimeric construction in 1994 by linking the *Ultrabithorax* homeodomain of *Drosophila* to the cleavage domain of the *Fok*I restriction endonuclease of a *Flavobacterium* (Kim & Chandrasegaran, 1994). Two years later, this group succeeded in using zinc-finger polypeptides (ZFP) as a substitute for a homeodomain, which led to zinc-finger nuclease (ZFN) (for a review, see Quétier, 2016). From 1996 to 2015, scientific progress has been made in order to constitute a whole set of engineered nucleases with the well-known ZFNs, transcription activator-like effector nucleases (TALENs), RNA-guided engineered nucleases (RGENs) and more recently the CRISPR/Cas systems. These engineered nucleases enabled to breed diverse plants since 2005 (cucumber, grapefruit, melon, rice, tomato, wheat).

The development of new breeding techniques (NBTs) raises several issues concerning regulatory capping of the NBT products. This chapter also aims to determine whether or not modifications using genome editing could be targets for industrial property. This regulatory dimension is crucial and even decisive for the adoption of these techniques by plant breeders. Whatever the result, it is also important to determine if plants generated by genome editing have to be considered or not by a new risk assessment approach. This chapter addresses the milestones of the reflection underlying the debate around these questions through:

- a scientific description of NBT, their evolution and development;
- an analysis of intellectual property issues linked to these techniques upstream (patent owners, licence policy) and downstream;
- an evaluation of challenges of NBT for breeding companies;
- an analysis of regulation issues linked to these NBT (are these products GMO or not?—what about the traceability and detectability of these products, which are sometimes impossible to distinguish from plants generated by conventional breeding).

2. INTELLECTUAL PROPERTY AND GMOs

2.1 The Major Principles Regarding the Intellectual Properties in the European Union

Intellectual property (IP) refers to the protection of creations of the mind, which have both a moral and a commercial value. IP law typically grants the author of intellectual creation exclusive rights for exploiting and benefiting from their creation. However, these rights, also called monopoly right of exploitation, are limited in scope, duration and geographical extent. In the field of plant biotechnologies, it is possible to distinguish two protection ways: the patent and the Proprietary Variety Certificate (PVC).

2.1.1 The UPOV Convention and the European Legislation

The plant variety protection was adopted with the International Convention for the Protection of New Varieties of Plants (UPOV) during a diplomatic conference in Paris on the 2nd December 1961. The UPOV Convention provides a *sui generis* form of intellectual property protection that has been specifically adapted for the process of plant breeding: the PVC (UPOV, 2015). UPOV rules have been transcribed into European Community law via the regulation 2100/94/EC of 22nd July 1994 which states that "it is appropriate that the implementation and application of this Community regime should not be carried out by the authorities of the Member States but by a Community Office with legal personality, the Community Plan Variety Office"; even if plant variety rights are granted also by national offices (European Commission, 2015).

The rules concerning the patent application within the EU are provided by the European Patent Convention (EPC), a multilateral treaty signed the 5th October 1973 in Munich. This treaty is completed by the directive

98/44/EC on the legal protection of biotechnological inventions. This treaty provides the creation of a European patent, which is not an unitary patent protection valid in all Member States but a group of national independent patents. The European patents are delivered by the Member States or by the European Patent Office (EPO). If a European patent is valid for one Member State, the patent owner should valid the patent in each country of the Union. The Regulation 1257/2012/EU simplifies the European procedure because it creates a "European patent with unitary effect", which means a European patent which benefits from unitary effect in the 28 participating Member States (European Commission, 2015).

2.1.2 Principle of Patent and Patentable Invention

Pursuant to the Article 52 EPC, "European patents shall be granted for any inventions, in all fields of technology, provided that they are new, involve an inventive step and are susceptible of industrial application". Article 53b EPC gives exceptions to patentability: European patents shall not be granted in respect of "inventions the commercial exploitation of which would be contrary to 'public order' or morality" and "plant varieties or essentially biological processes for the production of plants; this provision shall not apply to microbiological processes or the products thereof". Rule 26(4) EPC of the Guidelines for Examination gives the definition of plant varieties, which means "any plant grouping within a single botanical taxon of the lowest known rank, which grouping, irrespective of whether the conditions for the grant of a plant variety right are fully met, can be: (a) defined by the expression of the characteristics that results from a given genotype or combination of genotypes, (b) distinguished from any other plant grouping by the expression of at least one of the said characteristics, and (c) considered as a unit with regard to its suitability for being propagated unchanged" (EPO, 2016).

The patentability of biotechnological invention is stated by Rule 27(b) Article 52 EPC, because biotechnological inventions shall also be patentable if they concern "plants if the technical feasibility of the invention is not confined to a particular plant variety". Article 4.2 of the Directive 98/44/EC also states that "inventions which concern plants shall be patentable if the technical feasibility of the invention is not confined to a particular plant variety". Considering these two articles, plant varieties are excluded from patentability (they are then protected by PVC). Processes for genetic modification, processes for plant transformation, processes for transgenic plant production, gene and sequence of interest, and inventions which concern transgenic plant if technical feasibility is not confined to a particular plant variety, are thus patentable (European Commission, 2015; EPO, 2016).

2.1.3 Rights Protected by a PVC and a Patent

A patent is valid for 20 years since the application as long as the IP owner pays an annual fee called annuity. The PVC is valid for 25 years since the delivery, but it is valid for 30 years for trees, vines and potatoes. Once this period has passed, the invention or the plant variety becomes public property, and everyone is free to use it. Before this term, third parties, excepted by licence agreement, cannot use a patented invention. The PVC grants, like patent, the IP exclusive control over the propagating material, but unlike patent, the PVC grants free access (for research and not for production) to protected varieties as an initial source of variation for breeders. The research exception also helps to increase the efficiency of the programs designed to improve plants.

Furthermore, Regulation 2100/94/EC states that "for the purposes of safeguarding agricultural production, farmers are authorized to use for propagating purposes in the field, on their own holding the product of the harvest which they have obtained by planting, on their own holding, propagating material of a variety other than a hybrid or synthetic variety, which is covered by a Community plant variety right". This "farmer privilege" regulation first applied to IP without patent and has been extended to patent cases.

2.2 The Major Principles Concerning GMOs in the European Union

2.2.1 The Community Legislation Within the European Union

The European Commission is the main actor of the legal framework concerning GMOs. However, the EU must take care about the international market, its GMO policy shall comply with the World Trade Organization (WTO) rules. The building blocks of the GMO legislation are:

- Directive 2001/18/EC (12th March 2001) on the deliberate release into the environment of GMOs and repealing Council Directive 90/220/EEC (1990);
- Regulation 1830/2003/EC (22nd September, 2003) concerning traceability and labelling of GMOs and the traceability of food and feed products produced from GMOs.

The directive 90/220/CEE and then the directive 2001/18/EC gives the legal definition of a genetically modified organism: "A GMO means an organism, with the exception of human beings, in which the genetic material has been altered in a way that does not occur naturally by mating and/or natural recombination". It also makes the distinction between techniques leading or not to "genetic modifications". From a legal point of view, the techniques leading to genetic modifications are listed below:

- "recombinant nucleic acid techniques involving the formation of new combinations of genetic material by the insertion of nucleic acid molecules produced by whatever means outside an organism, into any virus, bacterial plasmid or other vector system and their incorporation into a host organism in which they do not naturally occur but in which they are capable of continued propagation;
- techniques involving the direct introduction into an organism of heritable material prepared outside the organism including microinjection, macroinjection and microencapsulation;
- cell fusion (including protoplast fusion) or hybridization techniques where live cells with new combinations of heritable genetic material are formed through the fusion of two or more cells by means of methods that do not occur naturally".

However, on condition that they do not involve the use of recombinant nucleic acid molecules or GMOs, techniques of in vitro fertilization, conjugation, transduction, transformation and polyploidy induction are not considered to result in "genetic modification" from a legal point of view. Furthermore, several genetic modification techniques yielding organisms are excluded from the directive, on the condition that they do not involve the use of recombinant nucleic acid molecules or GMOs:

- mutagenesis;
- cell fusion (including protoplast fusion) of plant cells of organisms which can exchange genetic material through traditional breeding methods.

This directive introduces an enormous gap between the scientific definition of a genetic modification and its legal definition in the European framework.

The protection of human health and the environment requires that due attention be given to controlling risks from the deliberate release into the environment of GMOs. For that reason, the EU has adopted regulations and administrative provisions related to the deliberate release and placing on the market of GMOs in accordance with the precautionary principle. The precautionary principle, detailed in Article 191 of the Treaty on the Functioning of the EU, provides that "where scientific data do not permit a complete evaluation of the risk, recourse to this principle may, for example, be used to stop distribution or order withdrawal from the market of products likely to be hazardous". This framework contributes to set up a common risk assessment method in the case of GMO use (under containment, confined and limited field trials). The measures set by this directive concerning the procedure for granting consent for the deliberate release and placing on the market of GMOs provide:

- authorization of release for 10 years and renewable;
- controls after placing on the market;
- common risks assessment method for the environment;
- common goals of monitoring;
- mechanism enabling the modification, the suspension or the termination of the release;
- public consultation by an information exchange system (Directive 90/220/ESC).

The regulation 1830/2003/EC sets the framework to ensure products containing GMOs, and food and animal feed derived from them, can be traced at all stages of the production and distribution chain. The rules do not cover human and veterinary drugs, which are authorized by the regulation 2309/93/ESC. Final consumer packaging or prepackaged products containing GMOs should be labelled, except if the GMO rate does not reach 0.9% and if the presence of GMOs in the product is involuntary or technically inevitable. Traceability has to be effective for 5 years (European Commission, 2015).

2.2.2 A Legislative Flexibility for the Member States in the European Union

A new European directive, the Directive 2015/412/EU from the 11th March 2015, allows flexibility for the Member States. It sets that "the Member States may restrict or prohibit the cultivation of GMOs in all or part of their territory as from the date of entry into force of the Union authorization and for the whole duration of the consent/authorization, provided that an established standstill period, during which the Commission was given the opportunity to comment on the proposed measures, has elapsed". This procedure is called the "opt-out" (EurLex, 2015). Since then, despite this flexibility, no qualified majority was reached at the EU level for new authorizations and the dead-lock remains.

2.2.3 Safety Evaluation of GMOs

The EU legal framework on GMOs provides that no GMO or genetically modified food and feed can be placed on the market before it has been granted an authorization in accordance with that legal framework. To this end, a scientific assessment is performed by the European Food Safety Authority (EFSA), in cooperation with the scientific bodies of the Member States. Where the results of the risk assessment show that the product does not pose a risk to health or to the environment under the proposed conditions of placing on the market/use, the Commission submits to the Member States, represented in the Standing Committee, a draft implementing decision of authorization, in accordance with the procedure set out in Regulation (EC) No 182/2011 (comitology procedure). Under this procedure, Member States vote under the "qualified majority" rule defined in the Treaty. In the Standing Committee, if Member States vote "Yes", the Commission adopts the draft decision. If they vote "No", or if the result of the vote is "No opinion" (no qualified majority either in favour or against is expressed), the Commission may submit the draft decision to another body representing the Member States at a higher level: The Appeal Committee. In the Appeal Committee, the Member States vote a second time on the draft decision tabled by the Commission. If the Member States vote "Yes", the Commission adopts the draft decision. If they vote "No", the Commission cannot adopt the draft decision. If the result of the vote is "No opinion", the Commission is required by the GMO legal framework and by the Charter of Fundamental Rights to adopt a decision on the application, and therefore in practice has little choice but to give the authorization (European Commission, 2015). However, there has been no authorization for cultivation in Europe since the adoption of this new framework.

3. GENOME EDITING WITH ENGINEERED NUCLEASES

3.1 Interaction of Engineered Nucleases With DNA and Genomic Consequences

Programmable nucleases produce site-specific DNA DSBs. These DSBs are then repaired by two endogenous mechanisms: homology-directed repair (HDR) with the presence of a gene-targeting vector or a homologous donor DNA, or nonhomologous end joining (NHEJ) without complementary DNA. These two mechanisms have several consequences: small insertions–deletions (INDELs) in the case of NHEJ repair, substitution, gene disruption, insertion, correction and chromosomal rearrangements in the case of HDR repair. Outcomes of genome editing using programmable nucleases are represented in Fig. 1 (Quétier, 2016).

In the field of biotechnology, three applications are particularly interesting. The first one is gene disruption—gene knockout (Fig. 1A). In the absence of DNA template, the error-prone NHEJ creates INDELs which

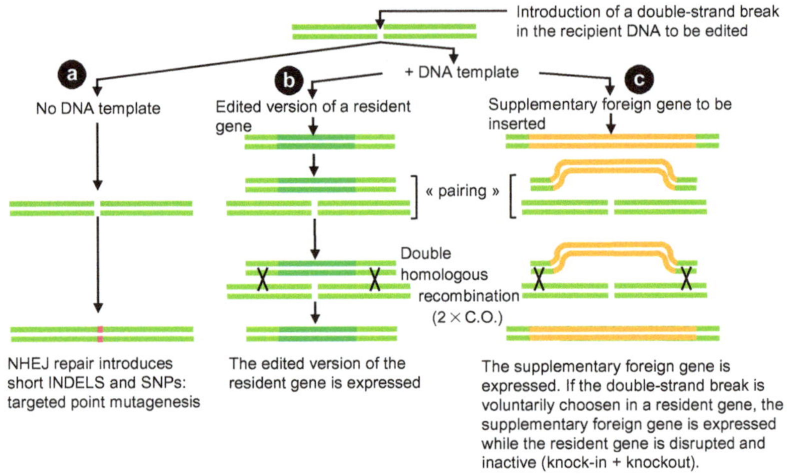

Fig. 1 The three classes of targeted genome editing: (A) in the absence of template DNA, targeted point mutations (SNPs, INDELs) are achieved through NHEJ, (B) with DNA template, targeted allele substitution is achieved by HDR and (C) with DNA template, targeted insertions of a supplementary gene are achieved by HDR. *Source: Reprinted with permission from Quétier, F. (2016). The CRISPR-Cas9 technology: Closer to the ultimate toolkit for targeted genome editing.* Plant Science, *242, 65–76.*

often cause frameshifts in the coding region, giving rise to the disruption of genetic information and resulting in gene knockout. The second one is gene correction (allele substitution), which takes place through HDR when an edited version of the resident gene, flanked on both ends by the homologous sequences in the resident genome, is added as a template (Fig. 1B). The third application is the insertion of a new gene in the resident genome, which follows exactly the same process as correction but involves the selection of a resident sequence devoid of any annotated information (genes, enhancers, noncoding RNAs). Conversely, choosing a gene to be inactivated as a target allows a double editing; the new gene is functional, while the resident gene is no more functional because of the induced disruption (Fig. 1C).

3.2 Different Kinds of Engineered Nucleases

3.2.1 Zinc-Finger Nucleases

In 1994 Kim and Chandrasegaran created the ZFN, a chimeric construction between a ZFP and the nuclease *Fok*I (Kim & Chandrasegaran, 1994). The structure of ZFN is represented in Fig. 2.

Each zinc-finger helix contains four variable amino acids at invariable locations, which specifically bind to three successive bases in the substrate

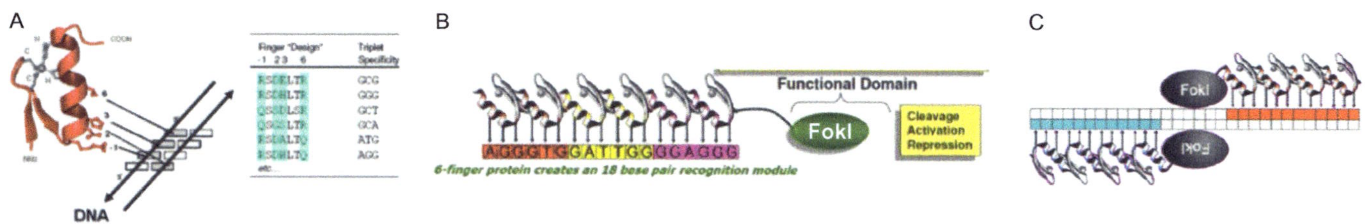

Fig. 2 Zinc-finger nucleases: (A) 3D structure of a zinc-finger showing the four amino acids which selectively bind to three adjacent DNA bases and a sample of the correspondence code, (B) structure of an array of zinc-fingers linked to a *Fok*I catalytic domain, which cuts one DNA strand, (C) spontaneous dimerization of two units (displaying different zinc-finger arrays) leads to a double-strands breaks targeted by a long specific recognition site. *Source: Adapted from EXZACT® Precision Technology, Dow Chemical http://www.exzactprecisiontechnology.com/exzact_slides.pdf*

A

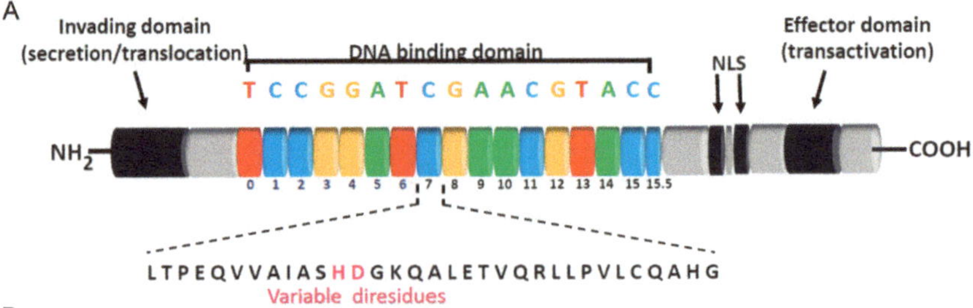

B

NG	NI	HD	NH	NN	N*
T	A	C	G	G/A	5mC

Fig. 3 Structure of a typical *Xanthomonas* TALE and the specific base recognition correspondence table: (A) the different domains of a typical 15.5 repeat TALE. *Black segments*: functional domains used in pathogenic bacteria, NLS are the nuclear localization signals. The DNA-binding domain here depicted is an array of 15.5 repeated units (the repeat No 15.5 is the N-terminal half part of a unit). The amino acid sequence of one repeat is enlarged to show the invariable residues (in *black*) and the variable diresidues which specifically recognize one nucleotide (*violet*). (B) The table of correspondence between the variable diresidues and the nucleotide which is recognized. *Source: Reprinted with permission from Quétier, F. (2016). The CRISPR-Cas9 technology: Closer to the ultimate toolkit for targeted genome editing.* Plant Science, *242, 65–76.*

DNA as shown in Fig. 2A. Tandem arrays of successive zinc-fingers can therefore target specific DNA. It was shown some years later that the table of correspondence between sets of four amino acids in zinc-fingers and the three bases to which they bind is not ambiguous, but is however degenerate, like the genetic code (Fig. 2B). Functional artificial ZFN can be constructed by tandem arrays of zinc-fingers fused to the *Fok*I coding DNA as represented in Fig. 2B. Because spontaneous dimerization of such ZFNs-*Fok*I occurs in vivo, dimers display 20–30 bp recognition sites (Fig. 3C). A public Zinc-finger Consortium made available a library of zinc-fingers, complemented with an in vivo bacterial testing to functionally select the best combination of arrayed zinc-fingers.

3.2.2 Transcription Activator-Like Effector Nucleases

It has been demonstrated between 2006 and 2009 by Schornack, Meyer, Römer, Jordan, and Lahaye (2006), Boch et al. (2009), Moscou and Bogdanove (2009) and Boch and Bonas (2010) that TALE (transcription activator-like effectors) proteins are responsible for the virulence of the bacteria *Xanthomonas* (pathogen of rice) by binding specifically to regulatory sequences in the host plant.

TALE proteins appeared to be similar to ZFP and can be fused to the nuclease *Fok*I to constitute TALEN. Interestingly, the TALE code is 2 amino acids for 1 nucleotide and seems less degenerated (Fig. 3).

Each of these TALEs proteins encompasses from 13.5 to 25.5 repeats (the last C-terminal unit is an N-terminal half of a repeat) of a 34–35 amino acid motif, each of which recognizes a single base-pair in the major DNA groove. The nucleotide specificity of each repeat domain is determined by the two amino acids at positions 12 and 13, which are called repeat variable diresidues (RVDs). Four different RVD modules—namely, Asn–Asn, Asn–Ile, His—Asp and Asn–Gly—are most widely used to recognize guanine, adenine, cytosine and thymine, respectively. Like ZFN, each monomer must bind to adjacent half-sites that are separated by spacers of 12–21 bp (Quétier, 2016). To generate TALEN, up to 20 RVDs are necessary, enabling a targeting of 30–40 bp in length, excluding spacers (Kim & Kim, 2014).

The OPEN source TALE (The Zinc Finger Consortium, 2014) provides a complete set of plasmids for building TALENs and transcription factors (TALE-TFs) for genome editing and transcription modulation applications.

Li, Liu, Spalding, Weeks, and Yang (2012) used TALEN to create disease-resistant varieties of rice. Many plants were successfully transformed since by using TALEN.

3.2.3 The CRISPR/Cas9 System as an Example of RGEN

Ishino, Shinagawa, Makino, Amemura, and Nakata (1987) reported the presence of an intriguing stretch of DNA, close to a bacterial protein gene and composed of short direct-repeats interspaced by short unique sequences (*protospacers*) in *Escherichia coli* genome. Actually, bacteria insert protospacers into their own genome to form a CRISPR (clustered regularly interspaced short palindromic repeats). These protospacers are coming from bacteriophages or plasmids. The Cas protein is an endonuclease involved in a bacterial defence mechanism against bacteriophages by DSB. During 2011 and 2012, Emmanuelle Charpentier's groups deciphered the molecular mechanism governing the accurate positioning of the DSB in DNA in the natural CRISPR/Cas9 system, which lead to the construction of an artificial CRISPR/Cas9 engineered to produce DSB at any position chosen along the DNA sequence (see Quétier, 2016). This DSB is induced in a specific way by a sgRNA. Accordingly, the CRISPR/Cas9 system is considered as a RGEN, a representation of which is shown in Fig. 4A.

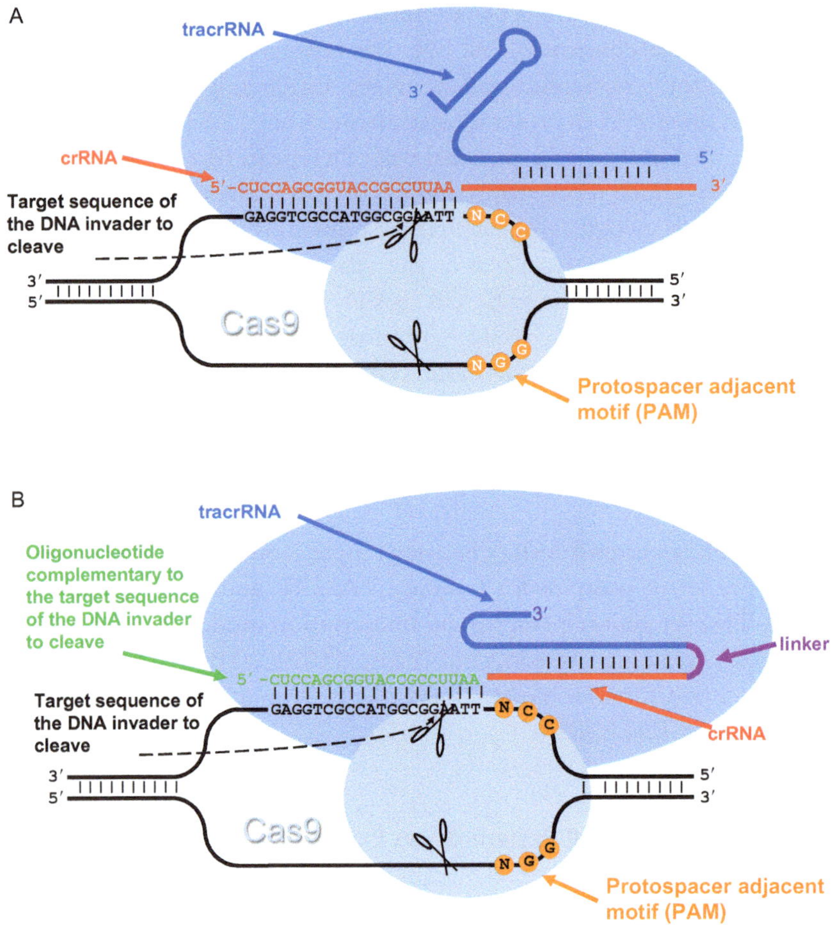

Fig. 4 Structure of CRISPR-Cas9: (A) the natural CRISPR/Cas9, driven by a guide RNA involving two partially hybridized RNA molecules (crRNA and tracrRNA), (B) an engineered CRISPR/Cas9 driven by a single molecule guide RNA obtained by a fusion between the crRNA and the tracrRNA by a short linker. *Source: Reprinted with permission from Quétier, F. (2016). The CRISPR-Cas9 technology: Closer to the ultimate toolkit for targeted genome editing.* Plant Science, *242, 65–76.*

In type II CRISPR systems, the transcription of a CRISPR unit leads to a RNA called crRNA, which binds by a complementary base-pairing in 3′ to another RNA in 5′, called tracrRNA (Fig. 4A). Once these two RNA are complexed, the new structure called sgRNA is blocked in the catalytic site of the Cas9 protein, to form an active DNA endonuclease, which is often termed dualRNA–Cas9. This complex can bind to DNA and recognize a 23-bp target DNA sequence that is composed of the 20 bp guide sequence in the crRNA (the protospacer) and the 5′-X_{20}NGG-3′ sequence known as protospacer adjacent motif (PAM), which is recognized by Cas9.

Natural CRISPR/Cas9 can be engineered simply by linking crRNA to the tracrRNA through a short linker and adding a 20mer-long oligoribonucleotide whose sequence is complementary to the target sequence to cleave (Fig. 4B).

3.3 Comparison of the Different Engineered Nucleases

By their mechanism, engineered nucleases seem to be the optimal solution thanks to their specificity. In terms of specificity, it is very important to assess their off-target activity, corresponding to an unwanted activity of the engineered nucleases. Thus, DSBs can be induced on other loci than the desired one, and these DSB are repaired by NHEJ which leads to INDELs. These off-target modifications can induce toxicity by inactivating vital genes by knockout. It is necessary to quantify off-target activity and to determine hot-spots sites, which are sites that might be prone to off-target activity (Hendel, Fine, Bao, & Porteus, 2015).

The comparison of the different engineered nucleases gives the following insights:

- technical feasibility: RGENs have the advantage to only need to synthetize one 22pb crRNA or sgRNA, while ZFN and TALEN require a heaviest protein engineering (up to 20 RVDs for TALEN with many homologous sequences and consequently a risk of in vivo recombination);
- activity and efficacy: TALEN and RGENs have the best activity with mutation rates of 1%–60% and 2%–79%, respectively, and success rates close to 100%;
- specificity: optimal for TALEN and RGEN.

The technology CRISPR/Cas9 is seen nowadays as the ultimate tool when TALEN seemed previously to be more performant. RGENs are now the most used because they are easy to obtain (in vitro RNA synthesis) and are active as monomer (Kim & Kim, 2014). Čermák et al. (2017) report targeted, specific modification of monocot and dicot genomes with geminivirus-based vectors for precise gene editing through homologous recombination and with reagents which are systematized for fast, modular cloning and accommodate diverse regulatory sequences to drive reagent expression.

3.4 Engineered Nucleases and Intellectual Property

Following the discovery of the first engineered nuclease began a race between academia and several companies. ZFN being known since 1996, the Pasteur Institute (France) developed several nucleases and did a technological transfer to its spin-off Cellectis (a French SME), created in 1999.

On the commercial level, the race to obtain tailored ZFNs grew up in 2006. Precision Bioscience, a US SME linked to Bayer Crop Science, obtained an exclusive licence from Duke University to exploit a technology called DNE™ and appeared on the market. Then, Dow AgroSciences acquired an exclusive licence from Sangamo Biosciences (USA) to commercialize an endonuclease called ExZactTM to be used for plants only. SMEs therefore use very similar technologies, based on closely related principles; competition became intense, and a court case soon opposed Cellectis and Precision Bioscience (Cellectis filed a suit against Precision Bioscience, accusing it of infringing two US Patents). On the academic level, The Zinc-finger Consortium made available its OPEN strategy which combined a library of zinc-fingers as well as a bacterial test to functionally select the optimized combination of zinc-fingers (The Zinc Finger Consortium, 2014). TALEN technology has been developed by a group affiliated with Iowa University and Minnesota University (Li et al., 2012). Cellectis acquired, in early 2011, the exclusive licence from Iowa University and from the Minnesota University to distribute TALEN (Quétier, 2016). For the CRISPR/Cas9 technology, the patent race opposes two major players to secure rights on the CRISPR/Cas9 system: the group lead by Jennifer Doudna from the University of California in USA and her chief collaborator, Emmanuelle Charpentier (University of Vienna, Austria) and the group headed by Feng Zhang from the Broad Institute of Harvard and MIT in United States. While Doudna's group may have published first and have received a number of awards in this field, it is Zhang who has been awarded the first patent on the basic CRISPR/Cas9 technology in 2014 (US Patent No. 8.697.359) (Smith-Willis & San, 2015). The EPO has announced on February 2017 it intends to grant its first CRISPR–Cpf1 patent to Zhang's group allowing the inventors' claims to an engineered Cpf1 system, including methods for use in genome editing in eukaryotic cells. The natural CRISPR–Cpf1 system is an RNA-guided system that does not require a tracrRNA component. Beside patents, several companies won licence agreements and are able to commercialize engineered nucleases (Kim & Kim, 2014).

Genome editing appears to be a real innovation drawing a great interest by the scientific community and companies, especially in RGEN. As it is evident with the patent race concerning this technology and the diversity of resources, engineered nucleases are used in many fields, in particular for crop improvement. This chapter has only considered genome editing, even if engineered nucleases can be used in several ways: DNA cut or genetic regulation by activating or blocking the transcription. This chapter also limits RGEN to the type II CRISPR/Cas9 system, while there are five types of CRISPR system (Makarova et al., 2011).

4. GENOME EDITING WITH ENGINEERED NUCLEASES FOR PLANT BREEDING

Genome editing with engineered nucleases has rapidly become the main tool for investigating gene function and for creating genetic variation using site-directed genomic alterations. When applied to economically important plant species, genome editing additionally provides a highly specific and efficient means to generate useful novel phenotypes (Wolt, Wang, & Yang, 2016). Indeed, once the function of a gene is known, it is possible to modify or transfect it. For example, a herbicide-tolerance or insect-resistance gene, abiotic factors-tolerance gene such as drought or water stress, could be inserted in a plant of interest, conferring the desired trait to the plant. The second option is to knock out a gene which confers agronomic advantages to the plant.

4.1 Methodology to Obtain Edited Plants

The process to obtain an edited plant (Fig. 5) consists in three steps:

1. transformation of a T_0 generation with an engineered nuclease;
2. quantitative and qualitative analysis of the T1 generation and its offspring;
3. characterization of the final product.

4.1.1 Transformation of a T1 Generation With an Engineered Nuclease

To produce the T1 generation, engineered nucleases can be delivered by several ways:

- plasmid DNA by electroporation;
- plasmid DNA by transfection via liposome;
- protoplast microinjection of mRNA;
- injection of purified proteins (preassembled sgRNA–Cas9 ribonucleoprotein, for example).

It is advisable to engineer several nucleases for a unique target and to test the off-target activity of each nuclease. ZFNs are the most compact member of the nuclease triad, and sequences encoding ZFNs can be packaged into small vectors, while small vector can be incompatible with TALENs because highly homologous TALE repeats often lead to unwanted recombination

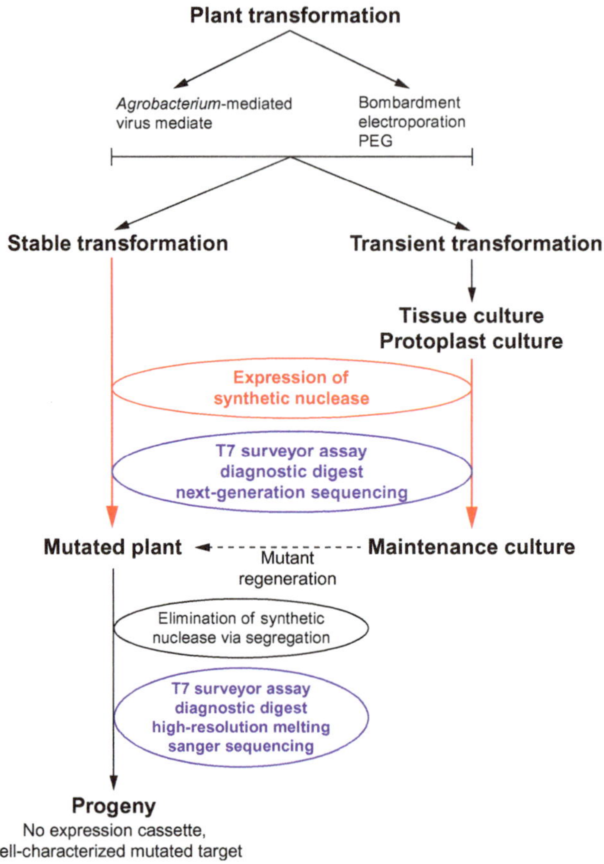

Fig. 5 Plant genome editing with engineered nucleases and controls associated. A T1 plant is obtained; it carries the expression cassette for a synthetic nuclease in the case of a stable transformation event, which is not the case for a transient expression event. Depending on the promoter used, the nuclease may be expressed in an induced, constitutive or tissue-specific manner. Mutations are induced in the meristematic or somatic tissue in the T1 generation. During T1 development, nuclease activity may be characterized in a qualitative manner using a T7 surveyor assay or a diagnostic digest. Mutations that enter the germ line of a T1 plant will be clonal in the T2 generation, resulting in a mutated plant. Mutated plants may then be screened for individuals that carry the mutation in a homozygous state. To eliminate the nuclease expression cassette, plants are further propagated. Plants without an expression cassette which are homozygous for the individual mutation may then be identified in the next generation. *Source: Reprinted with permission from Puchta, H., & Fauser, F. (2014). Synthetic nucleases for genome engineering in plants: Prospects for a bright future.* The Plant Journal, 78, *727–41.*

in cells (Kim & Kim, 2014). The protocol also depends on the used tissue and the presence of exogenous DNA or not. For knockout, *Arabidopsis thaliana* is typically transformed via the *Agrobacterium*-mediated flower-dip protocol, whereas other species such as *Nicotiana benthamiana* or *Solanum lycopersicum* rely on transformation of explants or electroporation. For knock-in, cotransformation of the synthetic endonuclease and the DNA template is required for HDR-mediated DSB by *Agrobacterium* transformation or by direct gene transfer. Plants are not easy to transform, and transformation efficiency is extremely low in the case of a stable transformation (Puchta & Fauser, 2014). Furthermore, isolation of protoplasts from plant tissue requires enzymatic digestion and removal of the cell wall, which is a heavy and time-consuming procedure. The challenge is then to regenerate a whole plant from the transformed protoplast; while species such as *N. benthamiana* and *A. thaliana* can be rather easily regenerated (Belhaj, Chaparro-Garcia, Kamoun, & Nekrasov, 2013), others may be more or less recalcitrant. To overcome this problem, it is better to use transient transformation in planta protocol, which should be applicable to all transformable plant species and tissues. It must be underlined that collateral unwanted genomic changes may occur when an in vitro culture step is used to regenerate an entire plant from protoplasts or tissue cultures, which induces somaclonal variations that should not be confused with true off-target errors.

Delivery of preassembled sgRNA–Cas9 ribonucleoproteins into plant cells is also possible, which cleave chromosomal target sites immediately after transfection and are degraded rapidly by endogenous proteases in cells. There are several advantages to use this technique. First, there is no need to optimize codon usage (see below) or find promoters that will express Cas9 and sgRNA. Second, it could remove the likelihood of inserting recombinant DNA in the host genome and it enables in vitro prescreening to guide the choice of highly active sgRNAs and genotyping of mutant clones via restriction fragment length polymorphism (RFLP) analysis. The incubation of ribonucleoproteins with protoplasts derived from *A. thaliana*, *Nicotiana attenuata*, *Oryza sativa* and *Lactuca sativa* in the presence of polyethylene glycol achieve targeted mutagenesis in regenerated plants at frequencies of up to 46% (Woo et al., 2015).

Using CRISPR/Cas9 in plants required rewriting its coding sequences to take into account the difference in the codon usage between bacteria and plants. This plant codon optimization has been done by Ma et al. (2015), who achieved to generate Cas9 whose 400 bp 5′ end terminal region displays a 62.5% GC content. For applications in eukaryotic organisms, the Cas9 protein is expressed in plants as a fusion to a nuclear localization signal (NLS) to ensure delivery into nuclei. Sometimes introns are added, and the promoter is also optimized. The key parameter for the achievement of genome editing with the lowest level of off-target events nowadays is clearly the design of the sgRNA. This step has been made possible by the complete genome sequencing of most of the crop and model plants. Several in silico facilities are available to get the best candidate sequences corresponding to a query sequence, of the basis of the PAM sequences database (there is roughly one PAM every 12 bp) and of the number of off-target events predicted for each candidate. Additional software to identify potential off-target sites in plant genomes has been developed in 2014. A standardized detailed practical procedure has been recently published by Li, Zhang, and Sheen (2015). In plants, sgRNAs (5′-X_{20}NGG-3′) are expressed using plant RNA polymerase III promoters, such as U6 and U3. These promoters have a defined transcription start nucleotide, which is "G" or "A", in the case of U6 or U3 promoters, respectively, which have to be taken into account during the design (Belhaj et al., 2013; Quétier, 2016).

4.1.2 From the Analysis of the T1 Generation and Its Offspring to the First Mutant

DSBs induced by engineered nucleases are repaired by HDR or NHEJ, but in both cases it is necessary to quantify and qualify the results in order to determine the on-target/off-target effects ratio. Indeed, HDR and NHEJ may occur at relatively equal frequencies (even if NHEJ-mediated events may predominate in certain conditions). Different techniques exist to assess on-target activity of genome editing. One method to quantitatively measure HDR-mediated outcomes is via an RFLP assay, by introducing synonymous SNPs that create a restriction enzyme site within the donor DNA. An alternative approach is to target a constitutively active reporter gene like GFP between the homology arms and to analyse fluorescence by flow cytometry. The best solution is the next-generation sequencing. These recent high-throughput sequencing technologies allow sequencing DNA and RNA much more quickly and cheaply than the previously used Sanger sequencing (Hendel et al., 2015). The scanning of mutations often associated with off-target effects is now better achieved by using the T7 endonuclease 1 (T7E1) assay which cleaves nonperfectly matched DNA as heteroduplex DNA (Quétier, 2016).

It is possible to remove some off-target mutations by meiosis purification when ZFN are used, provided that they are not too close to each other. TALEN bring on less off-target activity (Quétier, 2016). When a targeted mutagenesis event enters into the germline of T1 plants, these events may be either homozygous or heterozygous in the T2 offspring. Nevertheless, plants that harbour a mutated target may be easily genotyped via the qualitative assays mentioned earlier or by PCR-based high-resolution melting analysis. Plants that are homozygous for the individual mutation may then be identified in the next generation (Puchta & Fauser, 2014).

4.1.3 From a Plant With the Desired Modification to an Edited Plant

Only plants without a nuclease expression cassette are selected, and there are two situations depending on the transformation protocol (stable or transient). In the case of stable transformation, the resulting plant contains the nuclease-coding sequence within its genome. This cassette is not commonly linked to the edition target, and therefore the transgenic elements are readily removed by selection for null segregant lines to produce subsequent plant generations that have been edited but are devoid of the tools used for the process (Wolt et al., 2016). However, removing this DNA by segregation is not possible for species like grapevine, potato and banana which are propagated by vegetative multiplication. It is also possible to remove the expression cassette by two different methods used in transgenesis. The first one is the Cre-Lox system and the second tool takes advantage of a suicide gene, a powerful bacterial RNase, whose promoter is specifically active in the tapetum cells during the pollen formation. All these tools are nevertheless error-prone, the efficacy reaching ca. 90%. TALEN and CRISPR/Cas9 systems may be compatible with these tools in the future (Quétier, 2016). Plants resulting from protocols for transient expression do not have foreign DNA integrated into their genome. However, transfected plasmids are degraded in cells by endogenous nucleases, and the resulting small DNA fragments are sometimes inserted at both on-target and off-target sites in host cells (Woo et al., 2015).

Modifications by transgenesis are often carried out in nuclear genome, but stable chloroplast genome editing has been successively reported, from 2001 to 2011 for tomato, rape, carrot, cotton, soy, petunia, potato, lettuce, rice, cauliflower and cabbage. In spite of the numerous attempts which have been carried out by many research groups since the 1990s, no successful editing of a plant mitochondrial genome has been reported so far (Quétier, 2016).

4.2 Genome Editing in Plants With Engineered Nucleases From 2005 to Nowadays

The research group lead by Voytas (Gabriel & Voytas, 1997) in the United States pioneered the use of ZFNs for genome editing in plant cells in 2005 in tobacco protoplasts; they demonstrated that gene insertion frequencies may be enhanced by up to 10% in comparison with random integration (Wright et al., 2005). At the same time, Lloyd, Plaisier, Carroll, and Drews (2005) demonstrated that ZFNs may be used to generate INDELs in an artificially introduced restriction site in the *A. thaliana* genome via NHEJ. In 2009, these techniques were further improved and enabled to modify agronomic traits: Townsend et al. (2009) used ZFNs to induce SNPs via HDR in the surB loci in tobacco, establishing resistance to various herbicides. ZFNs have also been used for maize by Shukla et al., 2009. These ZFN-based technologies have been then used to modify plant species, such as *Arabidopsis*, soybean, etc., with a success in targeted editing ranged from 2% to 25% (Quétier, 2016).

Studies from 2010 to 2011 have demonstrated the huge potential use of TALEN for genetic engineering in plants. The first proof of concept has been given by Li et al. (2012) who have introduced a mutation into the promoter of the *OsWEET14* gene in rice, which makes it resistant to TALE proteins from *Xanthomonas*. Again, important contributions were made by Voytas and his group in 2013. They introduced targeted mutations within the ALS (acetolactate synthase) gene in up to 30% of transformed tobacco protoplasts using TALENs. They also performed a TALEN-mediated gene-targeted experiment in tobacco using a donor template that created an in-frame gene fusion between ALS and a YFP marker gene showing that frequencies were as high as 14%. In a second set of experiments, this group used a TALEN and a 322-bp homologous donor DNA differing by 6 bp from the ALS-coding sequence (Voytas, 2013). Even without selection, 4% of the regenerated calli showed evidence of targeted gene replacement (Puchta & Fauser, 2014). TALEN technology was used to induce modifications in the genome of many plants between 2011 and 2015: rice, barley, hexaploid wheat, tomato, soybean and maize (Quétier, 2016). Several studies have demonstrated the efficacy of CRISPR/Cas9 system for crop improvement since 2011 (see Bortesi & Fischer, 2015 for details) which shows its great agronomic potential. For example, Lawrenson et al. (2015) demonstrated the successful use of sgRNA–Cas9 genome editing to knockout the function of target genes in both *Hordeum vulgare* and *Brassica oleracea* (mono- and dicotyledonous), the transmission of the mutation to progeny plants in both species and the segregation of the transgenic edited locus (encoding the Cas9 and sgRNA) from the mutation, resulting in transgene-free plants that contain the desired mutation. In *H. vulgare*, the targeted gene was *HvPM19* (two copies out of the four existing in the genome) which is encoding an ABA-inducible plasma membrane protein, which in wheat acts as a positive regulator of grain dormancy, an important agronomic trait in cereals. The mutation was present in 23% of the T1 progeny. In *B. oleracea*, the targeted gene was *BolC.G4A.a*, which is an orthologue of the gene *GA4* from *A. thaliana*, whose loss-of-function mutants exhibit a dwarf structure and a reduced fruit dehiscence, important targets for crop improvement in Brassicas. The mutation was present in 10% of the T1 progeny (Lawrenson et al., 2015).

The necessity of innovating quickly in order to introduce novel traits in crops is heightened by increased world food demand and increasing use of crops as sources of renewable energy. However, the opportunity for transgenic crop innovation is limited by regulatory hurdles and continued public unease. New plant-breeding technologies, especially site-directed genome editing, are viable alternatives to transgenic crop production that provide new opportunities for innovation (Wolt et al., 2016).

5. THE STATUSES TO BE GIVEN TO PLANTS EDITED BY USING ENGINEERED NUCLEASES

5.1 Edited Plants and Intellectual Property

The French High Council for Biotechnology (Haut Conseil des Biotechnologies, HCB) published a report in 2013 addressing plant biotechnology and IP. Seed companies used patent for the fields related to Directive 98/44/EC: transgenic plants (even if the operators request double protection, PVC for variety and patents for trait) and isolated genes enabling a particular function to the plant. If HDR-based edited plants are also considered as transgenic plants because their genomes contain "foreign" DNA, they are therefore patentable. For patentability EPO requires that the production of plants to patent implies a human intervention exercising a determinant effect on the product, criteria allowing protecting processes like targeted mutagenesis. Plant displaying new NHEJ-based traits should then also be patentable (HCB, 2013). The debate concerning patentability for edited plants with engineered nucleases is actually settled. Indeed, even plants obtained by "essentially biological processes" are patented. This fact is illustrated by decisions G2/12 and G2/13 (known as "broccoli and tomato cases" http://xepc.eu/node/oj2012-130) from the Enlarged Board of Appeal, while the patentability of plants from conventional breeding should be excluded.

However, the European Parliament did not approve the decision taken by EPO and called on the European Commission, "as a matter of urgency, to clarify the scope and interpretation of Directive 98/44/EC, and in particular Articles 4, 12(3)(b) and 13(3)(b) thereof, in order to ensure legal clarity regarding the prohibition of the patentability of products obtained from essentially biological processes, and to clarify that breeding with biological material falling under the scope of a patent is permitted" by a resolution taken the 15th December 2015 (European Parliament, 2015).

In addition to statements from specialists related to the EU context, it is also pertinent to study the intellectual property of crops obtained by NBT in different countries.

- *Argentina*: A new seed law is being prepared. The law, if passed, would clarify the situation by allowing farm-saved seed of all varieties against compensation for the breeder, with the exception of small farmers who would be exempted, and establishing the breeder's privilege for all varieties including transgenic varieties (Le Buanec & Ricroch, 2014).
- *Brazil*: It would be possible to use a transgenic variety in a breeding program, but the new variety could be freely marketed only if the transgenic trait is not expressed or if expressed, at the expiration of the patent (Le Buanec & Ricroch, 2014).
- *China*: Transgenic varieties, after marketing authorization, must obtain a PVP certificate as plant varieties are not patentable as such (Le Buanec & Ricroch, 2014).
- *USA*: The plant breeder has the following options to protect his innovations:
    - for biotechnological inventions the "Utility Patent", like in Europe, but with no specific exemptions for plant breeding "breeder's privilege" or for farm-saved seed;
    - for plant varieties different solutions are available:
        - for asexually propagated varieties the plant patent acts with a scope of protection similar to the one of utility patent;

- for sexually propagated and tuber propagated varieties, either the PVP, with the breeders exception and a broad farm-saved seed exception, or the Utility Patent, with a limited research exception and no possibility for farm-saved seed. Given the obvious imbalance of the scope of variety protection between PVP and patent, it is not surprising that in the USA breeders are massively applying for patent protection for their varieties (Le Buanec & Ricroch, 2014).

5.2 Edited Plants and GMO Legislation

The consensus arising from scientific discussions shows that the degree of regulatory scrutiny on genome-edited crops will be determined by the nature of the DNA repair process used, the characteristics and intended use of the phenotype that is developed and the existing regulatory structures within the geopolitical region of release (Wolt et al., 2016).

5.2.1 Scientific and Legal Grounds on the Statuses of Edited Plants

For the scientific community, the nature of the DNA repair process (NHEJ or HDR) is the main factor to evaluate for edited plants, but the transformation protocol is also important to consider because it has consequences on the way the nuclease is expressed in plants and on the presence of an expression cassette (which is a transgene) in the edited plant. If no DNA sequence is added, the NHEJ repair process gives rise to SNPs and short INDELs, and the result is clearly equivalent to classical mutagenesis, whether obtained through natural/spontaneous processes or accelerated by chemical compounds (ethyl methanesulphonate, EMS) or physical irradiation (UV or γ radiations). Since these accelerating agents work randomly, treated plants usually display several mutagenized loci scattered all along the genome, with possible and unpredictable gene knockouts. Genome editing represents the obvious advantage of targeted and secured mutagenesis. NHEJ-based edited plants should therefore not be considered as GMOs regarding Directive 2001/18/CE. Plants whose genome has been edited with the use of foreign template DNA by HDR repair process should be considered as transgenic plants and should adhere to the current regulatory policy, whatever the origin of the template DNA (Quétier, 2016).

As the transgenic donor locus may be segregated from the targeted integration site, a plant may be obtained that carries only the designed change in the target without any additional transgenic sequences being inserted (Puchta & Fauser, 2014).

Plants resulting from protocols for transient expression do not have foreign DNA integrated into their genome. It is therefore questionable whether they may actually be considered as transgenic organisms regarding Directive 2001/18/CE. Hartung and Schiemann (2014) suggest that the fast-evolving development of engineered nucleases has to change our current understanding of transgenic organisms. In the case of HDR-based knock-in, the transgene is an exogenous DNA but the number of unwanted random integration events is therefore minimized in comparison with classical transgenesis approaches and leading likely to only one copy of the transgene within the genome.

Regulatory consideration of these various methodologies is an ongoing process as regulators throughout the world adhere to a case-by-case paradigm for decision-making. Therefore, until a significant body of regulatory decisions emerges, there will be uncertainty as to the regulatory classification of plants developed by genome editing. The general scientific sentiment of some regulators (not all) to date is that gene-edited crops should not uniformly be subject to the same regulatory standards as transgenic crops. Fortunately, progress is being made in shaping sensible and pragmatic approaches towards the application of genome editing for crop improvement, but at some point new product-based paradigms for regulation of new breeding technologies must emerge (see Ricroch, Ammann, & Kuntz, 2016). Based on the Podevin, Devos, Davies, & Nielsen (2012) work, Wolt et al. (2016) proposed a diagram (Fig. 6) for regulatory implications specific to various genome-editing techniques.

Techniques in Category 1 involve transient introduction of recombinant DNA that do not integrate into the host genome and produce genotypes that are indistinguishable from plants obtained through conventional plant breeding. Category 2 consists of stable introduction of recombinant DNA into the host genome followed by breeding selection for null segregants that results in phenotypes that are indistinguishable from phenotypes obtained through conventional plant breeding too. These categories are crossed with the types of site-directed modification: site-specific random mutations by NHEJ (SDN1) and site-specific mutations with DNA repair via homologous recombination (SDN2). Plant phenotypes developed by SDN1 methods represent simple point mutations and regulators should not consider crops developed by mutagenesis as GM crops. The regulatory opinions regarding plant phenotypes developed by SDN2 methods are not as clear, as the nature and extent of the edits used to effect the desired change in the phenotype obtained by the technique would influence opinions as to whether the phenotype represents a GM product. For instance, deletions

Method \ Category	Category 1 Transient expression resulting in site-specific DSB and repair	Category 2 Stable genomic introduction of rDNA with intermediate steps to generate transgene-free null segregants	Category 3 Stable genomic integration of recombinant DNA
SDN1* Site-directed random mutation involving NHEJ	Low	• Low for deletions • Case-by-case for addition • Higher as size of insertion increases	N/A
SDN2 Site-directed homologous repair involving one or very few nucleotides	Case-by-case	Case-by-case	N/A
SDN3 Site-directed transgene insertion	N/A	N/A	High, moderated for well-characterized insertion sites

Fig. 6 Relationship of site-directed genome approach to the anticipated degree of regulatory scrutiny of the plant phenotype obtained. *SDN*: site-directed nucleases. *Source: Reprinted with permission from Wolt, J. D., Wang, K., & Yang, B. (2016). The regulatory status of genome-edited crops.* Plant Biotechnology Journal, 14, *510–518.*

are viewed as less consequential than are additions. And in the case of additions, the greater the number of bases added, the greater could be the level of regulatory concern. Finally, Category 3 involves techniques that result in stable integration of recombinant DNA where genome editing is used to specifically target delivery of a transgene or multiple transgenes through insertion by homologous recombination (SDN3). Thus, they simply represent a refined technique to accomplish transgenesis and would be considered no differently than GM products by regulators. But because the technique can specifically target transgene delivery into the genome, plants developed using SDN3 methods should require less data for risk characterization than more conventional approaches to transgenesis according to EFSA (Wolt et al., 2016). Within the EU, the European Commission and the Member States could use this diagram to draw up an opinion. Although this diagram only considers ZFNs, it is applicable to all engineered nucleases (TALENs and RGENs) (Wolt et al., 2016).

5.2.2 State-of-the-Art of Regulation Adopted for Edited Plants Worldwide

In addition to advices from scientists related to the EU context, it is also pertinent to study the regulation of crops obtained by NBT in different countries.

- *Argentina*: The Argentinean authorities have so far not received any applications for authorization of crops obtained through NBTs. However, it is presumed that decisions concerning the regulatory classifications of NBTs will be possible on the basis of the current legislation (Lusser & Rodríguez, 2012).
- *Australia*: The 2001 Australian Gene Technology Regulations include a list of organisms that are not regarded as GMOs and Item 1 of this "schedule 1" (a mutant organism in which the mutational event did not involve the introduction of any foreign nucleic acid, which is non-homologous DNA, usually from another species) is specifically relevant in the context of NBTs. For example, '*A mutant organism in which the mutational event did not involve the introduction of any foreign nucleic acid; 'a whole animal, or a human being, modified by the introduction of naked recombinant nucleic acid'* are not GMOs (https://www.legislation.gov.au/Details/F2009C00287). The Office of Parliamentary Counsel has not publicly given general guidance for specific new techniques, but rather encourages developers to contact them with specific cases where the regulatory status is not clear (Lusser & Rodríguez, 2012). According to the Food Standards Australia New Zealand (FSANZ, 2013) changes introduced using techniques for targeted mutagenesis, including ODM and ZFN technology, would be typically small and definable and have predictable outcomes. Such techniques would therefore be similar to traditional mutagenic techniques used in conventional plant breeding and food derived from these plants should not be regarded as GM food.
- *Canada*: Under the Canadian legislation, crops obtained through biotechnology are treated in the same way as any other plants with new traits. Crops with novel traits have to pass safety assessments and an authorization process, which is product based. Therefore, the Canadian regulatory process does not need to be changed or specifically adapted for crops derived from NBTs (Lusser & Rodríguez, 2012). In 2015, a crop phenotype for herbicide resistance developed by genome editing was approved by Canadian regulators (Wolt et al., 2016).
- *Japan*: The GMO definition under Japanese law follows the definition of "living modified organisms resulting from modern biotechnology" in the Cartagena Protocol. Officials from the six ministries responsible for regulating GMOs collect information related to NBT crops and discuss and consider their classification as GMOs or non-GMOs on a case-by-case basis, but they have not reached conclusions so far.

- *USA*: Authorizations are given by several authorities, namely, US Food and Drug Administration (FDA), Environmental Protection Agency (EPA) and Department of Agriculture (USDA) on a case-by-case basis, according to the "coordinated framework" which relies on existing statuses for regulatory authority rather than on a national biosafety law as found elsewhere in the world. A "regulatory trigger" is based on the method (biotech use), but then the assessment is product based. The USDA has provided guidance to product developers as to the regulatory status of products of modern biotechnology, which have broad relevance to FDA and EPA as well through the US coordinated framework. The guidance to date suggests that SDN1 techniques would not be regulated articles under USDA statutes nor would approaches initiated with transgenesis but which in intermediary steps selected for the absence of the transgenic elements (null segregants). However, for site-directed methods involving targeted oligonucleotide insertions or substitutions, further case-specific determinations would be required (Wolt et al., 2016).
- *South Africa*: Experience of South Africa with NBTs is limited to some research activities. No regulatory decisions have been taken yet and no applications concerning NBTs have been received (Lusser & Rodríguez, 2012).

5.2.3 State-of-the-Art of Progress Concerning the EU and the Member States' Regulation

5.2.3.1 European Commission

The European Commission opinion, awaited since 2007, was due by the end of 2015, but procedure has been delayed. The European Commission's silence is hardly explicable because many organizations linked to European institutions brought key elements since 2007:

- Joint Research Center (JRC) in 2011 and 2012 (see also Lusser & Rodríguez, 2012)
- European Food Safety Authority (EFSA) in EFSA (2012a) and EFSA (2012b)
- European Academies Science Advisory Council (EASAC) in 2013, 2015, and 2017
- Scientific Advisory Mechanism (SAM) in 2016 and 2017

These advices are all convergent, recommending that edited plants with no foreign DNA should not be considered as GMOs regarding Directive 2001/18/CE. Furthermore, the evaluation process should consider product over process as a regulatory paradigm (Wolt et al., 2016). Enrico Brivio, EU Commission spokesperson, said that "reflection on NBTs is ongoing inside the Commission but the outcome is not yet clear". He added that "in that context we would like to invite to move away from a GMO-centered discussion, when it comes to innovation in plant reproductive materials", underlining that "we should not treat all new techniques as *hidden* GMOs" (Michalopoulos, 2016). In any case, before new measures to be adopted by the Commission they have to be presented to Member States.

In the USA, once a product is deregulated (i.e. approved) by the various agencies (USDA-APHIS, FDA, EPA), its commercialization is automatically authorized. There is no compulsory variety registration in the USA. Seeds do have to meet federal and state standards in terms of germination percentage, purity and free from noxious weeds. Approvals from US agencies simply assert that the biotech crops are not different from conventional versions of the same crop type. In EU, EFSA is an official and public agency without any regulatory power. They act as scientific consultants (the so-called GMO panel), but decisions regarding authorization of new agricultural biotech products are taken at various political or semipolitical levels: committees of agriculture ministries, EU Commission, etc. In the United States, when an agency (USDA-APHIS, FDA, EPA) makes a final determination authorizing the commercial release of an agricultural biotech product that authorization is final and takes effect without any further political review. In that sense, the USA approach differs from the EU approach in which the EFSA makes a recommendation to a political entity—the EU.

5.2.3.2 United Kingdom: An Approach Based on the Evaluation of the Final Product

In the United Kingdom, the House of Commons (one of the two chambers of the Parliament) gave an advice the 11th February 2015 on defaults of the EU GMO legislation and the consequences for future plant breeding techniques, products from these techniques and trade issues. This text was based on three previous analyses:

- the Advisory Committee on Releases to the Environment (ACRE) in July 2013 https://www.gov.uk/government/collections/genetically-modified-organisms-applications-and-consents
- the Biotechnology and Biological Science Research Council (BBSRC) in September 2014 http://www.bbsrc.ac.uk/documents/genetic-crop-improvement-position-statement-pdf/

- the Parliamentary Office of Science and Technology in February 2017 http://researchbriefings.parliament.uk/ResearchBriefing/Summary/POST-PN-0548

A process-based regulation also fails to keep pace with technology and therefore potentially acts as a block on innovation. Second, the current system assesses the risks posed by these products but fails to take account of their potentially significant benefits. The House of Commons fears that an approached appearing as too binding restricts the development of plant biotech in the United Kingdom. In that way, the Commons calls for a real debate at the European scale; Directive 2015/412/EU is a good initiative but appears inadequate. Finally, the Commons is committed to an approach based on the evaluation of the final product and not on the techniques used for the production of this product (House of Commons, 2015). The Parliamentary Office of Science and Technology (POST) confirmed this advice in February 2017 (http://researchbriefings.parliament.uk/ResearchBriefing/Summary/POST-PN-0548).

5.2.3.3 Sweden: Field Trial in the Absence of a Decision of the Commission

The Swedish Board of Agriculture (SBA) did not give an advice on legislation's concerns but on issues raised by researchers from Umeå and Uppsala University concerning field trials of certain CRISPR/Cas9-mutated *Arabidopsis*. This advice is therefore important since it authorized the trials in certain cases on the 16th November 2015, whereas the European Commission indicated in June 2015 to not give authorizations, while the issue was not treated. CRISPR/Cas9-mutated plants carrying the expression cassette inserted in their genome must obtain permission by the SBA for carrying out field trials as they fall within the remit for regulation by the European Directive. But CRISPR/Cas9-mutated plants in which expression cassette is removed by crossing are exempted because the SBA considers that the methods yielding these specific mutations do not involve the use of recombinant nucleic acid molecules. To refer to Fig. 6, the SBA considers Category 2—SDN1 as out of the Directive 2001/18/EC but Category 3—SDN1 as belonging to the Directive 2001/18/EC. Indeed, the changes induced by CRISPR/Cas9 in the plants are changes in the plant ´s genome with the building blocks of their own cells; the SBA believes that those plants are produced by mutagenesis, and mutagenesis is exempted from regulation according to Annex 1B of the Directive 2001/18/EC, despite the early involvement of a transgene which is later eliminated (Swedish Board of

Agriculture, 2015). Scandinavian perspectives on plant gene technology (Sweden and Finland) are reported by Eriksson et al. (2017).

5.2.3.4 Germany: A Legal Opinion

The Bundesamt für Verbraucherschutz und Lebensmittelsicherheit (BVL) gave an advice the 7th December 2015 on the legal classification of NBT, in particular CRISPR/Cas9 (published by BVL, 2016). The BVL concluded that plants which exhibit point mutations induced by means of ODM (oligonucleotide-directed mutagenesis) and CRISPR–Cas9 techniques do not constitute GMOs within the context of the Directive 2001/18/EC. To reach this conclusion, the BVL analysed the wording of Article 2(2) of the Directive. It cannot be derived from the wording of Article 2(2) that the GMO definition is only based on the process by which the genetic modification is achieved. Thus, the exclusion of natural events is directly related to the genetic material and not to the technology used to carry out the modification. For point mutations induced by CRISPR–Cas9, the technique cannot be differentiated from point mutations that have arisen as a result of natural or induced mutagenesis. On the other hand, the sgRNAs used to carry out the CRISP/Cas9 technique are recombinant nucleic acids in the sense of the Directive (BVL, 2016). This legal opinion is unclear because it only considers "point mutations" that seems to be SDN1 but maybe SDN2 too (referring to Fig. 6), but it is not clearly mentioned which repair system is considered (Sprink, Eriksson, Schiemann, & Hartung, 2016). SDN2 may be authorized when the size of the insertion is not too important.

5.2.3.5 France: Scientific Reports

In the context of a referral procedure launched by the French ministries of Agriculture and of Ecology, the High Council for Biotechnology (HCB) published a report the 4th February 2016 on its first step of reflections about the status of edited plants. HCB advices are similar as those of its European counterparts. The HCB first proposal is that regulation should be based on the characteristics of the products obtained rather than the techniques used to generate them. The HCB also raised the question of assessment if one organism cannot be differentiated from another one obtained by a process which is not subject to assessment (e.g. modifications generated by SDN1 techniques can be identical to those obtained by conventional mutagenesis or conventional crossbreeding, NBTs simply save time when creating a variety). New plant varieties obtained by a process which is not subject to assessment according to Annex 1B of Directive 2001/18/EC (i.e. not a "GMO") must therefore simply be subject to the conventional IP and marketing authorization regulations. The situation for SDN2 techniques is the same as SDN1 on all points, except that breeders could choose to introduce molecular signatures for traceability in order to support ownership claims. In plant breeding, using negative segregation to remove a genetic modification event, of whatever origin (SDN3), is the standard procedure. After molecular confirmation that the modification has been removed, the resulting plant should be exempt from risk assessment and could be considered as a plant obtained by conventional breeding. But if an exogenous gene is not removed, the most logical solution might be to use GMO assessment principles. Thus, referring to Fig. 6, Category 1, Category 2, SDN1 and SDN2 techniques are not considered as producing GMOs (HCB, 2016).

The same advice was elaborated by the French Academy of Agriculture and the French Academy of Technologies, and endorsed by the Union of European Academies of Agriculture, which stated that targeted mutagenesis techniques could be excluded from the Directive 2001/18/EC (Académie d'Agriculture de France, 2016).

Furthermore, the French National Assembly requested a referral procedure of the Parliamentary Office for Scientific and Technological Assessment (Office Parlementaire d'Evaluation des Choix Scientifiques et Technologiques, OPECST) the 11th February 2015 on the same issue. The OPECST stated that Category 1, SDN1 and SDN2 techniques are not considered as producing GMOs (OPECST, 2017).

5.3 Key Elements Taking Part the Debate on Genome Editing in Plants

5.3.1 Societal Issues

The area planted in Bt corn in four countries (Spain, Portugal, Czech Republic and Slovakia) in Europe was 136,337 ha in 2016. This area corresponds to 1/1000 of Utilized Agricultural Land (UAL). The share of Bt corn with respect to the total corn area is 1.55%. Even if the majority of farmers do not cultivate GM crops, over a half of German farmers and almost a half of Czech and UK farmers indicated that they would be keen on adopting GM herbicide-tolerant oilseed rape (Lucht, 2015); over a third of Spanish, French and Hungarian farmers would be interested in adopting GM herbicide-tolerant maize. However, the current state of the GMO issue within the EU is influenced by some NGOs that played a role in the 1990s in framing GMOs as threat to biodiversity, farmer autonomy and food safety (Lucht, 2015). Therefore, people have a higher sensibility about GM foods: only 37% of the public at large responded that GM foods are generally safe to eat, whereas 88% of scientists interviewed recognized GM foods as generally safe (Pew Research Center, 2015, cited by Wolt et al., 2016). Transgenic technologies continue to elicit considerable public misunderstanding and mistrust despite 19 years of commercial use and over 185.1 million ha in production globally in 2016. Europeans seem to have in general a rather positive attitude with respect to technological development, but the application of biotechnology in GM foods is a difficult societal and political case. In 2010, 23% of respondents from the EU 27 states thought that GM food should be supported, while 61% disagreed with this view. The fact is that consumers do not perceive the benefits of GM crops. Indeed, the vast majority of GM crops cultivated today have phenotypes in order to reduce the technological itinerary for farmers and to increase yields, but have no perceptible influence on food characteristics or quality (Wolt et al., 2016).

A step can be crossed for a societal acceptance of GM crops if the consumer can perceive benefaction of new varieties, in particular if GM food is cheaper. Another factor could be biofortification, meaning genetic modifications to improve micronutrient content and availability as ω3 fatty acids, for example. But for the time being, consumers worry about possible negative effects of GM food on their health even if these perceived risks are scientifically unfounded. In this context, there is a lack of plant biotech research in Europe, which is also the consequence of numerous destructions of GM crop field trials. Concerning NBTs, anti-GMO NGOs are warning that plants obtained by genome editing are "hidden GMO", which will have heavy consequences for NBT products to come to market because even if the European Commission concludes that these products are not GMOs, public acceptance seems compromised. That is the reason why some organizations want these products to also be labelled, despite the fact that these products can be indistinguishable from products obtained through conventional plant breeding (HCB, 2016). In addition, NGOs opposing genetic engineering, including Greenpeace and Friends of the Earth Europe, are warning that some of the NBTs allow more radical changes to plant genomes than currently used genetic engineering methods, and that therefore the precautionary principle should be applied (Lucht, 2015).

5.3.2 Technical Issues Raised by Off-Target Effects and Technical Improvement

Off-target effects are sometimes noticed and represent one of the main challenges for risk assessment. However, when the following requirement are combined:

- the double-nickase version of the CRISPR–Cas9, which provides a 40-bp specific targeting instead of the usual 20-bp one, is used (Ran et al., 2013),
- the sgRNAs are designed according to the exact sequence of the plant variety or line to be edited,
- the two short PAM sequences (which specify the exact nucleotide where the DSB will take place) are correctly chosen (see Quétier, 2016),

then no off-target events occur.

Libraries and dedicated softwares for this double nickase technology have been provided with open access. It is highly recommended that journals' reviewers be more stringent in checking the material and methods to see whether or not the correct protocol to avoid off-target events has been used.

Interestingly, simple nickases, which are now known as "dead-Cas9" or "d-Cas9" give rise to a powerful class of genome engineering enzymes (Horlbeck et al., 2016), the most exciting of them being:

- the DNA base editing enzyme (using d-Cas9) which can change a chosen DNA A-T base-pair into a DNA G-C base-pair (Gaudelli et al., 2017),
- the RNA base editing enzyme (using d-Cas13) which can change a chosen RNA A-base into an RNA I-base, that cell machineries read G (Cox et al., 2017), which surfaced very recently.

6. CONSEQUENCES AND PERSPECTIVES FOR THE EU ON THE STATUTES ACCORDED TO EDITED PLANTS

The decision to grant or not GMO status to edited plants concern two kinds of actors, those who produce and those who use seeds, such as seeders and farmers. Innovations and improvements can only be disseminated if regulatory costs are acceptable with regard to the size of the market targeted. The authorization costs of a GM plant are approximately € 30–40 million per trait in the EU, having as consequences a centralization of innovation by a couple of actors, which is preventing plant biotechnologies to deliver its full promises. Currently, breeders only market traits that can be sold at very large scale in order to get a return on their investment (HCB, 2016). Furthermore, the registration time of a GM crop in EU legislation is 51 months on average, while it takes at least 15–20 months less than that in the USA, Brazil and Canada. Due to these unfavourable conditions, major plant biotechnology companies have moved research and development activities away from Europe. It was the case for the large multinational company BASF, in 2012, which closed its European plant biotech research facilities to focus on the American and Asian markets, and abandoned the development of GM crop varieties for Europe (Lucht, 2015).

"This dynamics of EU regulation may be seen as an ongoing 'Schumpeterian' chain of public choices: the calculus of consent drives politicians more than a science-based approach to law-making" according to Tagliabue (2017). Political attitudes towards GM crops are not uniform across all European countries. Rather, EU Member States can be separated into three categories depending on their acceptance of plant biotech: "adopters", "conflicted" and "opposed" Member States (Lucht, 2015). With this separation, the debate on NBTs is becoming more complex because the European Commission has to consider all positions: some Member States authorize Bt corn cultivation, others do not. A lot of researchers and seed companies hope that edited plants would ease public acceptance, and that NBT could reinvigorate the EU market. Furthermore, UE has to keep in mind that from 2007 through 2011, more than 35% of genome editing publications emanated from Europe (Wolt et al., 2016). At the end, crop improvement belongs to farmers and genome editing can contribute to optimize or improve current agronomic practices, and to develop new products and new market opportunities (HCB, 2016).

Consequences of patentability of edited plants are far more critical. Even if patents seem necessary for seed companies, their use could lead to two phenomena. The locking of access to certain genetic resources for certain actors, in particular small breeders and farmers since they do not have all technical means to benefit from breeder exception. Furthermore, they will not always be able to know if a given plant or trait is the object of a patent, and may be unable to distinguish natural or obtained through essentially biological process. The second fear is the constitution of oligopoly because the costs of licence may cause concentration phenomenon in seed industry, which could be as important as the one triggered by the costs for transgenesis (HCB, 2016).

7. CONCLUSION

Genome editing represents surely a great innovation in plant breeding, the use of engineered nucleases enabling a time gain, a cost decrease and targeted modification in the genome. Based on a literature review, the main use of CRISPR systems is to achieve improved yield performance, biofortification, biotic and abiotic stress tolerance, with rice being the most studied crop (Ricroch, Clairand, & Harwood, 2017).

Since the first transgenic plant placed on the market in 1996, the "GMO" issue is prone to controversy in the EU. Nevertheless, technology is constantly evolving, new techniques appear rapidly as it is the case for genome editing with engineered nucleases. The European agriculture needs NBT because agricultural production depends on the evolving of biotechnologies to adapt varieties to the needs of the markets. Besides agriculture, research and seed industry in the EU also need these new techniques. However, innovations and improvement can only be diffused if regulatory costs and the period of registration are acceptable regarding the size of the targeted market. The regulatory evaluation of these plants is a crucial issue for the EU, which has not given an answer regarding the status of NBT since 2007. In the European context, associating a product to the term "GMO" is not insignificant as it is a synonym for ban. Thus the questioning on NBTs is directly shifted from their evaluation to their potential ban through a classification as GMOs. A first approach is based on current regulation: the Directive 2001/18/EC. However, there is no consensus for a unique interpretation. Furthermore, an approach based only on techniques used for plant production leads to a regulatory case-by-case situation, which is not practicable. On the other hand, a large scientific consensus is rapidly growing in favour of a product-based approach, which is more realistic and would be a real breakthrough. This approach adopted by North American regulators involves regulating gene edited crops varieties based on the trait instead of the method used to generate it. Some European researchers suggested as to how these new plants may be regulated on a product basis (Davison & Ammann, 2017; Ricroch et al., 2016). This is also the approach defended by some European organizations. Indeed, when genome editing is used for plant breeding, the outcome of genetic modifications can be assessed, but it is not always possible to know which technique was used.

Based on current legislation, taking into accounts scientific advices and the JRC's advice in 2012, edited plants should not be considered as GMOs when engineered nucleases are used for targeted mutagenesis, but should be considered as GMOs when engineered nucleases are used for gene insertion. During this debate, an interesting phenomenon is observed. Due to the silence of the European Commission and thanks to Directive 2015/412/EU (opt-out possibility), Member States took part in the NBT debate. The United Kingdom adopted a hostile approach criticizing EU policy for not being adapted to the market and is in favour of adopting a product-based approach. Sweden did not wait for EU agreement to conduct field trials of edited plant and authorized trials which used engineered nucleases to carry out targeted mutagenesis. Finally, France and Germany gave advices which are in line with the idea that plants which have no foreign DNA (expression cassette, inserted DNA) and point mutations should not be considered as GMOs.

The European Commission decision is still under discussion. Since seed industry and plant biotech research are important for the EU, the categorization of gene-edited plants as non-GM would be a serious advantage for the EU. If European Commission classes gene-edited crops as GMOs, the EU should import gene-edited crops or food produced from or using gene-edited crops, the European seed industry and research will probably not be able to catch up with the technological gap accumulated over the past few decades.

REFERENCES

Académie d'Agriculture de France (2016). Avis sur la réglementation des mutagénèses ciblées en amélioration des plantes. https://www.academie-agriculture.fr/publications/academie-communique/avis/13-juillet-2016-avis-sur-la-reglementation-des-mutageneses. In English: http://ueaa.info/wp-content/uploads/2016/11/French-Opinion-Targeted-Mutagenesis_Apendix_AAF-AT_2016.pdf.

Belhaj, K., Chaparro-Garcia, A., Kamoun, S., & Nekrasov, V. (2013). Plant genome editing made easy: Targeted mutagenesis in model and crop plants using the CRISPR/Cas system. *Plant Methods*, *9*, 39.

Boch, J., & Bonas, U. (2010). Xanthomonas AvrBs3 family-type III effectors: Discovery and function. In N. K. VanAlfen, G. Bruening, & J. E. Leach (Eds.), *Book Series: Annual review of phytopathology: Vol. 48* (pp. 419–436). Annual Reviews.

Boch, J., Scholze, H., Schornack, S., Landgraf, A., Hahn, S., Kay, S., et al. (2009). Breaking the code of DNA binding specificity of TAL-Type III effectors. *Science*, *326*, 1509–1512.

Bortesi, L., & Fischer, R. (2015). The CRISPR/Cas9 system for plant genome editing and beyond. *Biotechnology Advances*, *33*, 41–52.

Bundesamt für Verbaucherschutz und Lebensmittelsicherheit (BVL). (2016). *Opinion on the legal classification of new plant breeding techniques, in particular ODM and CRISPR-Cas9.*

Čermák, T., Curtin, S. J., Javier, G.-H., Čegan, R., Kono, J. Y., Konečná, E., et al. (2017). A multipurpose toolkit to enable advanced genome engineering in plants. *The Plant cell*, *29*, 1196–1217.

Colleaux, L., D'Auriol, L., Betermier, M., Cottarel, G., Jacquier, A., Galibert, F., et al. (1986). Universal code equivalent of a yeast mitochondrial intron reading frame is expressed into E. coli as a specific double strand endonuclease. *Cell*, *44*(4), 521–533.
Colleaux, L., D'Auriol, L., Galibert, F., & Dujon, B. (1988). Recognition and cleavage site of the intron encoded omega transposase. *Proceedings of the National Academy of Sciences of the USA*, *85*(16), 6022–6026.
Cox, D. B. T., Gootenberg, J. S., Abudayyeh, O. O., Franklin, B., Kelner, M. J., Joung, J., et al. (2017). RNA editing with CRISPR-Cas13. *Science*, *358*(6366), 1019–1027.
Davison, J., & Ammann, K. (2017). New GMO regulations for old: Determining a new future for EU crop biotechnology. *GM Crops & Food*, *8*, 12–14.
EFSA. (2012a). Scientific opinion addressing the safety assessment of plants developed through cisgenesis and intragenesis. *EFSA Journal*, *10*(2), 2561.
EFSA. (2012b). Scientific opinion addressing the safety assessment of plants developed using Zinc Finger Nuclease 3 and other Site-Directed Nucleases with similar function. *EFSA Journal*, *10*(10), 2943.
Eriksson, D., Brinch-Pedersen, H., Chawade, A., Holme, I., Hvoslef-Eide, T. A. K., Ritala, A., et al. (2017). Scandinavian perspectives on plant gene technology: Applications, policies and progress. *Physiologia Plantarum*. http://onlinelibrary.wiley.com/doi/10.1111/ppl.12661/full.
European Academies Science Advisory Council (EASAC). (2013). *Planting the future: Opportunities and challenges for using crop genetic improvement technologies for sustainable agriculture, EASAC policy report 21*. http://www.easac.eu/biosciences/reports-and-statements/detail-view/article/planting-the.html.
European Academies Science Advisory Council (EASAC). (2015). *New breeding techniques*. http://www.easac.eu/biosciences/reports-and-statements/detail-view/article/easac-statem-2.html.
European Academies Science Advisory Council (EASAC). (2017). *Genome editing: Scientific opportunities, public interests, and policy options in the EU*. http://www.easac.eu/biosciences/reports-and-statements/detail-view/article/genome-editi.html.
European Commission (Eur-Lex). (2015). *Review of the decision-making process on GMOs in the EU*. http://eur-lex.europa.eu/legal-content/en/TXT/?uri=CELEX%3A52015DC0176.
European Parliament. (2015). *Joint motion for a resolution on patents and plant breeders rights—RC-B8-1394/2015*. http://www.europarl.europa.eu/sides/getDoc.do?pubRef=-//EP//NONSGML+TA+P8-TA-2015-0473+0+DOC+PDF+V0//EN.
European Patent Office (EPO). (2016). *The European Patent Convention*. http://www.epo.org/law-practice/legal-texts/epc.html.
Food Standards Australia New Zealand (FSANZ). (2013). *New plant breeding techniques*. http://www.foodstandards.gov.au/publications/Documents/New%20Plant%20Breeding%20Techniques%20Workshop%20Report.pdf.
Gabriel, A., & Voytas, D. (1997). DNA on the move—Keystone Symposium: Transposition- and site-specific recombination, Santa Fe, NM, USA, 1–7 March 1997. *Trends in Genetics*, *13*(7), 258–259.
Gaudelli, M. N., Komor, A. C., Rees, H. A., Packer, M. S., Badran, A. H., Bryson, D. I., et al. (2017). Programmable base editing of A · T to G · C in genomic DNA without DNA cleavage. *Nature*, *551*(7681), 464–471.
Hartung, F., & Schiemann, J. (2014). Precise plant breeding using new genome editing techniques: Opportunities, safety and regulation in the EU. *The Plant Journal*, *78*, 742 752.
Haut Conseil des Biotechnologies (HCB) (2013) Recommandation sur l'évolution de la propriété industrielle dans le domaine des biotechnologies végétales. HCB-2013.06.12. http://www.hautconseildesbiotechnologies.fr/en/avis/recommandation-sur-levolution-propriete-industrielle-dans-domaine-biotechnologies-vegetales.

Haut Conseil des Biotechnologies (HCB) (2016) Nouvelles techniques. New plant breeding techniques. Première étape de la réflexion du HCB Introduction générale". 20 janvier 2016. http://www.hautconseildesbiotechnologies.fr/fr/article/publications-hcb.

Hendel, A., Fine, E. J., Bao, G., & Porteus, M. H. (2015). Quantifying on- and off-target genome editing. *Trends in Biotechnology*, 33, 132–140.

Horlbeck, M. A., Gilbert, L. A., Villalta, J. E., Adamson, B., Pak, R. A., & Chen, Y. (2016). Compact and highly active next-generation libraries for CRISPR-mediated gene repression and activation. *eLife*, *2016*(5), e19760. https://doi.org/10.7554/eLife.19760.

House of Commons. (2015). *Advanced genetic techniques for crop improvement: Regulation, risk and precaution.* https://www.parliament.uk/business/committees/committees-a-z/commons-select/science-and-technology-committee/news-parliament-2015/gm-government-response-publsihed-15-16/.

Ishino, Y., Shinagawa, H., Makino, K., Amemura, M., & Nakata, A. (1987). Nucleotide-sequence of the IAP gene responsible for alkaline-phosphatase isozyme conversion in *E. coli* and identification of the gene product. *Journal of Bacteriology*, *169*, 5429–5433.

Lusser, M., Parisi, C., Plan, D., & Rodríguez-Cerezo, E. (2011). *New plant breeding techniques. State-of-the-art and prospects for commercial development.* JRC technical report EUR 24760 EN. (European Commission Joint Research Centre, Rome, 2011). Joint Research Center (JRC).

Kim, Y. G., & Chandrasegaran, S. (1994). Chimeric restriction endonuclease. *Proceedings of the National Academy of Sciences of the United States of America*, *91*, 883–887.

Kim, H., & Kim, J.-S. (2014). A guide to genome engineering with programmable nucleases. *Nature Reviews. Genetics*, *15*, 321–334.

Lawrenson, T., Shorinola, O., Stacey, N., Li, C., Østergaard, L., Patron, N., et al. (2015). Induction of targeted, heritable mutations in barley and *Brassica oleracea* using RNA-guided Cas9 nuclease. *Genome Biology*, *16*, 258. https://doi.org/10.1186/s13059-015-0826-7.

Le Buanec, B., & Ricroch, A. (2014). Intellectual property protection of plant innovation. In A. Ricroch, S. Chopra, & S. Fleischer (Eds.), *Plant biotechnology—Experience and future prospects.* Springer.

Li, T., Liu, B., Spalding, M. H., Weeks, D. P., & Yang, B. (2012). High efficiency TALEN-based gene editing produces disease resistance rice. *Nature Biotechnology*, *30*, 390–392.

Li, J. F., Zhang, D., & Sheen, J. (2015). Targeted plant genome editing via the CRISPR/Cas9 technology. In J. M. Alonso & A. N. Stepanova (Eds.), *Plant functional genomics* (pp. 239–255). New York: Springer.

Lloyd, A. A., Plaisier, C. L., Carroll, D., & Drews, G. N. (2005). Targeted mutagenesis usingzinc-finger nucleases in Arabidopsis. *Proceedings of the National Academy of Sciences of the United States of America*, *102*, 2232–2237.

Lucht, J. (2015). Public acceptance of plant biotechnology and GM crops. *Viruses*, *7*, 4254–4281.

Lusser, M., & Rodríguez, C. E. (2012). In *Comparative regulatory approaches for new plant breeding techniques. JRC workshop proceedings: Institute for Prospective technological studies.*

Ma, X., Zhang, Q., Zhu, Q., Liu, W., Chen, Y., Qiu, R., et al. (2015). A robust CRISPR-Cas9 system for convenient, high efficiency multiplex genome editing in monocot and dicot plants. *Molecular Plant*, *8*, 1274–1284.

Makarova, S., Haft, K. H., Barrangou, D., Brouns, R. J. J., Charpentier, E., et al. (2011). Evolution and classification of the CRISPR-Cas systems. *Nature Reviews. Microbiology*, *9*(6), 467–477.

Michalopoulos, S. (2016). *Fresh EU-US trade spat brewing over new plant breeding techniques.* EurActiv.com.

Moscou, M. J., & Bogdanove, A. J. (2009). A simple cipher governs DNA recognition by TAL effectors. *Science*, *326*, 1501.

Office Parlementaire d'Evaluation des Choix Scientifiques et Technologiques (OPECST). (2017). *Les enjeux économiques, environnementaux, sanitaires et éthiques des biotechnologies à la lumière des nouvelles pistes de recherche*. https://www.senat.fr/notice-rapport/2016/r16-507-1-notice.html.

Podevin, N., Devos, Y., Davies, H. V., & Nielsen, K. M. (2012). Transgenic or not? No simple answer! *EMBO Reports*, *13*, 1057–1061.

Puchta, H., Dujon, B., & Hohn, B. (1993). Homologous recombination in plant cells is enhanced by in vivo double strand breaks into DNA by a specific endonuclease. *Nucleic Acids Research*, 21, 5034–5040.

Puchta, H., Dujon, B., & Hohn, B. (1996). Two different but related mechanisms are used in plants for the repair of genomic double-strand breaks by homologous recombination. *Proceedings of the National Academy of Sciences of the United States of America*, *93–10*, 5055–5060.

Puchta, H., & Fauser, F. (2014). Synthetic nucleases for genome engineering in plants: Prospects for a bright future. *The Plant Journal*, *78*, 727–741.

Quétier, F. (2016). The CRISPR-Cas9 technology: Closer to the ultimate toolkit for targeted genome editing. *Plant Science*, *242*, 65–76.

Ran, F. A., Hsu, P. D., Wright, J., Agarwala, V., Scott, D. A., & Zhang, F. (2013). Genome engineering using the CRISPR-Cas9 system. *Nature Protocols*, *8*, 2281–2308.

Ricroch, A., Ammann, K., & Kuntz, M. (2016). Editing EU legislation to fit plant genome editing. *EMBO Reports*, *17*, 1365–1369.

Ricroch, A., Clairand, P., & Harwood, W. (2017). Use of CRISPR systems in plant genome editing: Toward new opportunities in agriculture. *Emerging Topics in Life Sciences*, *1*, 169–182.

Schornack, S., Meyer, A., Römer, P., Jordan, T., & Lahaye, T. (2006). Gene-for-gene-mediated recognition of nuclear-targeted AvrBs3-like bacterial effector. *Journal of Plant Physiology*, 163, 256–272.

Scientific Advisory Mechanism High Level Group (SAM HLG) of EU. (2016). *Scoping paper: New techniques in agricultural biotechnology. https://ec.europa.eu/research/sam/pdf/meetings/hlg_sam_052016_scoping_paper_agribiotechnology.pdf.*

Scientific Advisory Mechanism High Level Group (SAM HLG) of EU. (2017). *Explanatory note on new techniques in agricultural biotechnology*. https://ec.europa.eu/research/sam/pdf/topics/explanatory_note_new_techniques_agricultural_biotechnology.pdf.

Shukla, V. K., Doyon, Y., Miller, J. C., DeKelver, R. C., Moehle, E. A., Worden, S. E., et al. (2009). Precise genome modification in the crop species *Zea mays* using zinc-fingernuclease. *Nature*, *459*, 437–441.

Smith-Willis, H., & San, M. B. (2015). Revolutionizing genome editing with CRISPR/Cas9: Patent battles and human embryosle. *Cell and Gene Therapy Insights*, *1*, 253–262.

Sprink, T., Eriksson, D., Schiemann, J., & Hartung, F. (2016). Regulatory hurdles for genome editing: process- vs product-based approaches in different regulatory contexts. *Plant Cell Report*, *35*, 1493–1506.

Swedish Board of Agriculture. (2015). *CRISPR/Cas9 mutated Arabidopsis*. https://www.upsc.se/documents/Information_on_interpretation_on_CRISPR_Cas9_mutated_plants_Final.pdf.

Tagliabue, G. (2017). The EU legislation on "GMOs" between nonsense and protectionism: An ongoing Schumpeterian chain of public choices. *GM Crops & Food*, 35–51.

The Zinc Finger Consortium. (2014). http://www.zincfingers.org/scientific-background.htm.

Townsend, J. A., Wright, D. A., Winfrey, R. J., Fu, F., Maeder, M. L., Joung, J. K., et al. (2009). High frequency modification of plant genes using engineered zinc finger nucleases. *Nature*, *459*(7245), 442–445.

UPOV. (2015). *Mission statement.* http://www.upov.int/overview/en/.

Voytas, D. F. (2013). Plant genome engineering with sequence-specific nucleases. *Annual Review of Plant Biology, 64*, 327–350.

Wolt, J. D., Wang, K., & Yang, B. (2016). The regulatory status of genome-edited crops. *Plant Biotechnology Journal, 14*, 510–518.

Woo, J. W., Kim, J., Kwon, S. I., Corvalán, C., Woo, C. S., Kim, H., et al. (2015). DNA-free genome editing in plants with preassembled CRISPR-Cas9 ribonucleoproteins. *Nature Biotechnology, 33*, 1162–1164.

Wright, D. A., Townsend, J. A., Winfrey, R. J., Irwin, P. A., Rajapogal, J., Lonosky, P. M., et al. (2005). High-frequency homologous recombination in plant mediated by zinc-finger nuclease: Recombination and zinc-finger nucleases. *The Plant Journal, 44*, 693–706.

FURTHER READING

Puchta, H., & Fauser, F. (2016). The regulatory status of genome-edited crops. *Plant Biotechnology Journal, 14*, 510–518.

CHAPTER NINE

Epigenetics, Epigenomics and Crop Improvement

Aliki Kapazoglou*, Ioannis Ganopoulos†, Eleni Tani‡,1, Athanasios Tsaftaris§,¶

*Institute of Olive Tree, Subtropical Crops and Viticulture, Hellenic Agricultural Organisation "DEMETER" (HAO-Demeter), Lykovrisi, Athens, Greece
†Institute of Plant Breeding and Genetic Resources, Hellenic Agricultural Organization DEMETER (ex NAGREF), Thermi, Greece
‡Laboratory of Plant Breeding and Biometry, Agricultural University of Athens, Athens, Greece
§Institute of Applied Biosciences (INAB), CERTH, Thermi-Thessaloniki, Greece
¶Perrotis College, American Farm School, Thessaloniki, Greece
[1]Corresponding author: e-mail address: etani@aua.gr

Contents

Advances in Botanical Research, Volume 86
ISSN 0065-2296
https://doi.org/10.1016/bs.abr.2017.11.007

Abstract

Epigenetics refers to heritable alterations in chromatin architecture that do not involve changes in the underlying DNA sequence but profoundly affect gene expression and impact cellular function. Epigenetic regulation is attained by specific mechanisms involving DNA methylation, histone posttranslational modifications and the action of noncoding (nc) RNAs which lead to open or closed chromatin states associated with gene activation or gene silencing, respectively. Over the past two decades extensive investigations have provided a wealth of information on epigenetic regulation at specific loci both in model and crop plants and the effect it may have on various aspects of plant development such as proper vegetative growth, successful reproduction and viability, effects on yield, and efficiency in coping with stress. In recent years, the rapid progress of high-throughput technologies has led to the unveiling of epigenetic landscapes at genome-wide scale (epigenomes) exemplified by the deciphering of the full methylomes, at single base resolution, of the model plant Arabidopsis and crop plants such as rice and tomato. An increasing number of epigenomes are now being investigated on crops of high economic value. Transgenerational natural or induced epigenetic variation can be a new source of phenotypic diversity especially for species with low genetic variation. The comparison of different epigenomes arising from different genotypes/tissues/cell types/environmental conditions can offer valuable information for the development of biomarkers paving the way to what is nowadays termed plant epibreeding. This review will attempt a comprehensive presentation of the progress in plant epigenetics both at small scale (single locus) and large scale (epigenome-wide) during development and in response to environmental stress, focusing on agronomically important crops and the impact that epigenetics, epigenomics and the new emerging field of epibreeding may have on crop improvement.

1. EPIGENETIC REGULATION AND PLANT DEVELOPMENT

Epigenetics refers to stable and heritable changes in gene expression that occur without changes in the underlying DNA sequence (Fujimoto et al., 2012). Epigenetic regulation during plant development and in response to environmental conditions is attained by three epigenetic mechanisms, namely, DNA methylation, posttranslational histone modifications and the action of noncoding RNA molecules which are either small RNAs (small interfering RNAs, siRNAs and microRNAs, miRNAs) or long noncoding RNAs (long ncRNAs). The activation of these pathways leads to changes in chromatin architecture that impacts gene expression. Open (active) chromatin form or closed (repressive) chromatin conformation is associated with gene activation or gene silencing, respectively, and governs proper onset of gene expression programmes in distinct cell types, tissues and under different developmental or environmental stimuli (Allis & Jenuwein, 2015; Berger, 2007; Kouzarides, 2007) (Fig. 1). Dynamic transitions from open to closed chromatin forms and interactions among the epigenetic mechanisms are necessary to ensure proper cellular function at all stages of development and at different environmental conditions (Berger, 2007; Du, Johnson, Jacobsen, & Patel, 2015; Forderer, Zhou, & Turck, 2016; Jarillo, Piñeiro, Cubas, & Martínez-Zapater, 2009; Van Oosten, Bressan, Zhu, Bohnert, & Chinnusamy, 2014).

Early and extensive investigations in Arabidopsis had provided a plethora of information concerning epigenetic mechanisms in plant growth and development that were then followed by intensive investigation in crop plants like rice, maize, barley, tomato, soybean, both at the locus-specific and genome-wide levels (in depth reviews: Gallusci, Hodgman, Teyssier, & Seymour, 2016; Kapazoglou, Papaefthimiou, & Tsaftaris, 2012; Kapazoglou & Tsaftaris, 2011; Niederhuth & Schmitz, 2017; Tsaftaris, Kapazoglou, & Darzentas, 2012; Van Oosten et al., 2014).

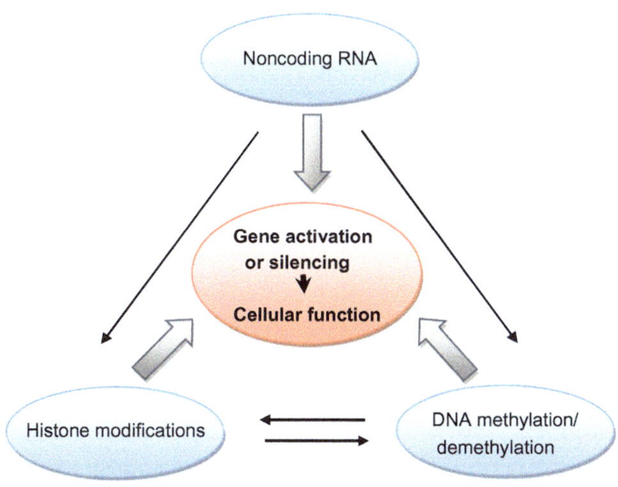

Fig. 1 The three epigenetic mechanisms affecting chromatin architecture: DNA methylation, histone modifications and the action of noncoding long or small RNA molecules. The three epigenetic mechanisms can lead to changes in chromatin structure without altering the underlying DNA sequence, resulting in open (permissive) or closed (repressive) forms associated with gene activation or gene silencing, respectively. Ultimately, alterations in chromatin architecture govern gene expression programmes and regulate cellular function. The three epigenetic mechanisms exist in a dynamic state, affecting one another in an interactive manner.

1.1 DNA Methylation

DNA methylation is a well-conserved epigenetic mark in eukaryotes that involves the addition of a methyl group on the 5′carbon of the cytosine base to form 5-methylcytosine. In plants, DNA methylation occurs in all three sequence contexts: the symmetric CG and CHG contexts (H = C, A or T) and the asymmetric CHH context (Feng, Jacobsen, & Reik, 2010). It is an instrumental epigenetic mark with a vital role in silencing deleterious transposon insertions, defending genome stability and regulating global gene expression across all developmental stages and environmental conditions (Niederhuth & Schmitz, 2017). DNA methylation in promoter regions mostly leads to gene silencing, whereas DNA methylation in the gene body is associated with gene expression. A dynamic interplay between DNA methylation and demethylation accomplished through specific enzymes is critical for proper cellular regulation during plant development (Van Oosten et al., 2014). Two antagonizing classes of enzymes, cytosine-5 DNA methyltransferases and DNA demethylases, govern the distribution and abundance of DNA methylation in plant genomes. DNA METHYLTRANSFERASE1 (MET1) family establishes and maintains CG methylation (Chan, Henderson, & Jacobsen, 2005), whereas CHROMOMETHYLASES CMT2 and CMT3 in cooperation with DOMAINS REARRANGED METHYLASE DRM2 shape the non-CG methylation pattern of the plant genome (Stroud et al., 2014). CMT2 and CMT3, in coordination with H3K9 methylation, methylate non-CG and CHG sites, respectively (Du et al., 2012; Matzke & Mosher, 2014; Stroud et al., 2014). *DRM2* regulates non-CG methylation and is being guided to specific targets by an RNA-directed DNA methylation pathway that involves 24-nt siRNAs (Law & Jacobsen, 2010; Niederhuth & Schmitz, 2017; Van Oosten et al., 2014). Early studies in Arabidopsis demonstrated that DNA methylation plays a vital role in all aspect of development, as single and double DNA methyltransferase mutants resulted in severe defects such as arrest of vegetative development, shorter stature, failure to enter the floral transition stage, extremely late flowering, abnormal embryo and endosperm development, reduced leaf size or complete sterility (Cokus et al., 2008; Law & Jacobsen, 2010; Van Oosten et al., 2014). In rice, mutation of the CG methylase *MET1* causes genome-wide hypomethylation, deregulation of gene expression and seedling lethality (Hu et al., 2014). The antagonizing actions of DNA methylation and demethylation dynamically regulate diverse processes in plant development. Four DNA demethylase genes have been identified in *A. thaliana*; *DEMETER (DME)*, *REPRESSOR OF SILENCING 1*

(*ROS1*), *DEMETER LIKE 2* (*DML2*) and *DEMETER LIKE 3* (*DML3*). The DNA glycosylase (or demethylase) enzymes remove 5-methylcytosine from its targets through a base excision repair mechanism and replace it with an unmethylated cytosine (Gehring et al., 2006; Morales-Ruiz et al., 2006). In Arabidopsis, AtDME was originally characterized as an epigenetic regulator required for maternal allelelic expression of the MEDEA (MEA) gene, encoding a repressive H3K27 methyltransferase, in the central cell of the embryo and endosperm (Choi et al., 2002; Kohler & Villar, 2008). Proper embryo and endosperm development depends on the expression of the maternal allele and silencing of the paternal allele (imprinting) of certain components of the Polycomb group Repressive complex2 (PRC2). In particular, *MEA* (*MEDEA*), *FERTILIZATION-INDEPENDENT ENDOSPERM* (*FIE*) and *FERTILIZATION-INDEPENDENT SEED2* (*FIS2*) are maternally expressed in the early stages of embryo and endosperm development. This expression is regulated, in part, by promoter demethylation of the maternal *MEA* promoter by DME (Gehring et al., 2006; Van Oosten et al., 2014). Tight regulation of PRC2 components is essential for preventing abnormal central cell proliferation before fertilization by repressing the expression of target genes, such as the Type-I MADS-box transcription factor *PHERES1* (*PHE1*) (Kohler et al., 2003; Makarevich, Villar, Erilova, & Kohler, 2008). DME was later found to be responsible for all of the active DNA demethylation taking place in the central cell preferentially targeting small, AT-rich and nucleosome-depleted transposable elements (TEs) leading to proper endosperm development (Ibarra et al., 2012). DME also demethylates similar sequences in the vegetative cell of the male gametophyte, whereas its suppression in the vegetative cell causes reduced small RNA-directed DNA methylation of transposons in the companion sperm cell, indicating movement of small RNAs between the two cell types (Calarco et al., 2012; Ibarra et al., 2012; Schoft et al., 2011). *DML2* and *DML3* are expressed in vegetative and reproductive tissues, and they prevent DNA hypermethylation at specific genomic regions (Van Oosten et al., 2014).

The ROS1 DNA demethylase was found to target genomic regions, particularly TEs, with distinct chromatin modifications and close to protein-coding genes suggesting spreading of demethylation and presumably gene activation to nearby genes (Tang, Lang, Zhang, & Zhu, 2016). In addition, other recent work has demonstrated that *ROS1* gene expression itself is methylation-sensitive and acts as an epigenetic rheostat sensing the methylation levels of the genome via a regulatory circuit (Lei et al., 2015; Williams, Pignatta, Henikoff, & Gehring, 2015). In rice, ROS1 was shown to play a key role in seed development, as downregulation of rice *ROS1* leads to wrinkled seeds, and a null *ROS1a* mutation results in abnormal early endosperm development and nonviable seed (La et al., 2011; Ono et al., 2012). Downregulation of a wheat *DME* led to reduced expression of endosperm glutens, a powerful immune reactant in celiac disease patients (Wen et al., 2012). A barley *DME* was shown to be associated with seed development and the response to drought (Kapazoglou, Drosou, Argiriou, & Tsaftaris, 2013). Finally, an orthologue of Arabidopsis ROS1 in tomato, termed SlDML2, was shown to have a vital role in regulation of fruit ripening (Liu, Feng, Li, & He, 2015; Liu, How-Kit, et al., 2015).

The first reports on global DNA methylation profiling of endosperm and embryo genomes demonstrated widespread reduction of DNA methylation in the endosperm, particularly at regions corresponding to TEs and small RNAs (Gehring, Bubb, & Henikoff, 2009; Hsieh et al., 2009; Li et al., 2012; Zemach, Kim, et al., 2010; Zemach, McDaniel, Silva, & Zilberman, 2010). Genome-wide DNA methylation studies in rice, maize, tomato, soybean and common bean (Crampton, Sripathi, Hossain, & Kalavacharla, 2016; Eichten et al., 2013; Li, Eichten, et al., 2014; Li, Yan, et al., 2014; Li et al., 2012; Song et al., 2013; Xing et al., 2015; Zemach, Kim, et al., 2010; Zemach, McDaniel, et al., 2010; Zhong et al., 2013) have provided important information with respect to global DNA methylation in order to understanding further the function of DNA methylation in vegetative, seed and fruit development (discussed in more detail in later sections).

1.2 Histone Modifications

Histone modification is another epigenetic mechanism with a crucial role in diverse aspects of plant development (Forderer et al., 2016; Fujimoto et al., 2012; Jiang & Kohler, 2012; Kapazoglou et al., 2012; Kapazoglou & Tsaftaris, 2011). Histone modifications refer to chemical alterations taking place on the N-tails of nucleosomal histones and include methylation, acetylation, phosphorylation, ubiquitination, biotinylation and sumoylation on specific lysine, arginine, serine and threonine residues (Engelhorn, Blanvillain, & Carles, 2014; Gan, Huang, & Ito, 2013; Grimanelli & Roudier, 2013). These chemical modifications affect chromatin architecture leading to open (active) or closed (repressive) chromatin states which regulate transcriptional activity and are inherited by daughter cells (Allis & Jenuwein, 2015; Berger, 2007). Usually, trimethylation of histone 3 on lysine 4 (H3K4me3) and acetylation of histone 3 or histone 4 result in gene expression, whereas trimethylation on lysine 27 of histone 3 (H3K27me3), dimethylation of histone 3 on lysine 9 (H3K9me2) and general deacetylation lead to gene silencing. Histone modifications influence multiple facets of developmental processes like flowering, seed development and responses to external stimuli (Banerjee & Roychoudhury, 2017; Gallusci et al., 2016; Gehring & Satyaki, 2017; Kim & Sung, 2014; Lämke & Bäurle, 2017; Pandey, Sharma, Pankaj Sahu, & Prasad, 2016; Yamamuro, Zhu, & Yang, 2016). An excellent example of the way histone modifications may act interactively in a synergistic or antagonistic manner for ensuring proper control of a major developmental process is the regulation of the *FLOWERING LOCUS C* (*FLC*), a repressor of flowering (Dean & Whittaker, 2017). The transition from the vegetative to the floral state is a crucial developmental change in a plants' life cycle. To ensure reproductive success, plants have to flower under favourable temperature conditions and appropriate day lengths. In temperate plants a long period of cold (the cold of winter), known as vernalization, is required before flowering in the spring (Amasino, 2010; Sung & Amasino, 2006). An elaborate system involving a series of histone repression and activation marks is employed by the shoot meristematic cell on *FLC* chromatin to ensure proper flowering time. Prior to vernalization, in the fall, the chromatin around the *FLC* gene is enriched with the active histone marks H3K4me3, H3K4me36 and H2Bub1 (H2B monoubiquitylation), resulting in high expression of the *FLC* gene and consequent repression of the floral transition. Permissive chromatin marking is accomplished through the action of chromatin remodeling complexes such as Compass-c and PAF1-c, with H3K4me3 and H3K36me3 methyltransferase activity, respectively, and RAD6/BRE1-c with histone H2B monoubiquitination activity (Kim & Sung, 2014). During vernalization, in the winter, the levels of the active histone marks are reduced, and *FLC* chromatin is enriched with chromatin repressive marks, such as H3K27me3, through the action of the POLYCOMB GROUP REPRESSIVE COMPLEX2 (PRC2) with methyltransferase activity, and *FLC* expression is now reduced (Bastow, Mylne, Lister, & Lippman, 2004; Chanvivattana et al., 2004).

1.3 Noncoding RNA

Notably, the PRC2 complex is recruited to *FLC* chromatin through a long ncRNA, COLDAIR, transcribed from the first intron of *FLC* (Heo & Sung, 2011). In the spring season, *FLC* is further suppressed through the continuous action of PRC2-c and the combinatorial action of a PRC1-like complex which mediates enrichment with repressive H2Aub, and associates with H3K27me3, contributing to chromatin compaction and stabilizing the repressive state. By this intricate regulatory network, *FLC* is repressed and flowering is promoted in the spring when temperature is optimal and days are longer. Analogous but not identical epigenetic mechanisms are involved in vernalization in other species. For example, in cereals, a flowering time regulator, *VERNALIZATION1* (*VRN1*) which is activated through active H3K4me3 marks, silences the flowering repressor *VRN2* and promotes flowering under the most favourable conditions (Oliver, Finnegan, Dennis, Peacock, & Trevaskis, 2009).

A massive number of small RNAs have been identified and associated with all aspects of plant development and in response to stress, either by locus-specific or large-scale genome-wide studies both in Arabidopsis and crop plants. Due to space limitations this review will not be able to contain the full extent of progress regarding small RNA investigations (for excellent reviews, the reader is referred to Banerjee, Sirohi, Ansari, & Gill, 2017; Erdmann & Gehring, 2016; Kamthan, Chaudhuri, Kamthan, & Datta, 2015; Kumar, Verma, & Trivedi, 2017; Lewsey et al., 2016; Li, Castillo-González, Yu, & Zhang, 2017; Li & Zhang, 2016; Martinez & Kohler, 2017; Rajeev Kumar, Safia, & Sathishkumar, 2017; Rodrigues & Miguel, 2017; Tang, Wen, Lin, & Wang, 2017; Yu, Liu, Li, & Yu, 2017; Zhang, 2015; Zhang & Chen, 2017).

2. EPIGENETICS AND ENVIRONMENTAL STRESS

Numerous studies have shown that a plethora of abiotic stresses such as cold, salinity, drought, plant density, osmotic stress and excess or limitation of minerals, control genes that encode epigenetic regulators (Fang, Liu, Thorn, Duan, & Tian, 2014; Li, Eichten, et al., 2014; Li, Yan, et al., 2014; Su et al., 2015; Tani, Polidoros, Nianiou-Obeidat, & Tsaftaris, 2005; Tsaftaris, Kafka, & Polidoros, 1997). Under stress, DNA methylation, histone posttranslational modifications (PTMs), chromatin remodeling and mechanisms such as RNA-interference can rapidly regulate gene expression (Bocchini et al., 2015; Boyko & Kovalchuk, 2008; Chen, Luo, Wang, & Wu, 2010; Goll & Bestor, 2005; Kim et al., 2008; Liu, Feng, et al., 2015; Liu, How-Kit, et al., 2015; Lukens & Zhan, 2007). After stress, this type of modification can be 'memorized' by plant somatic cells and can be utilized as an epigenetic mark that can be inherited transgenerationally, thus the same epigenetic modification will occur when the progenies will face stressful conditions (Bruce, Matthes, Napier, & Pickett, 2007; Feng et al., 2012; Grativol, Hemerly, & Ferreira, 2012; Hauser, Aufsatz, Jonak, & Luschnig, 2011; Mathieu, Reinders, Čaikovski, Smathajitt, & Paszkowski, 2007; Migicovsky, Yao, & Kovalchuk, 2014; Tricker, López, Gibbings, Hadley, & Wilkinson, 2013). However, this process appears to function mainly through the female gamete (Wibowo et al., 2016). This epigenetic plasticity that is transmitted through generations is a key factor in the plant immediate response to stress and its long-term adaptation and can have a great impact on breeding programs (Fortes & Gallusci, 2017; Mirouze & Paszkowski, 2011; Peng & Zhang, 2009).

Different types of abiotic stresses can cause different types of epigenetic regulation. For example, it was shown that under some abiotic stresses such as chilling, planting density, cutting and successive rounds of subculture, a decline of DNA methylation levels occurs, whereas under salt stress an increase is generally found (Peng & Zhang, 2009). Due to space limitations some representative examples will be given below, showing the epigenetic control of stress adaptation with special emphasis on cold and drought stress.

2.1 Epigenetic Control of Adaptation to Low Temperatures

In roots of maize seedlings, chilling decreases the amount of DNA methyltransferase which results in 10% reduction of DNA methylation (Steward, Ito, Yamaguchi, Koizumi, & Sano, 2002). Similar hypomethylation events of approximately 10% of loci also took place in the tobacco genome under several abiotic stresses including aluminium, salt and cold stress. Such hypomethylation events might enhance alteration of chromatin structure, thus inducing transcriptional activation of stress-responsive genes (Choi & Sano, 2007). Interestingly, under cold stress, the change of methylation in corn seedlings seems to be organ and site specific: methylation is decreased in roots, whereas no methylation change arises in shoots. Moreover, demethylation takes place only in Ac/Ds transposons, and not in other genes (Steward et al., 2002). Therefore, demethylation under stress may induce the activation of Ac/Ds transposons. However, in *Antirrhinum majus*, the activation of Tam3 transposons is the cause of decrease in methylation status under low temperatures rather that the result of this change (Hashida, Kitamura, Mikami, & Kishima, 2003; Hashida et al., 2006). In previous studies methylation mapping has revealed genome-wide demethylation in maize roots subjected to cold stress mainly at core regions where linker regions are not influenced (Steward et al., 2002).

Recent studies have highlighted the importance of epigenetic alterations in response to chilling during dormancy break and subsequent fruit set in apple. Under ideal high chill conditions, total methylation was decreased, which leads to a reinitiation of active growth (bud break) and a subsequent fruit set (Kumar, Rattan, & Singh, 2016).

Interestingly, Zhang et al. (2016) have demonstrated that the reduced flavour quality of ripe tomato fruits results from greatly reduced levels of transcripts for some key volatile synthesis enzymes associated with consumer liking, a phenomenon that is tightly linked to major changes in the methylation status of promoters of key genes that take part in ripening, quality and flavour volatile synthesis, under the postharvest chilling storage temperatures.

Although most of the studies refer to epigenetic control in response to cold stress, some studies have focused on the epigenetic regulation of warm and high temperatures. It is documented that posttranscriptional gene silencing (PTGS), along with the formation of double-stranded RNA (dsRNA) siRNAs biogenesis are key players that are involved in transgenerational epigenetic regulation in response to heat stress. Notably, the duration of heat treatment has diverse effects on the epigenetic mechanisms, suggesting the complexity in the epigenetic regulation of heat stress (Liu, Feng, et al., 2015; Liu, How-Kit, et al., 2015; Zhong et al., 2013).

2.2 Epigenetic Events and Adaptation to Drought

In many cases, the abundance in transcript levels of drought-responsive genes was found to be correlated with changes in histone modification and nucleosome density. In fact under drought stress several histone alterations can occur including acetylation, methylation, phosphorylation and SUMOylation (Kim, Sasaki, Ueda, Sako, & Seki, 2015). Moreover, results by several groups implied that epigenetic responses are depended on the intensity of the drought stress. For example, upon severe drought conditions, epigenetic responses such as histone modifications and nucleosome loss in the same gene region were more profound, compared to moderate drought conditions (Kim et al., 2012; Kim et al., 2008; Kiriga, Yu, & Bill, 2016).

Many reports have documented that drought stress response is memorized via histone alteration on several drought stress-induced genes (Ding, Bellizzi, Ning, Meyers, & Wang, 2012). Trimethylation at lysine 4 on histone 3 (H3K4me3) is a good epigenetic mark of gene induction and stress memory (Ding, Avramova, & Fromm, 2011; Ding et al., 2012; Oh, Park, & van Nocker, 2008; Zhang, Bernatavichute, Cokus, Pellegrini, & Jacobsen, 2009). In Arabidopsis, H3K4me3 is modified by the histone methyltransferase trithorax-like 1 (*ATX1*) and subsequently activates *NCED3*, which encodes a key enzyme in the ABA biosynthesis pathway, under drought stress conditions (Ding et al., 2011; van Dijk et al., 2010). Osmotic stress in Arabidopsis also causes histone modification leading to an increase in phosphorylated histone H3 threonine 3 (H3T3ph) and a subsequent transcriptional activation of the underlying DNA sequences (Wang et al., 2015). The increase of H3T3ph levels in pericentromeric regions is thought to provide a condensed heterochromatic structure and proper chromosome organization.

An excellent review by Han and Wagner (2014) discusses each of the different epigenetic mechanisms that affect gene expression upon drought stress and how these mechanisms contribute to water stress responses, highlighting that PTMs of histones are critical for successful water stress responses in plants.

In addition, DNA methylation plays an important role in the regulation of plant responses to abiotic stresses. Indeed drought stress induces extensive alterations of DNA methylation patterns in Arabidopsis (Colaneri & Jones, 2013; Tricker, Gibbings, Rodríguez López, Hadley, & Wilkinson, 2012; Wibowo et al., 2016; Yong-Villalobos et al., 2015), barley (Chwialkowska, Nowakowska, Mroziewicz, Szarejko, & Kwasniewski, 2016) or poplar (Liang et al., 2014). Recent studies suggested that ABA can regulate gene expression through DNA methylation and histone acetylation status (Sah, Reddy, & Li, 2016). For example, in response to drought stress, an increase was documented in the histone acetylation level of *RD20*, *RD29A* and *RD29B* genes thought to be drought-induced. Moreover, studies by Khraiwesh et al. (2010) and Gohlke et al. (2013) showed that ABA represses gene expression in *Physcomitrella patens* (a moss) and Arabidopsis, respectively, via DNA methylation of gene promoters in ABA-repressive genes.

In the case of poplar, drought stress and also nutrient deficiency cause variation of DNA methylation at repetitive elements, which subsequently controls adjacent gene expression (Liang et al., 2014). Additionally, when maize seedlings were subjected to various abiotic stresses such as salt, heat and UV radiation, and when rice plants were subjected to Pi deficiency, TEs regulated the expression of many stress-inducible genes (Makarevitch et al., 2015). In the latter case, however, TEs were hypermethylated, an event that most likely prevented potentially deleterious activity of TEs near genes that are highly induced upon stress.

Finally, many epigenetic mechanisms can function together in order to provide abiotic stress adaptation in plants. Studies in three rice cultivars grown under drought and salt stress have shown that methylation levels were significantly different in the three rice cultivars, and many methylated regions were associated with differential expression of genes important for abiotic stress response. Moreover, hypermethylated regions were positively correlated with small RNA abundance (Garg, Narayana Chevala, Shankar, & Jain, 2015).

3. EPIGENETIC CHANGES AND BIOTIC STRESS

There are fewer studies regarding epigenetic regulation under biotic stress compared to abiotic stress. However, it was recently documented that chromatin modifications are implicated in the plant immune against both necrotrophic and biotrophic pathogens (De-La-Peña, Rangel-Cano, & Alvarez-Venegas, 2012; Ding & Wang, 2015; Soyer et al., 2014). On the other hand, pathogenic infection can influence the levels of methylation in plants by two different ways: by hypermethylation on the genome-wide level and hypomethylation of resistance-related genes (Peng & Zhang, 2009).

Expression of several *R* genes (*SUPPRESSOR OF npr1-1, CONSTITUTIVE1* (*SNC1*) and *RESISTANCE TO PERONOSPORA PARASITICA4*) was shown to be controlled by Arabidopsis E3 ubiquitin ligase genes, HISTONE MONOUBIQUITINATION1 (HUB1) and HUB2 (Zou, Yang, Shi, Dong, & Hua, 2014). Moreover, HUB1 and HUB2 mediate histone 2B (H2B) monoubiquitination at the *SNC1 R* gene locus and regulate its expression, indicating that H2B monoubiquitination and histone modification at the *R* gene locus play a critical role in biotic responses in plants. Another study conducted with Arabidopsis showed that histone deacetylase HDA19 negatively controls basal defence mediated by the SA-dependent signaling pathway, thus loss of HDA19 induces the upregulation of SA biosynthetic genes and renders resistance to the virulent *Pseudomonas syringae* Pst DC3000 strain (Choi et al., 2012). In rice, Li et al. (2013) showed that Jumonji C protein may remove dimethylated or trimethylated histone H3K27me2/3 demethylase from marked silent or repressed defence-related genes and increase their basal and induced expression during pathogen infection.

On the other hand, when plants are exposed to different forms of biotic stresses, a plethora of stress-induced differentially methylated regions (DMRs) alter their level of methylation which is often coupled to differential gene expression (Dowen et al., 2012). Some pathogen response genes contain plenty of repeats in their promoter regions and are negatively regulated by DNA methylation. After pathogen invasion, these promoter loci may undergo a decrease in DNA methylation so they can be more easily triggered at the transcriptional level (Le et al., 2014; Sahu et al., 2013; Yu et al., 2013).

Epigenetic inheritance helps plants to have both long- and short-term adaptation to unfavorable abiotic and biotic environmental conditions. However, few cases of transgenerational epigenetic marks influencing defence responses to biotic stresses have been documented, mainly regarding model plants. An in-depth understanding of the way that epigenetic mechanisms contribute to plant defence and the concomitant pathogen reaction for their own benefit could help in utilizing new crop protection tools (Zhu, Shan, Ayliffe, & Wang, 2015).

4. EPIGENETICS AND GRAFTING

Grafting is an ancient agricultural practice employed widely in crops such as woody fruit trees, grapes and vegetables to improve plant performance (Goldschmidt, 2014). It was originally used to avoid the detrimental effects of soil borne pathogens infecting the roots of susceptible plants by utilizing resistant rootstocks grafted onto the plant of interest (Louws, Rivard, & Kubota, 2010). Beneficial rootstocks provide resistance to pathogens, tolerance to abiotic stress, and enhanced yield and fruit quality. Graft-induced epigenetic changes concerning mainly DNA methylation and small RNA expression have been documented in grafted plants as compared to the seed plants, in Arabidopsis, and in *Solanaceae* and *Cucurbitaceae* species (Avramidou et al., 2014; Molnar et al., 2010; Wu et al., 2013). Through genome-wide methylation profiling with methylation-sensitive amplified polymorphisms or via locus-specific bisulphite sequencing (see below), substantial alterations in DNA methylation were detected in grafted *Solanaceae* after reciprocal interspecies grafting (Wu et al., 2013). Specifically, the global methylation pattern of tomato, eggplant and pepper scions was found significantly altered both in the CG and CHG context upon grafting, and these graft-induced DNA methylation changes were inherited in the self-pollinated progeny indicating that grafting can induce epigenetic changes that can be stably transferred to subsequent generations (Wu et al., 2013).

An investigation involving *Cucurbitaceae* interspecies heterografting revealed significant increase in DNA methylation levels in the scions of melon and cucumber that were grafted onto pumpkin rootstocks (Avramidou et al., 2014). Transport of small noncoding RNAs (siRNAs and miRNAs) across graft segments has also been reported. Migration of small RNAs of homografted and heterografted plants has been demonstrated in Arabidopsis, *Solanaceae* and *Cucurbits* and often the mobile RNA molecule has a transcriptional or PTGS effect (Haroldsen et al., 2012; Melnyk, Molnar, & Baulcombe, 2011; Molnar et al., 2010). MiRNAs act as posttranscriptional gene regulators and target a wide range of transcription factors related to developmental processes and responses to abiotic and biotic stresses as well as enzymes involved in metabolic pathways (Barrera-Figueroa, Wu, & Liu, 2013; Li & Zhang, 2016). In a recent study, *Cucurbitaceae* intraspecies grafting induced significant differences in DNA methylation in scions of grafted plants and correlated with changes in fruit morphology and metabolic profile (A. Xanthopoulou, personal communication). In addition a series of miRNAs were differentially expressed in scions of grafted plants as compared to the shoots of seed plants. Taken together these data indicate that grafting induces epigenetic changes associated to plant performance and may significantly affect important agricultural traits. Exploitation of this induced epigenetic variation and differential epimarking could lead to development of epimolecular markers for generation of superior grafts and selection of compatible grafting partner combination. Given the growing interest in improving vegetables and other crops through grafting this endeavour acquires additional value.

5. EPIGENETICS AND HYBRID VIGOUR

Hybrid vigour or heterosis is defined as the phenomenon whereby a progeny exhibits phenotypic superiority over its parents with regard to traits such as growth rate, reproductive success and yield (Lippman & Zamir, 2007). Generation of hybrids is a practice used in agriculture for hundreds of years and has achieved increased yields in crops such as maize and rice. Likewise, studies with Arabidopsis hybrids have shown heterosis phenotype for a number of traits including vegetative biomass and seed yield (Groszmann et al., 2014; Groszmann et al., 2015; Ni et al., 2009). The molecular basis of heterosis is an active field of investigation but not fully understood, as yet. Many interactive attributes have been associated with heterosis, including gene expression alterations, metabolic activity and epigenetic regulation (Greaves et al., 2015; Shen et al., 2017; Varriale, 2017; Zhang et al., 2014; Zhu, Greaves, Dennis, & Peacock, 2017). Comparative transcriptome analysis between hybrids and their parents indicated that a vast array of genes have altered levels of expression as compared to the average levels of expression of their parents (nonadditive expression) (Chen, 2013; Meyer et al., 2012; Zhu et al., 2017). During the last several years genome-wide-based studies have investigated the epigenetic mechanisms involved in heterosis (Greaves et al., 2015). In Arabidopsis, significant changes in genome-wide DNA methylation associated with gene expression have been evidenced in hybrids (Shen et al., 2012). Similarly, global epigenetic and transcriptional changes were reported in rice hybrids and maize hybrids as compared to their parents (He et al., 2013; He et al., 2010). Furthermore, noncoding small RNAs, which play crucial roles in plant development and in response to stress, have been associated with heterosis (Groszmann et al., 2011; Ng, Lu, & Chen, 2012; Shen et al., 2017; Zhang et al., 2014). Small RNA profiling of an F1 super-hybrid in rice identified distinct miRNAs and trans-acting siRNAs which were differentially expressed in the hybrid (Zhang et al., 2014). A great number of the targets of these small RNAs was found to be transcription factors associated with a series of developmental processes, including the auxin signaling pathway. Additionally, a number of the differentially expressed small RNAs and their target genes were significantly enriched in quantitative trait loci related to important agronomic traits, such as plant architecture and grain yield, implying their contribution to the superior performance of hybrid rice (Zhang et al., 2014). Similarly, genome-wide DNA methylation was increased in a *Brassica napus* F1 hybrid and was particularly enriched in sequences containing TEs, whereas global small RNA expression was also found to be enhanced. Interestingly, gene expression was altered for hormone-related genes (auxin and salicylic acid) in the hybrid as compared to the parental lines (Shen et al., 2017).

Finally, another recent report has demonstrated the implication of the other epigenetic system, histone modifications, in the molecular mechanism of heterosis in Arabidopsis hybrids at the early stages of seedling development (Zhu et al., 2017). Alterations in two gene activating marks, H3K4me3 and H3K9ac, and two repressive marks, H3K27me3 and H3K9me2, were evidenced in a Ler/C24 hybrid at specific loci in the genome and were accompanied by expression changes in a small subset of genes. It was suggested that altered levels of histone modifications and gene activity at early seedling stages may play an essential role in biomass increase at the later stages of development (Zhu et al., 2017).

6. MAIN EXPERIMENTAL TECHNIQUES FOR DNA METHYLATION AND HISTONE MODIFICATION PROFILING IN PLANTS

6.1 Whole-Genome Bisulphite Sequencing

Whole-genome bisulphite sequencing, WGBS (BS-seq; MethylC-seq), is a next-generation sequencing (NGS) technology used to determine the DNA methylation status of single cytosines by treating the DNA with sodium bisulphite before sequencing. Sodium bisulphite is a chemical compound that converts unmethylated cytosines into uracil (Gu et al., 2011). Genomic DNA is purified and sheared into fragments. The fragmented DNAs are end-repaired; adenine bases are added to the 3′ end (A-tailing) of the DNA fragments, and methylated adapters are ligated to the DNA fragments (Urich, Nery, Lister, Schmitz, & Ecker, 2015). Sodium bisulphite treatment is followed by construction of a library with size-selected fragments, PCR amplification, and sequence determination. The major advantage of WGBS is that it can decipher methylation state at single base resolution determining the methylation mark at every CpG site, including regions of low CpG density, such as 'gene deserts', and distal regulatory elements. WGBS was firstly used in plants for decoding the Arabidopsis methylome (Cokus et al., 2008). It enabled the single base mapping of cytosine methylation at all three sequencing contexts, CpG, CHG and CHH as well as the determination of bulk methylation levels within each context. Furthermore, it unraveled the genome-wide methylation pattern in wild-type and methylation-related mutants and revealed specific methylation sites associated with gene expression (Cokus et al., 2008). WGBS unraveled the global methylation pattern of the maize genome, identifying a highly methylated genome and a specific 'CHH island' upstream of transcriptional start sites (Gent et al., 2013; Regulski et al., 2013). In addition to global pattern identification, differential methylation in specific regions or even at particular loci was determined using bioinformatics tools. For instance, Hsieh et al. (2009) compared Arabidopsis endosperm and embryo methylomes and reported that virtually the entire endosperm genome is demethylated, coupled with extensive local non-CpG hypermethylation of siRNA-targeted sequences. Lu et al. (2015) performed WGBS of maize embryo and endosperm, and the results revealed hypomethylation in the endosperm compared to the embryo.

6.2 Reduced Representation Bisulphite Sequencing

Reduced representation bisulphite sequencing (RRBS) was originally developed for purposes of deciphering the mammalian methylome at low costs (Meissner et al., 2008). It is based on *Msp*1 restriction enzyme digestion and size selection of *Msp*I-digested fragments (40 and 220 bps) for bisulphite conversion and sequence. This translates to coverage of 85% of CGIs (CpG islands), mostly in promoters, which comprises only 1%–3% of the mammalian genome and thereby it achieves sequencing of a much smaller portion of the genome (Gu et al., 2011). Although RRBS is more cost-effective than WGBS as it relies on the enrichment of specific CpG regions, it may provide low coverage at intergenic regions. RRBS has been widely used in analysing methylation patterns in large-scale samples. RRBS has been very useful in profiling plants genomes, such as oak populations (Platt, Gugger, Pellegrini, & Sork, 2015) and *Brassica rapa* (Chen et al., 2015). Platt et al. (2015) suggested that CpG methyl polymorphisms as a whole are involved in local adaptation, either directly or through linkage to regions under selection. Furthermore, Chen et al. (2015) found that DNA methylation in *B. rapa* indicated that 52.4% of CG sites were present as ^{5m}CG, with 31.8% of CHG and 8.3% of CHH. They proposed that genic regions of single copy genes had significantly higher methylation compared to those of two or three copy genes.

Enrichment-based technologies have emerged as alternative sequencing methods to bisulphite treatment. The basis of methylated DNA immunoprecipitation sequencing (MeDIP-seq) is the utilization of a suitable antibody to precipitate fragments of DNA containing a methylated cytosine (Zhao, Whyte, Hopkins, Kirk, & Prather, 2014). Based on the specific antibody, methylated cytosines in CpG context or total cytosines can be targeted, and the technique can be successfully applied to low (160 ng) DNA concentrations (Weng, Huang, & Yan, 2009). Other related methods are MBD-seq (methyl-binding domain sequencing) (Clark et al., 2012) and MethylCap-seq (methyl capture sequencing) (Taiwo et al., 2012), which make use of a methyl CpG-binding domain protein to precipitate DNA fragments containing a high density of methylated CpG sites. The advantage of these approaches resides in the fact that the fraction of the genome to be sequenced is much reduced providing high coverage at low cost. The major disadvantage is that it is only possible to detect methylation sites to within 150–200 bases (Weng et al., 2009). In addition, a decrease in sequence reads may be due to either a decrease in methylation or a reduction in read coverage compromising accurate quantification of methylation levels.

6.3 Chromatin Immunoprecipitation Techniques

Chromatin immunoprecipitation (ChIP) is a technique used for assaying protein–DNA binding in vivo. In principle, genomic DNA is sheared by sonication into small fragments, preferentially in the range of a 200- to 800-base pairs. The proteins are then linked to the DNA by gentle formaldehyde treatment. An antibody specific to the protein of interest is used to immunoprecipitate the protein–DNA assembly and enrich the DNA fragments associated with the protein. When the epigenetic landscape of histone modifications are to be determined, antibodies raised against the particular histone mark (e.g. H327K3me or H34Kme3) are used. The immunoprecipitated DNA is released by acid treatment and amplified by PCR (Park, 2008, 2009).

6.3.1 ChIP-PCR

The immunoprecipitated DNA is amplified and quantified by real-time PCR (RT-PCR), via TaqMan or Syber Green technologies, with specific primers in order to analyse a specific region of the genome that was associated with a particular histone mark.

6.3.2 ChIP-seq

Progress in NGS technology has allowed the combination of ChIP and NGS known as ChIP-seq. ChIP-seq combines ChIP and direct sequencing technology for genome-wide analysis aiming to reveal the genome-wide distribution of histone marks. The immunoprecipitated DNA fragments are sequenced directly using a variety of sequence platforms, the most frequently used being the Illumina Genome Analyzer (Park, 2009).

7. FUTURE TRENDS: THIRD-GENERATION SEQUENCING

Recently, in plant epigenetics research emerging third-generation sequencing technologies (Schadt, Turner, & Kasarskis, 2010), including single-molecule real-time sequencing (SMRT) and Oxford Nanopore technology, have been adopted.

7.1 Single-Molecule Real-Time Sequencing

This technology was developed by Pacific Biosystems and entails direct detection of base modifications through monitoring of DNA polymerase activity during the incorporation of different fluorescently labeled nucleotides into complementary DNA strands (Eid et al., 2009). The direct detection of various base modifications involves the measurement of the kinetics variation in the time between base incorporations. The advantages of this technology over second-generation sequencing include minimizing chemical modification during library preparation, no requirement for DNA amplification, generating longer reads with average read length of 3000 bp and enabling the detection of different types of epigenetic modifications (Eid et al., 2009). Recently, Liang et al. (2017) reported global profiling of 6mA sites at single-nucleotide resolution in the genome of Arabidopsis using SMRT. They found that 6mA sites are widely distributed across the Arabidopsis genome and enriched over the pericentromeric heterochromatin regions.

7.2 Nanopore Sequencing

In nanopore sequencing, single-stranded DNA is pulled by a phage DNA polymerase through a bacterial pore in single-nucleotide steps, and the ion current through the pore is recorded (Laszlo et al., 2013). C can be distinguished from 5mC and 5hmC based on differences in the current traces. Although encouraging results have been produced using nanopore sequencing for DNA methylation profiling of a single locus (for the detection of 5mC and 5hmC), the application of this method to genome-wide DNA methylation profiling has yet to be further established. Third-generation sequencing evolution resulted to new discoveries of different epigenetic modifications and potentially novel functions of these epigenetic marks in gene expression. Despite many advantages of third-generation sequencing, the higher error rate, cost and lower throughput than second-generation sequencing technologies result in a limited broad application (Rhoads & Au, 2015). The throughput and accuracy must be substantially improved before applying these approaches to studies involving complex genomes. Hence, significant contributions for the characterization of functions and mechanisms of methylation in plant systems have been established, applying high-throughput sequencing to evaluate cytosine methylation. Bisulphite sequencing is the current gold standard for genome-wide mapping of the methylome, offering largely unbiased, base-level resolution maps of methylation. Alternative methods, usually offering higher coverage over some regions of the genome, either represent strong biases in the portions of the genome sequenced or are of low resolution. Plummeting sequencing costs (Wetterstrand, 2013) suggest that the advantages offered by these methods over bisulphite sequencing are unlikely to overcome these disadvantages.

8. REFERENCE METHYLOMES

To date, whole-genome bisulphite sequencing (WGBS-seq) or high-performance liquid chromatography (HPLC) has been used to analyse the methylomes of about 90 various plant species (for review, see Vidalis et al., 2016). These species include angiosperms (flowering plants), gymnosperms, ferns, and nonvascular plants, which diverged nearly 500 million years ago and thus provide a decent coverage of the plant kingdom taxa. Vidalis et al. (2016) analysed and reviewed the methylomes of about 90 diverse plant species using WGBS-seq or HPLC. A reference genome sequence results in a paradigm shift in accessing genetic variation and understanding their roles in plant biology. Nowadays, cytosine methylation patterns can be determined at the single nucleotide level by WGBS-seq and mapping to a reference genome, resulting in a reference methylome. Additionally to these interspecific data, deep genome and methylome sequencing has been performed for over 1000 natural Arabidopsis accessions from all over the world (Kawakatsu et al., 2016), as well as for several experimentally derived populations in Arabidopsis, *Zea mays* and *Glycine max* (for review, see Vidalis et al., 2016).

In order to search for methylation variations, completed sequenced methylome data can be used applying an identical method calculating methylation levels. Loci of commonly occurring DMRs can be identified based on the reference genome and later used for genotyping the methylation states of samples of interest by low coverage sequencing or sequencing of PCR amplicons for target regions. Moreover, reference methylomes from different species can be used to establish pan-methylomes, providing insights into epigenetic variation within a species.

9. EPIALLELES AS A SOURCE OF PHENOTYPIC VARIATION

Epigenetic changes, such as DNA methylation and histone modifications and the action of small RNAs, can lead to changes in gene expression without affecting the underlying DNA sequence. Stably inherited epigenetic changes could result in the formation of epialleles and lead to phenotypic variation (Fig. 2).

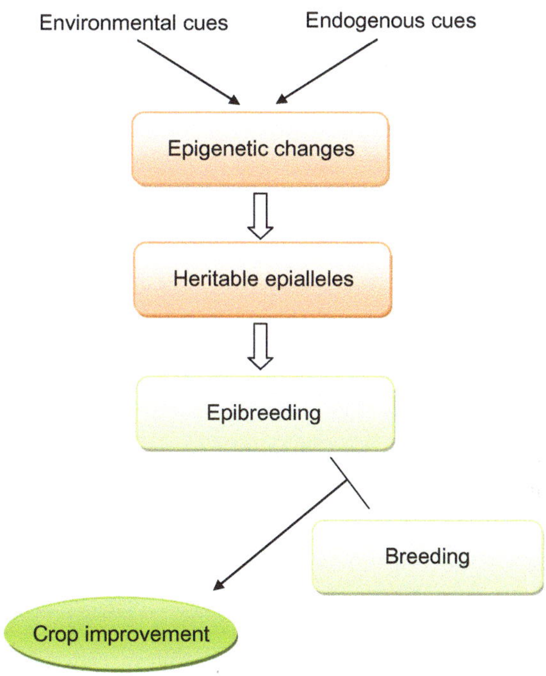

Fig. 2 External and internal stimuli may lead to epigenetic changes such as DNA methylation, histone modifications and the action of noncoding RNAs that alter chromatin structure resulting in changes in gene expression. If heritable, these changes can lead to stable epialleles impacting phenotypic variation and can be exploited for the selection of desired genotypes in epibreeding programmes. Ultimately the combinatorial application of epibreeding and breeding could develop efficient strategies for crop improvement.

9.1 Natural Epigenetic Variants Impacting Phenotypic Variation and Agronomical Traits

Transgenerational epigenetic stability is a source of phenotypic variation and can impact important agricultural traits. The most classical example of heritable epigenetic variation leading to phenotypic alteration is the *peloric* mutant in *Linaria vulgaris,* first described by Carl Linnaeus in 1749. Wild-type *L. vulgaris* plants exhibit flowers with bilateral symmetry, whereas the *peloric* mutant displays radial floral symmetry (Linnaeus, 1749). The molecular basis of this change in floral phenotype was discovered by Cubas, Vincent, and Coen (1999) who demonstrated that the mutation was due to promoter hypermethylation and subsequent gene silencing of a *CYCLOIDEA* homologue encoding a transcription factor involved in organ boundary formation and known to regulate floral asymmetry in Antirrhinum (Cubas et al., 1999; Zhang & Hsieh, 2013). This epimutation has been stably inherited and maintained for nearly 300 years.

In rice, several epialleles have been described affecting agronomical characters. The rice *epi-d1* allele associated with the *DWARF1* (*D1*) gene, encoding the α-subunit of a GTP-binding protein, leads to a dwarf phenotype (Miura et al., 2009). The *epi-d1* allele is correlated with DNA hypermethylation and histone-repressive marks in the promoter of the *D1* gene and consequently silencing of the *D1* gene. *Epi-d1* is a metastable mutation and has been used as breeding material in Japan for many decades (Fujimoto et al., 2012). Likewise, the *epi-df* epimutation affecting the *OsFIE* gene is linked to a heritable dwarf phenotype, in rice (Zhang et al., 2012). OsFIE is a member of the Polycomb group complex (PcG) with histone H3K27 trimethyl (H3K27me3) methyltransferase activity with crucial role in vegetative and seed development (Kohler et al., 2003). In the rice epimutant the promoter of the *OsFIE* gene is marked with chromatin activation marks such as H3K4me3, leading to *OsFIE* ectopic expression and a dwarf phenotype (Zhang et al., 2012). Another rice epiallele with agricultural relevance is associated with the quantitative trait locus WEALTHY FARMERS PANICLE (WFP) encoding the transcriptional factor SQUAMOSA-BINDING PROTEIN-LIKE 14 (OsSPL14). Increased expression level of *OsSPL14* is positively correlated with rice panicle branching and higher grain yield (Miura et al., 2010). In the Nipponbare rice cultivar which exhibits low amount of panicle branching and decreased grain yield, increased levels of DNA methylation were evidenced in the *OsSPL14* promoter region and reduced *OsSPL14* gene expression. Conversely, in the ST-12 line which shows increased degree of branching and higher grain yield, decreased promoter methylation and enhanced *OsSPL14* expression were observed (Miura et al., 2010).

In tomato, a ripening mutant known as *cnr* (colourless nonripening) exhibits an impaired phenotype with respect to fruit ripening. The underlying gene of the mutation was found to encode a SQUAMOSA-BINDING PROTEIN (SBP)-box transcription factor. The *cnr* epiallele harboured hypermethylation in the promoter of the *CNR-SBP* gene leading to repression of *CNR* expression and fruit maturation defects (Manning et al., 2006). This phenotype can be reversed by the application of a DNA methylation inhibitor, 5-azacytidine, an inhibitor of DNA methyltransferase enzymes (Martel, Vrebalov, Tafelmeyer, & Giovannoni, 2011). Another MADS-box transcription factor which is a major regulator of fruit ripening, RIN (RIPENING INHIBITOR), was found to interact with CNR in the wild type, but its binding was inhibited in the *cnr* epimutant. Upon 5-azacytidine application, the promoter was demethylated and binding was allowed, in accordance to the notion that DNA hypomethylation facilitates interaction between transcription factors and their promoter targets. Moreover, deciphering of the tomato methylomes in different fruit developmental stages revealed that a series of ripening regulators such as *CNR*, the pectinase polygalacturonase *PG2A* (*POLYGALACTURONASE2A*) and the gene encoding the rate-limiting enzyme in fruit carotenoid synthesis *PSY1* (*PHYTOENE SYNTHASE1*) harbour hypermethylated promoters prior to fruit ripening and the genes are silenced (Zhong et al., 2013). Promoters of postripening genes are demethylated and gene expression concurrent with demethylation lead to proper fruit maturation. Interestingly, the RIN factor, which binds the promoter of many ripening-related genes, fails to interact with these genes in the *cnr* epimutant suggesting complex interactions between fruit ripening-related genes during fruit development.

Finally, investigations in recent years have shown that epigenetic modifications are associated with alleles related to fruit quality. The way natural epialleles may impact important characters related to product quality is exemplified by the case of a metabolic epi-QTL in tomato and its association with vitamin E accumulation in tomato fruit (Quadrana et al., 2014). Fine mapping of the major quantitative loci for vitamin E content ($mQTL_{9\text{-}2\text{-}6}$) identified the causal gene, *VTE3(1)*, encoding a 2-methyl-6-phytylquinolmethyltransferase which catalyses the final steps of the biosynthesis of γ- and α-tocopherols (the main forms of vitamin E). *VTE3(1)* expression is associated with differential methylation of a SINE-retrotransposon residing in the promoter of the gene. It was demonstrated that differential expression of *VTE3(1)* alleles correlates with DNA methylation differences in the promoter region and leads to variation in vitamin E content (Quadrana et al., 2014). Furthermore, in apple and pear it was demonstrated that skin colour changes were associated with increased methylation levels in the promoter region of the transcription factor gene *MYB10,* leading to MYB10 suppression and inhibition of anthocyanin accumulation (El-Sharkawy, Liang, & Xu, 2015; Wang et al., 2013).

9.2 Induced Epigenetic Changes and Their Impact on Agricultural Traits

Apart from naturally occurring epimutations, induced epigenetic changes have been attempted in order to achieve epigenetic modulation leading to desired phenotypes. Akimoto et al. (2007) treated germinating rice seeds with the DNA methylation inhibitor, 5-aza-deoxycytidine, and surviving progeny was propagated in the field for over a decade. One of the propagated lines, designated as Line-2, displayed acquired resistance to the bacterial pathogen *Xanthomonas oryzae* pv. *oryzae* (Akimoto et al., 2007). Specifically, it was demonstrated that the promoter of the gene *Xa21G* which confers resistance to the pathogen was substantially demethylated and the *Xa21G* gene highly expressed in the resistant Line2. On the contrary, the *Xa21G* promoter was densely methylated and the gene was silenced in the susceptible wild-type plants. Notably, the erasure of methylation marks on the *Xa21G* promoter in the resistant lines was maintained for nine generations (Akimoto et al., 2007). Equally, in another study regarding rapeseed epigenetic modulation, it was demonstrated that subjection of *B. napus* seedlings to 5-azacytidine resulted in novel Brassica lines that contained increased amounts of seed protein and increased linoleic acid content (Amoah et al., 2012). Transgenerational stability of epialleles is essential for their utilization in crop breeding programmes. In *B. napus,* stable epigenotypes with improved energy use efficiency were achieved through recurrent phenotypic selection of isogenic *B. napus* lines. Notably, crosses between these epigenetic variable but genetically identical lines gave rise to hybrids with a 5% increase in yield (Hauben et al., 2009). Furthermore, much progress in understanding epigenetic regulation of numerous plant processes, and in creating epialleles, has arisen from the development of the so-called epi-Recombinant Inbred Lines (epiRILs). EpiRILs are recombinant inbred populations generated from parental strains that differ only in DNA methylation. EpiRILs are generated by crossing DNA hypomethylation mutants such as *met1*, *ddm1*, to wild-type plants. The resulting lines are homozygous differing only in their DNA methylation pattern and affect a range of characters such as plant height, flowering time, yield and salt stress tolerance (Johannes et al., 2009; Reinders et al., 2009). The transgenerational stability of these induced methylation marks was considerably high to allow the identification of quantitative trait loci for some phenotypes using DMRs as genetic markers (Cortijo et al., 2014; Teixeira et al., 2009). Similarly, linkage maps related to complex traits of agricultural significance were constructed using stable natural epigenetic variability of *B. napus* lines (Long et al., 2011). Generation of epiRILs in Arabidopsis has been a straightforward task due to the existence of global methylation mutants such as *met1* and *ddm1*. Nevertheless, developing epiRIL lines in crop species such as maize or rice is a much more challenging endeavour due to the lack of respective DNA methylation mutants and the large abundance of transposon sequences in their genomes.

10. EPIGENETIC VARIATION AS A POTENTIAL BIOMARKER TOOL

Over the past several years, much effort has focused on exploring the potential of differential epigenetic marks towards developing biomarker tools. Epigenetic chromatin marks, such as DNA methylation, histone modifications and small RNAs, affect all aspects of plant growth, development and response to environmental cues; therefore, it can be envisaged that these marks could be utilized as predictors for desired agronomical characters. Presently, studies on developing epibiomarker tools have focused on DNA methylation as it is the best-studied and most stably maintained epimark. On the other hand, studying of histone modifications is technically more difficult, and knowledge on its transgenerational inheritance is obscure (Niederhuth & Schmitz, 2017; Rodriguez-Lopez & Wilkinson, 2015). An excellent example of how variation in DNA methylation marking could be exploited in genotype selection is the case of epialleles related to the *MANTLED* locus regulating proper fruit development in the African oil palm, *Elaeis guineensis*, producer of about 50% of the world's edible vegetable oil (Ong-Abdullah et al., 2015). For many decades somaclonal variation has hindered the micropropagation of elite plants as the oil palm clones gave rise to spoilt 'mantled' fruit with massive losses in yield and negative impact for the environment (Mgbeze & Iserhienrhien, 2014). Recently, epigenome-wide association studies delineated the molecular basis of the faulty variation and attributed it to a DNA methylation change in the *DEFICIENS* gene, a regulator of sexual organ fate (Ong-Abdullah et al., 2015). Specifically, DNA hypomethylation of a LINE retrotransposon, known as *Karma,* located in the intron of the *DEFICIENS* gene is associated with premature termination, nonfunctional protein, parthenocarpy, and subsequent mantled fruit phenotype (*Bad Karma* epiallele). Conversely, dense methylation near the *Karma* splice site (termed the *Good Karma* epiallele) leads to proper intron splicing and synthesis of functional DEFICIENS protein and predicts normal fruit set. By using a diagnostic DNA methylation-based assay, hypomethylated plants can be easily identified at the early plantlet stage and be discarded, circumventing the need for transplantation and a 6-year growth period before phenotypic assessment. Thus, this epigenetic strategy, utilizing DNA methylation as a biomarker, could enable selection of healthy plants at early stages, minimize losses and transplantation costs, and save valuable land space from deforestation.

In another study, information on epigenetic differences among plants was used to predict plant height in Arabidopsis. These authors showed that epigenetic variation in methylation profile among individuals accounted for more than 65% of phenotypic variance in plant height (Hu, Morota, Rosa, & Gianola, 2015).

It is expected that in the near future many more assays utilizing epimolecular markers will emerge. Epigenetic molecular markers based on the distribution of small RNAs to predict vigour relationships in maize heterosis and on DNA methylation patterns to predict phenotypes of agricultural relevance in maize breeding programmes are currently underway (Slotkin, 2016). Finally, it is anticipated that DNA methylation could serve as a useful marker for molecular taxonomy of plant species, epigenotyping, and as a diagnostic tool for food quality assessments (Rodriguez-Lopez & Wilkinson, 2015). Methylation status could be used for detecting fraudulent labeling on plant products and thus contribute to protecting authenticity and geographical origin of crops and their products in the same manner that the HRM (high-resolution melting) technology has been used in recent years (Ganopoulos, Argiriou, & Tsaftaris, 2011).

11. CONCLUSIONS

Over the past two decades extensive epigenetic investigations involving locus-specific and genome-wide approaches both in model and crop plants have provided an explosive amount of information concerning the epigenetic regulation of all aspects of plant growth and development from vegetative to reproductive development and response to environmental stressors. It is well established, nowadays, that the contribution of heritable epigenetic variation to phenotypic plasticity is not only considerable but also it may be as important as the widely accepted genetic variation and its effect on variable phenotypes. The list of plant methylomes is ever growing, and comparative genome-wide methylome profiles under different environmental conditions and in contrasting phenotypes have been deciphered in major crops in just a period of a few years. Substantial progress has been achieved relating to the epigenetic mechanisms underlying the processes of grafting and heterosis which have drastic impact on crop improvement. Reports on heritable, natural or induced epialleles, affecting important agricultural traits, are ever increasing including plants of high agronomic value such as rice, tomato, rapeseed, maize, and it is expected that the list will grow in coming years. Much effort has focused in exploiting the potential of epigenetic variation and differential epimarking to develop efficient assays for selecting superior genotypes. Epigenotyping based on epigenome-wide studies will allow association of differential epimarking to desired phenotypes and thorough examination of the currently unexplored epigenetic variants will reveal novel cryptic diversity, especially for species with low genetic variation. The knowledge arising from epigenetic variation will allow the exploitation of differential epigenetic marking towards useful epibiomarker tools for superior genotype selection at early stages in breeding programmes, leading to what is nowadays called epibreeding. Collectively, breeding and the emerging area of epibreeding will pave the way for developing optimal strategies for crop improvement. In view of the adverse effects of global climate change, utilizing genetic and epigenetic variation for purposes of generating improved crop varieties is an undertaking of paramount importance.

REFERENCES

Akimoto, K., Katakami, H., Kim, H.-J., Ogawa, E., Sano, C. M., Wada, Y., et al. (2007). Epigenetic inheritance in rice plants. *Annals of Botany*, *100*(2), 205–217.

Allis, C. D., & Jenuwein, T. (2015). The molecular hallmarks of epigenetic control. *Nature Reviews Genetics*, *17*(8), 487–500.

Amasino, R. M. (2010). Seasonal and developmental timing of flowering. *Plant Journal*, *61*(6), 1001–1013.

Amoah, S., Kurup, S., Lopez, C. M. R., Welham, S. J., Powers, S. J., Hopkins, C. J., et al. (2012). A hypomethylated population of Brassica rapa for forward and reverse epigenetics. *BMC Plant Biology*, *12*(1), 193.

Avramidou, E., Kapazoglou, A., Aravanopoulos, F. A., Xanthopoulou, A., Ganopoulos, I., Tsaballa, A., et al. (2014). Global DNA methylation changes in Cucurbitaceae interspecies grafting. *Crop Breeding and Applied Biotechnology*, *15*(2), 112–116.

Banerjee, A., & Roychoudhury, A. (2017). Epigenetic regulation during salinity and drought stress in plants: Histone modifications and DNA methylation. *Plant Gene*, *11*, 199–204.

Banerjee, S., Sirohi, A., Ansari, A. A., & Gill, S. S. (2017). Role of small RNAs in abiotic stress responses in plants. *Plant Gene*, *11*, 180–189.

Barrera-Figueroa, B. E., Wu, Z., & Liu, R. (2013). Abiotic stress-associated microRNAs in plants: Discovery, expression analysis, and evolution. *Frontiers in Biology*, *8*(2), 189–197.

Bastow, R., Mylne, J. S., Lister, C., & Lippman, Z. (2004). Vernalization requires epigenetic silencing of FLC by histone methylation. *Nature*, *427*(6970), 164–167.

Berger, S. L. (2007). The complex language of chromatin regulation during transcription. *Nature*, *447*(7143), 407–412.

Bocchini, M., Bartucca, M. L., Ciancaleoni, S., Mimmo, T., Cesco, S., Pii, Y., et al. (2015). Iron deficiency in barley plants: Phytosiderophore release, iron translocation, and DNA methylation. *Frontiers in Plant Science*, *6*, 514.

Boyko, A., & Kovalchuk, I. (2008). Epigenetic control of plant stress response. *Environmental and Molecular Mutagenesis*, *49*(1), 61–72.

Bruce, T. J. A., Matthes, M. C., Napier, J. A., & Pickett, J. A. (2007). Stressful "memories" of plants: Evidence and possible mechanisms. *Plant Science*, *173*(6), 603–608.

Calarco, J. P., Borges, F., Donoghue, M. T., Van Ex, F. D. R., Jullien, P. E., Lopes, T., et al. (2012). Reprogramming of DNA methylation in pollen guides epigenetic inheritance via small RNA. *Cell*, *151*(1), 194–205.

Chan, S. W. L., Henderson, I. R., & Jacobsen, S. E. (2005). Gardening the genome: DNA methylation in Arabidopsis thaliana. *Nature Reviews. Genetics*, *6*(5), 351–360.

Chanvivattana, Y., Bishopp, A., Schubert, D., Stock, C., Moon, Y.-H., Sung, Z. R., et al. (2004). Interaction of Polycomb-group proteins controlling flowering in Arabidopsis. *Development*, *131*(21), 5263–5276.

Chen, Z. J. (2013). Genomic and epigenetic insights into the molecular bases of heterosis. *Nature Reviews. Genetics*, *14*(7), 471–482.

Chen, X., Ge, X., Wang, J., Tan, C., King, G. J., & Liu, K. (2015). Genome-wide DNA methylation profiling by modified reduced representation bisulfite sequencing in Brassica rapa suggests that epigenetic modifications play a key role in polyploid genome evolution. *Frontiers in Plant Science*, *6*, 836.

Chen, L.-T., Luo, M., Wang, Y.-Y., & Wu, K. (2010). Involvement of Arabidopsis histone deacetylase HDA6 in ABA and salt stress response. *Journal of Experimental Botany*, *61*(12), 3345–3353.

Choi, Y., Gehring, M., Johnson, L., Hannon, M., Harada, J. J., Goldberg, R. B., et al. (2002). DEMETER, a DNA glycosylase domain protein, is required for endosperm gene imprinting and seed viability in Arabidopsis. *Cell*, *110*(1), 33–42.

Choi, C.-S., & Sano, H. (2007). Abiotic-stress induces demethylation and transcriptional activation of a gene encoding a glycerophosphodiesterase-like protein in tobacco plants. *Molecular Genetics and Genomics*, *277*(5), 589–600.

Choi, S.-M., Song, H.-R., Han, S.-K., Han, M., Kim, C.-Y., Park, J., et al. (2012). HDA19 is required for the repression of salicylic acid biosynthesis and salicylic acid-mediated defense responses in Arabidopsis. *The Plant Journal*, *71*(1), 135–146.

Chwialkowska, K., Nowakowska, U., Mroziewicz, A., Szarejko, I., & Kwasniewski, M. (2016). Water-deficiency conditions differently modulate the methylome of roots and leaves in barley (Hordeum vulgare L.). *Journal of Experimental Botany*, *67*(4), 1109–1121.

Clark, C., Palta, P., Joyce, C. J., Scott, C., Grundberg, E., Deloukas, P., et al. (2012). A comparison of the whole genome approach of MeDIP-seq to the targeted approach of the Infinium Human Methylation450 BeadChip® for methylome profiling. *PLoS One*, 7(11), e50233.

Cokus, S. J., Feng, S., Zhang, X., Chen, Z., Merriman, B., Haudenschild, C. D., et al. (2008). Shotgun bisulfite sequencing of the Arabidopsis genome reveals DNA methylation patterning. *Nature*, *452*(7184), 215–219.

Colaneri, A. C., & Jones, A. M. (2013). Genome-wide quantitative identification of DNA differentially methylated sites in Arabidopsis seedlings growing at different water potential. *PLoS One*, *8*(4), e59878.

Cortijo, S., Wardenaar, R., Colomé-Tatché, M., Gilly, A., Etcheverry, M., Labadie, K., et al. (2014). Mapping the epigenetic basis of complex traits. *Science*, *343*(6175), 1145–1148.

Crampton, M., Sripathi, V. R., Hossain, K., & Kalavacharla, V. (2016). Analyses of methylomes derived from Meso-American common bean (Phaseolus vulgaris L.) using MeDIP-seq and whole genome sodium bisulfite-sequencing. *Frontiers in Plant Science*, 7, 447.

Cubas, P., Vincent, C., & Coen, E. (1999). An epigenetic mutation responsible for natural variation in floral symmetry. *Nature*, *401*(6749), 157–161.

Dean, C., & Whittaker, C. (2017). The FLC locus: A platform for discoveries in epigenetics and adaptation. *Annual Review of Cell and Developmental Biology*, *33*(1), 555–575.

De-La-Peña, C., Rangel-Cano, A., & Alvarez-Venegas, R. (2012). Regulation of disease-responsive genes mediated by epigenetic factors: Interaction of Arabidopsis–Pseudomonas. *Molecular Plant Pathology, 13*(4), 388–398.

Ding, Y., Avramova, Z., & Fromm, M. (2011). The Arabidopsis trithorax-like factor ATX1 functions in dehydration stress responses via ABA-dependent and ABA-independent pathways. *The Plant Journal, 66*(5), 735–744.

Ding, B., Bellizzi, M. d. R., Ning, Y., Meyers, B. C., & Wang, G.-L. (2012). HDT701, a histone H4 deacetylase, negatively regulates plant innate immunity by modulating histone H4 acetylation of defense-related genes in rice. *The Plant Cell, 24*(9), 3783–3794.

Ding, B., & Wang, G.-L. (2015). Chromatin versus pathogens: The function of epigenetics in plant immunity. *Frontiers in Plant Science, 6*, 675.

Dowen, R. H., Pelizzola, M., Schmitz, R. J., Lister, R., Dowen, J. M., Nery, J. R., et al. (2012). Widespread dynamic DNA methylation in response to biotic stress. *Proceedings of the National Academy of Sciences of the United States of America, 109*(32), E2183–E2191.

Du, J., Johnson, L. M., Jacobsen, S. E., & Patel, D. J. (2015). DNA methylation pathways and their crosstalk with histone methylation. *Nature Reviews. Molecular Cell Biology, 16*(9), 519–532.

Du, J., Zhong, X., Bernatavichute, Y. V., Stroud, H., Feng, S., Caro, E., et al. (2012). Dual binding of chromomethylase domains to H3K9me2-containing nucleosomes directs DNA methylation in plants. *Cell, 151*(1), 167–180.

Eichten, S. R., Briskine, R., Song, J., Li, Q., Swanson-Wagner, R., Hermanson, P. J., et al. (2013). Epigenetic and genetic influences on DNA methylation variation in maize populations. *The Plant Cell, 25*(8), 2783–2797.

Eid, J., Fehr, A., Gray, J., Luong, K., Lyle, J., Otto, G., et al. (2009). Real-time DNA sequencing from single polymerase molecules. *Science, 323*(5910), 133–138.

El-Sharkawy, I., Liang, D., & Xu, K. (2015). Transcriptome analysis of an apple (Malusxdomestica) yellow fruit somatic mutation identifies a gene network module highly associated with anthocyanin and epigenetic regulation. *Journal of Experimental Botany, 66*(22), 7359–7376.

Engelhorn, J., Blanvillain, R., & Carles, C. C. (2014). Gene activation and cell fate control in plants: A chromatin perspective. *Cellular and Molecular Life Sciences, 71*, 3119–3137.

Erdmann, R. M., & Gehring, M. (2016). Comprehensive profiling of small RNAs within the developing seed of Arabidopsis thaliana. In *BS biochemistry and chemistry, 2008* (p. 78). Madison, WI: University of Wisconsin-Madison.

Fang, H., Liu, X., Thorn, G., Duan, J., & Tian, L. (2014). Expression analysis of histone acetyltransferases in rice under drought stress. *Biochemical and Biophysical Research Communications, 443*(2), 400–405.

Feng, S., Jacobsen, S. E., & Reik, W. (2010). Epigenetic reprogramming in plant and animal development. *Science, 330*(6004), 622–627.

Feng, Q., Yang, C., Lin, X., Wang, J., Ou, X., Zhang, C., et al. (2012). Salt and alkaline stress induced transgenerational alteration in DNA methylation of rice ('Oryza sativa'). *Australian Journal of Crop Science, 6*(5), 877–883.

Forderer, A., Zhou, Y., & Turck, F. (2016). The age of multiplexity: Recruitment and interactions of Polycomb complexes in plants. *Current Opinion in Plant Biology, 29*, 169–178.

Fortes, A., & Gallusci, P. (2017). Plant stress responses and phenotypic plasticity in the Epigenomics era: Perspectives on the grapevine scenario, a model for perennial crop plants. *Frontiers in Plant Science, 8*, 82.

Fujimoto, R., Sasaki, T., Ishikawa, R., Osabe, K., Kawanabe, T., & Dennis, E. S. (2012). Molecular mechanisms of epigenetic variation in plants. *International Journal of Molecular Sciences, 13*(8), 9900–9922.

Gallusci, P., Hodgman, C., Teyssier, E., & Seymour, G. B. (2016). DNA methylation and chromatin regulation during fleshy fruit development and ripening. *Frontiers in Plant Science, 7*, 807.

Gan, E. S., Huang, J., & Ito, T. (2013). Functional roles of histone modification, chromatin remodeling and microRNAs in Arabidopsis flower development. *International Review of Cell and Molecular Biology*, *305*, 115–161.

Ganopoulos, I., Argiriou, A., & Tsaftaris, A. (2011). Microsatellite high resolution melting (SSR-HRM) analysis for authenticity testing of protected designation of origin (PDO) sweet cherry products. *Food Control*, *22*(3), 532–541.

Garg, R., Narayana Chevala, V. V. S., Shankar, R., & Jain, M. (2015). Divergent DNA methylation patterns associated with gene expression in rice cultivars with contrasting drought and salinity stress response. *Scientific Reports*, *5*, 14922.

Gehring, M., Bubb, K. L., & Henikoff, S. (2009). Extensive demethylation of repetitive elements during seed development underlies gene imprinting. *Science*, *324*(5933), 1447–1451.

Gehring, M., Huh, J. H., Hsieh, T.-F., Penterman, J., Choi, Y., Harada, J. J., et al. (2006). DEMETER DNA glycosylase establishes MEDEA polycomb gene self-imprinting by allele-specific demethylation. *Cell*, *124*(3), 495–506.

Gehring, M., & Satyaki, P. (2017). Endosperm and imprinting, inextricably linked. *Plant Physiology*, *173*(1), 143–154.

Gent, J. I., Ellis, N. A., Guo, L., Harkess, A. E., Yao, Y., Zhang, X., et al. (2013). CHH islands: De novo DNA methylation in near-gene chromatin regulation in maize. *Genome Research*, *23*(4), 628–637.

Gohlke, J., Scholz, C.-J., Kneitz, S., Weber, D., Fuchs, J., Hedrich, R., et al. (2013). DNA methylation mediated control of gene expression is critical for development of crown gall tumors. *PLoS Genetics*, *9*(2), e1003267.

Goldschmidt, E. E. (2014). Plant grafting: New mechanisms, evolutionary implications. *Frontiers in Plant Science*, *5*, 727.

Goll, M. G., & Bestor, T. H. (2005). Eukaryotic cytosine methyltransferases. *Annual Review of Biochemistry*, *74*, 481–514.

Grativol, C., Hemerly, A. S., & Ferreira, P. C. G. (2012). Genetic and epigenetic regulation of stress responses in natural plant populations. *Biochimica et Biophysica Acta (BBA)-Gene Regulatory Mechanisms*, *1819*(2), 176–185.

Greaves, I. K., Gonzalez-Bayon, R., Wang, L., Zhu, A., Liu, P.-C., Groszmann, M., et al. (2015). Epigenetic changes in hybrids. *Plant Physiology*, *168*(4), 1197–1205.

Grimanelli, D., & Roudier, F. (2013). Epigenetics and development in plants: Green light to convergent innovations. *Current Topics in Developmental Biology*, *104*, 189–222.

Groszmann, M., Gonzalez-Bayon, R., Greaves, I. K., Wang, L., Huen, A. K., Peacock, W. J., et al. (2014). Intraspecific Arabidopsis hybrids show different patterns of heterosis despite the close relatedness of the parental genomes. *Plant Physiology*, *166*(1), 265–280.

Groszmann, M., Gonzalez-Bayon, R., Lyons, R. L., Greaves, I. K., Kazan, K., Peacock, W. J., et al. (2015). Hormone-regulated defense and stress response networks contribute to heterosis in Arabidopsis F1 hybrids. *Proceedings of the National Academy of Sciences of the United States of America*, *112*(46), E6397–E6406.

Groszmann, M., Greaves, I. K., Albertyn, Z. I., Scofield, G. N., Peacock, W. J., & Dennis, E. S. (2011). Changes in 24-nt siRNA levels in Arabidopsis hybrids suggest an epigenetic contribution to hybrid vigor. *Proceedings of the National Academy of Sciences of the United States of America*, *108*(6), 2617–2622.

Gu, H., Smith, Z. D., Bock, C., Boyle, P., Gnirke, A., & Meissner, A. (2011). Preparation of reduced representation bisulfite sequencing libraries for genome-scale DNA methylation profiling. *Nature Protocols*, *6*(4), 468–481.

Han, S.-K., & Wagner, D. (2014). Role of chromatin in water stress responses in plants. *Journal of Experimental Botany*, *65*(10), 2785–2799.

Haroldsen, V. M., Szczerba, M. W., Aktas, H., Lopez-Baltazar, J., Odias, M. J., Chi-Ham,-C. L., et al. (2012). Mobility of transgenic nucleic acids and proteins within grafted rootstocks for agricultural improvement. *Frontiers in Plant Science*, *3*, 39.

Hashida, S.-N., Kitamura, K., Mikami, T., & Kishima, Y. (2003). Temperature shift coordinately changes the activity and the methylation state of transposon Tam3 in Antirrhinum majus. *Plant Physiology*, *132*(3), 1207–1216.

Hashida, S.-N., Uchiyama, T., Martin, C., Kishima, Y., Sano, Y., & Mikami, T. (2006). The temperature-dependent change in methylation of the Antirrhinum transposon Tam3 is controlled by the activity of its transposase. *The Plant Cell*, *18*(1), 104–118.

Hauben, M., Haesendonckx, B., Standaert, E., Van Der Kelen, K., Azmi, A., Akpo, H., et al. (2009). Energy use efficiency is characterized by an epigenetic component that can be directed through artificial selection to increase yield. *Proceedings of the National Academy of Sciences of the United States of America*, *106*(47), 20109–20114.

Hauser, M.-T., Aufsatz, W., Jonak, C., & Luschnig, C. (2011). Transgenerational epigenetic inheritance in plants. *Biochimica et Biophysica Acta (BBA)-Gene Regulatory Mechanisms*, *1809*(8), 459–468.

He, G., Chen, B., Wang, X., Li, X., Li, J., He, H., et al. (2013). Conservation and divergence of transcriptomic and epigenomic variation in maize hybrids. *Genome Biology*, *14*(6), R57.

He, G., Zhu, X., Elling, A. A., Chen, L., Wang, X., Guo, L., et al. (2010). Global epigenetic and transcriptional trends among two rice subspecies and their reciprocal hybrids. *The Plant Cell Online*, *22*(1), 17–33.

Heo, J. B., & Sung, S. (2011). Vernalization-mediated epigenetic silencing by a long intronic noncoding RNA. *Science*, *331*(6013), 76–79.

Hsieh, T.-F., Ibarra, C. A., Silva, P., Zemach, A., Eshed-Williams, L., Fischer, R. L., et al. (2009). Genome-wide demethylation of Arabidopsis endosperm. *Science*, *324*(5933), 1451–1454.

Hu, L., Li, N., Xu, C., Zhong, S., Lin, X., Yang, J., et al. (2014). Mutation of a major CG methylase in rice causes genome-wide hypomethylation, dysregulated genome expression, and seedling lethality. *Proceedings of the National Academy of Sciences of the United States of America*, *111*(29), 10642–10647.

Hu, Y., Morota, G., Rosa, G. J., & Gianola, D. (2015). Prediction of plant height in Arabidopsis thaliana using DNA methylation data. *Genetics*, *201*(2), 779–793.

Ibarra, C. A., Feng, X., Schoft, V. K., Hsieh, T.-F., Uzawa, R., Rodrigues, J. A., et al. (2012). Active DNA demethylation in plant companion cells reinforces transposon methylation in gametes. *Science*, *337*(6100), 1360–1364.

Jarillo, J. A., Piñeiro, M., Cubas, P., & Martínez-Zapater, J. M. (2009). Chromatin remodeling in plant development. *International Journal of Developmental Biology*, *53*(8–10), 1581–1596.

Jiang, H., & Kohler, C. (2012). Evolution, function, and regulation of genomic imprinting in plant seed development. *Journal of Experimental Botany*, *63*(13), 4713–4722.

Johannes, F., Porcher, E., Teixeira, F. K., Saliba-Colombani, V., Simon, M., Agier, N., et al. (2009). Assessing the impact of transgenerational epigenetic variation on complex traits. *PLoS Genetics*, *5*(6), e1000530.

Kamthan, A., Chaudhuri, A., Kamthan, M., & Datta, A. (2015). Small RNAs in plants: Recent development and application for crop improvement. *Frontiers in Plant Science*, *6*, 208.

Kapazoglou, A., Drosou, V., Argiriou, A., & Tsaftaris, A. (2013). The study of a barley epigenetic regulator, HvDME, in seed development and under drought. *BMC Plant Biology*, *13*(1), 172.

Kapazoglou, A., Papaefthimiou, D., & Tsaftaris, A. S. (2012). Histone modifiers in cereals. In C.-H. Shen (Ed.), *Histones: Class, structure and function*. NY: Nova Science Publisher, Inc.

Kapazoglou, A., & Tsaftaris, A. (2011). Epigenetic chromatin regulators as mediators of abiotic stress responses in cereals. In *Abiotic stress in plants-mechanisms and adaptations*. InTech.
Kawakatsu, T., Huang, S.-s. C., Jupe, F., Sasaki, E., Schmitz, R. J., Urich, M. A., et al. (2016). Epigenomic diversity in a global collection of Arabidopsis thaliana accessions. *Cell*, *166*(2), 492–505.
Khraiwesh, B., Arif, M. A., Seumel, G. I., Ossowski, S., Weigel, D., Reski, R., et al. (2010). Transcriptional control of gene expression by MicroRNAs. *Cell*, *140*(1), 111–122.
Kim, J.-M., Sasaki, T., Ueda, M., Sako, K., & Seki, M. (2015). Chromatin changes in response to drought, salinity, heat, and cold stresses in plants. *Frontiers in Plant Science*, *6*, 114.
Kim, D.-H., & Sung, S. (2014). Genetic and epigenetic mechanisms underlying vernalization. *The Arabidopsis Book*, *12*, e0171.
Kim, J.-M., To, T. K., Ishida, J., Matsui, A., Kimura, H., & Seki, M. (2012). Transition of chromatin status during the process of recovery from drought stress in Arabidopsis thaliana. *Plant and Cell Physiology*, *53*(5), 847–856.
Kim, J. M., To, T. K., Ishida, J., Morosawa, T., Kawashima, M., Matsui, A., et al. (2008). Alterations of lysine modifications on the histone H3 N-tail under drought stress conditions in Arabidopsis thaliana. *Plant and Cell Physiology*, *49*(10), 1580–1588.
Kiriga, W. J., Yu, Q., & Bill, R. (2016). Breeding and genetic engineering of drought-resistant crops. *International Journal of Agriculture and Crop Sciences*, *9*(1), 7–12.
Kohler, C., Hennig, L., Spillane, C., Pien, S., Gruissem, W., & Grossniklaus, U. (2003). The Polycomb-group protein MEDEA regulates seed development by controlling expression of the MADS-box gene PHERES1. *Genes and Development*, *17*, 1540–1553.
Kohler, C., & Villar, C. B. R. (2008). Programming of gene expression by Polycomb group proteins. *Trends in Cell Biology*, *18*, 236–243.
Kouzarides, T. (2007). Chromatin modifications and their function. *Cell*, *128*(4), 693–705.
Kumar, G., Rattan, U. K., & Singh, A. K. (2016). Chilling-mediated DNA methylation changes during dormancy and its release reveal the importance of epigenetic regulation during winter dormancy in apple (Malus x domestica Borkh.). *PLoS One*, *11*(2), e0149934.
Kumar, S., Verma, S., & Trivedi, P. K. (2017). Involvement of small RNAs in phosphorus and sulfur sensing, signaling and stress: Current update. *Frontiers in Plant Science*, *8*, 285.
La, H., Ding, B., Mishra, G. P., Zhou, B., Yang, H., del Rosario Bellizzi, M., et al. (2011). A 5-methylcytosine DNA glycosylase/lyase demethylates the retrotransposon Tos17 and promotes its transposition in rice. *Proceedings of the National Academy of Sciences of the United States of America*, *108*(37), 15498–15503.
Lämke, J. R., & Bäurle, I. (2017). Epigenetic and chromatin-based mechanisms in environmental stress adaptation and stress memory in plants. *Genome Biology*, *18*(1), 124.
Laszlo, A. H., Derrington, I. M., Brinkerhoff, H., Langford, K. W., Nova, I. C., Samson, J. M., et al. (2013). Detection and mapping of 5-methylcytosine and 5-hydroxymethylcytosine with nanopore MspA. *Proceedings of the National Academy of Sciences of the United States of America*, *110*(47), 18904–18909.
Law, J. A., & Jacobsen, S. E. (2010). Establishing, maintaining and modifying DNA methylation patterns in plants and animals. *Nature Reviews. Genetics*, *11*(3), 204–220.
Le, T.-N., Schumann, U., Smith, N. A., Tiwari, S., Au, P. C. K., Zhu, Q.-H., et al. (2014). DNA demethylases target promoter transposable elements to positively regulate stress responsive genes in Arabidopsis. *Genome Biology*, *15*(9), 458.
Lei, M., Zhang, H., Julian, R., Tang, K., Xie, S., & Zhu, J.-K. (2015). Regulatory link between DNA methylation and active demethylation in Arabidopsis. *Proceedings of the National Academy of Sciences of the United States of America*, *112*(11), 3553–3557.
Lewsey, M. G., Hardcastle, T. J., Melnyk, C. W., Molnar, A., Valli, A. N., Urich, M. A., et al. (2016). Mobile small RNAs regulate genome-wide DNA methylation. *Proceedings of the National Academy of Sciences of the United States of America*, *113*(6), E801–E810.

Li, S., Castillo-González, C., Yu, B., & Zhang, X. (2017). The functions of plant small RNAs in development and in stress responses. *The Plant Journal*, *90*(4), 654–670.
Li, T., Chen, X., Zhong, X., Zhao, Y., Liu, X., Zhou, S., et al. (2013). Jumonji C domain protein JMJ705-mediated removal of histone H3 lysine 27 trimethylation is involved in defense-related gene activation in rice. *The Plant Cell*, *25*(11), 4725–4736.
Li, Q., Eichten, S. R., Hermanson, P. J., Zaunbrecher, V. M., Song, J., Wendt, J., et al. (2014). Genetic perturbation of the maize methylome. *The Plant Cell*, *26*(12), 4602–4616.
Li, H., Yan, S., Zhao, L., Tan, J., Zhang, Q., Gao, F., et al. (2014). Histone acetylation associated up-regulation of the cell wall related genes is involved in salt stress induced maize root swelling. *BMC Plant Biology*, *14*(1), 105.
Li, C., & Zhang, B. (2016). MicroRNAs in control of plant development. *Journal of Cellular Physiology*, *231*(2), 303–313.
Li, X., Zhu, J., Hu, F., Ge, S., Ye, M., Xiang, H., et al. (2012). Single-base resolution maps of cultivated and wild rice methylomes and regulatory roles of DNA methylation in plant gene expression. *BMC Genomics*, *13*(1), 300.
Liang, Z., Shen, L., Cui, X., Bao, S., Geng, Y., Yu, G., et al. (2017). DNA N6-adenine methylation in Arabidopsis thaliana. *Mechanisms of Development*, *145*, S137–S138.
Liang, D., Zhang, Z., Wu, H., Huang, C., Shuai, P., Ye, C.-Y., et al. (2014). Single-base-resolution methylomes of populus trichocarpa reveal the association between DNA methylation and drought stress. *BMC Genetics*, *15*(1), S9.
Linnaeus, C. (1749). Peloria. Diss. Ac., Uppsala (Amoenitates Acad.)
Lippman, Z. B., & Zamir, D. (2007). Heterosis: Revisiting the magic. *Trends in Genetics*, *23*(2), 60–66.
Liu, J., Feng, L., Li, J., & He, Z. (2015). Genetic and epigenetic control of plant heat responses. *Frontiers in Plant Science*, *6*, 267.
Liu, R., How-Kit, A., Stammitti, L., Teyssier, E., Rolin, D., Mortain-Bertrand, A., et al. (2015). A DEMETER-like DNA demethylase governs tomato fruit ripening. *Proceedings of the National Academy of Sciences of the United States of America*, *112*(34), 10804–10809.
Long, Y., Xia, W., Li, R., Wang, J., Shao, M., Feng, J., et al. (2011). Epigenetic QTL mapping in Brassica napus. *Genetics*, *189*(3), 1093–1102.
Louws, F. J., Rivard, C. L., & Kubota, C. (2010). Grafting fruiting vegetables to manage soilborne pathogens, foliar pathogens, arthropods and weeds. *Scientia Horticulturae*, *127*(2), 127–146.
Lu, X., Wang, W., Ren, W., Chai, Z., Guo, W., Chen, R., et al. (2015). Genome-wide epigenetic regulation of gene transcription in maize seeds. *PloS One*, *10*(10), e0139582.
Lukens, L. N., & Zhan, S. (2007). The plant genome's methylation status and response to stress: Implications for plant improvement. *Current Opinion in Plant Biology*, *10*(3), 317–322.
Makarevich, G., Villar, C. B. R., Erilova, A., & Kohler, C. (2008). Mechanism of PHERES1 imprinting in Arabidopsis. *Journal of Cell Science*, *121*, 906–912.
Makarevitch, I., Waters, A. J., West, P. T., Stitzer, M., Hirsch, C. N., et al. (2015). Transposable elements contribute to activation of maize genes in response to abiotic stress. *PLOS Genetics*, *11*(1), e1004915. https://doi.org/10.1371/journal.pgen.1004915.
Manning, K., Tör, M., Poole, M., Hong, Y., Thompson, A. J., King, G. J., et al. (2006). A naturally occurring epigenetic mutation in a gene encoding an SBP-box transcription factor inhibits tomato fruit ripening. *Nature Genetics*, *38*(8), 948–952.
Martel, C., Vrebalov, J., Tafelmeyer, P., & Giovannoni, J. J. (2011). The tomato MADS-box transcription factor RIPENING INHIBITOR interacts with promoters involved in numerous ripening processes in a COLORLESS NONRIPENING-dependent manner. *Plant Physiology*, *157*(3), 1568–1579.

Martinez, G., & Kohler, C. (2017). Role of small RNAs in epigenetic reprogramming during plant sexual reproduction. *Current Opinion in Plant Biology*, *36*, 22–28.

Mathieu, O., Reinders, J., Čaikovski, M., Smathajitt, C., & Paszkowski, J. (2007). Transgenerational stability of the Arabidopsis epigenome is coordinated by CG methylation. *Cell*, *130*(5), 851–862.

Matzke, M. A., & Mosher, R. A. (2014). RNA-directed DNA methylation: An epigenetic pathway of increasing complexity. *Nature Reviews. Genetics*, *15*(6), 394–408.

Meissner, A., Mikkelsen, T. S., Gu, H., Wernig, M., Hanna, J., Sivachenko, A., et al. (2008). Genome-scale DNA methylation maps of pluripotent and differentiated cells. *Nature*, *454*(7205), 766.

Melnyk, C. W., Molnar, A., & Baulcombe, D. C. (2011). Intercellular and systemic movement of RNA silencing signals. *The EMBO Journal*, *30*(17), 3553–3563.

Meyer, R. C., Witucka-Wall, H., Becher, M., Blacha, A., Boudichevskaia, A., Dörmann, P., et al. (2012). Heterosis manifestation during early Arabidopsis seedling development is characterized by intermediate gene expression and enhanced metabolic activity in the hybrids. *The Plant Journal*, *71*(4), 669–683.

Mgbeze, G., & Iserhienrhien, A. (2014). Somaclonal variation associated with oil palm (Elaeis guineensis Jacq.) clonal propagation: A review. *African Journal of Biotechnology*, *13*(9), 989–997.

Migicovsky, Z., Yao, Y., & Kovalchuk, I. (2014). Transgenerational phenotypic and epigenetic changes in response to heat stress in Arabidopsis thaliana. *Plant Signaling & Behavior*, *9*(2), e27971.

Mirouze, M., & Paszkowski, J. (2011). Epigenetic contribution to stress adaptation in plants. *Current Opinion in Plant Biology*, *14*(3), 267–274.

Miura, K., Agetsuma, M., Kitano, H., Yoshimura, A., Matsuoka, M., Jacobsen, S. E., et al. (2009). A metastable DWARF1 epigenetic mutant affecting plant stature in rice. *Proceedings of the National Academy of Sciences of the United States of America*, *106*(27), 11218–11223.

Miura, K., Ikeda, M., Matsubara, A., Song, X.-J., Ito, M., Asano, K., et al. (2010). OsSPL14 promotes panicle branching and higher grain productivity in rice. *Nature Genetics*, *42*(6), 545–549.

Molnar, A., Melnyk, C. W., Bassett, A., Hardcastle, T. J., Dunn, R., & Baulcombe, D. C. (2010). Small silencing RNAs in plants are mobile and direct epigenetic modification in recipient cells. *Science*, *328*(5980), 872–875.

Morales-Ruiz, T., Ortega-Galisteo, A. P., Ponferrada-Marin, M. I., Martinez-Macias, M. I., Ariza, R. R., & Roldan-Arjona, T. (2006). DEMETER and REPRESSOR OF SILENCING 1 encode 5-methylcytosine DNA glycosylases. *Proceedings of the National Academy of Sciences of the United States of America*, *103*(18), 6853–6858.

Ng, D. W., Lu, J., & Chen, Z. J. (2012). Big roles for small RNAs in polyploidy, hybrid vigor, and hybrid incompatibility. *Current Opinion in Plant Biology*, *15*(2), 154–161.

Ni, Z., Kim, E.-D., Ha, M., Lackey, E., Liu, J., Zhang, Y., et al. (2009). Altered circadian rhythms regulate growth vigour in hybrids and allopolyploids. *Nature*, *457*(7227), 327–331.

Niederhuth, C. E., & Schmitz, R. J. (2017). Putting DNA methylation in context: From genomes to gene expression in plants. *Biochimica et Biophysica Acta (BBA)-Gene Regulatory Mechanisms*, *1860*(1), 149–156.

Oh, S., Park, S., & van Nocker, S. (2008). Genic and global functions for Paf1C in chromatin modification and gene expression in Arabidopsis. *PLoS Genetics*, *4*(8), e1000077.

Oliver, S. N., Finnegan, E. J., Dennis, E. S., Peacock, W. J., & Trevaskis, B. (2009). Vernalization-induced flowering in cereals is associated with changes in histone methylation at the VERNALIZATION1 gene. *PNAS*, *106*(20), 8386–8391.

Ong-Abdullah, M., Ordway, J. M., Jiang, N., Ooi, S. E., Kok, S.-Y., Sarpan, N., et al. (2015). Loss of Karma transposon methylation underlies the mantled somaclonal variant of oil palm. *Nature, 525*(7570), 533–537.

Ono, A., Yamaguchi, K., Fukada-Tanaka, S., Terada, R., Mitsui, T., & Iida, S. (2012). A null mutation of ROS1a for DNA demethylation in rice is not transmittable to progeny. *The Plant Journal, 71*(4), 564–574.

Pandey, G., Sharma, N., Pankaj Sahu, P., & Prasad, M. (2016). Chromatin-based epigenetic regulation of plant abiotic stress response. *Current Genomics, 17*(6), 490–498.

Park, P. J. (2008). Epigenetics meets next generation sequencing. *Epigenetics, 3*, 318–332.

Park, P. J. (2009). ChIP-seq: Advantages and challenges of a maturing technology. *Nature Reviews. Genetics, 10*, 669–680.

Peng, H., & Zhang, J. (2009). Plant genomic DNA methylation in response to stresses: Potential applications and challenges in plant breeding. *Progress in Natural Science, 19*(9), 1037–1045.

Platt, A., Gugger, P. F., Pellegrini, M., & Sork, V. L. (2015). Genome-wide signature of local adaptation linked to variable CpG methylation in oak populations. *Molecular Ecology, 24*(15), 3823–3830.

Quadrana, L., Almeida, J., Asís, R., Duffy, T., Dominguez, P. G., Bermúdez, L., et al. (2014). Natural occurring epialleles determine vitamin E accumulation in tomato fruits. *Nature Communications, 5*, 4027.

Rajeev Kumar, S., Safia, & Sathishkumar, R. (2017). Small RNAs: Master regulators of epigenetic silencing in plants. In N. Rajewsky, S. Jurga, & J. Barciszewski (Eds.), *Plant epigenetics. RNA technologies* (pp. 89–106). Cham, Switzerland: Springer.

Regulski, M., Lu, Z., Kendall, J., Donoghue, M. T. A., Reinders, J., Llaca, V., et al. (2013). The maize methylome influences mRNA splice sites and reveals widespread paramutation-like switches guided by small RNA. *Genome Research, 23*(10), 1651–1662.

Reinders, J., Wulff, B. B., Mirouze, M., Marí-Ordóñez, A., Dapp, M. l., Rozhon, W., et al. (2009). Compromised stability of DNA methylation and transposon immobilization in mosaic Arabidopsis epigenomes. *Genes & Development, 23*(8), 939–950.

Rhoads, A., & Au, K. F. (2015). PacBio sequencing and its applications. *Genomics, Proteomics & Bioinformatics, 13*(5), 278–289.

Rodrigues, A. S., & Miguel, C. l. M. (2017). The pivotal role of small non-coding RNAs in the regulation of seed development. *Plant Cell Reports, 36*, 653–667. https://doi.org/10.1007/s00299-017-2120-5.

Rodriguez-Lopez, C. M., & Wilkinson, M. J. (2015). Epi-fingerprinting and epi-interventions for improved crop production and food quality. *Frontiers in Plant Science, 6*, 397.

Sah, S. K., Reddy, K. R., & Li, J. (2016). Abscisic acid and abiotic stress tolerance in crop plants. *Frontiers in Plant Science, 7*, 571.

Sahu, P. P., Pandey, G., Sharma, N., Puranik, S., Muthamilarasan, M., & Prasad, M. (2013). Epigenetic mechanisms of plant stress responses and adaptation. *Plant Cell Reports, 32*(8), 1151–1159.

Schadt, E. E., Turner, S., & Kasarskis, A. (2010). A window into third-generation sequencing. *Human Molecular Genetics, 19*(R2), R227–R240.

Schoft, V. K., Chumak, N., Choi, Y., Hannon, M., Garcia-Aguilar, M., Machlicova, A., et al. (2011). Function of the DEMETER DNA glycosylase in the Arabidopsis thaliana male gametophyte. *Proceedings of the National Academy of Sciences of the United States of America, 108*(19), 8042–8047.

Shen, H., He, H., Li, J., Chen, W., Wang, X., Guo, L., et al. (2012). Genome-wide analysis of DNA methylation and gene expression changes in two Arabidopsis ecotypes and their reciprocal hybrids. *The Plant Cell, 24*(3), 875–892.

Shen, Y., Sun, S., Hua, S., Shen, E., Ye, C. Y., Cai, D., et al. (2017). Analysis of transcriptional and epigenetic changes in hybrid vigor of allopolyploid Brassica napus uncovers key roles for small RNAs. *The Plant Journal*, *91*(5), 874–893.

Slotkin, R. K. (2016). Plant epigenetics: From genotype to phenotype and back again. *Genome Biology*, *17*(1), 57.

Song, Q.-X., Lu, X., Li, Q.-T., Chen, H., Hu, X.-Y., Ma, B., et al. (2013). Genome-wide analysis of DNA methylation in soybean. *Molecular Plant*, *6*(6), 1961–1974.

Soyer, J. L., El Ghalid, M., Glaser, N., Ollivier, B., Linglin, J., Grandaubert, J., et al. (2014). Epigenetic control of effector gene expression in the plant pathogenic fungus Leptosphaeria maculans. *PLOS Genetics*, *10*(3), e1004227.

Steward, N., Ito, M., Yamaguchi, Y., Koizumi, N., & Sano, H. (2002). Periodic DNA methylation in maize nucleosomes and demethylation by environmental stress. *The Journal of Biological Chemistry*, *277*(40), 37741–37746.

Stroud, H., Do, T., Du, J., Zhong, X., Feng, S., Johnson, L., et al. (2014). Non-CG methylation patterns shape the epigenetic landscape in Arabidopsis. *Nature Structural & Molecular Biology*, *21*(1), 64–72.

Su, L.-C., Deng, B., Liu, S., Li, L.-M., Hu, B., Zhong, Y.-T., et al. (2015). Isolation and characterization of an osmotic stress and ABA induced histone deacetylase in Arachis hygogaea. *Frontiers in Plant Science*, *6*, 512.

Sung, S., & Amasino, R. M. (2006). Molecular genetic studies of the memory of winter. *Journal of Experimental Botany*, *57*(13), 3369–3377.

Taiwo, O., Wilson, G. A., Morris, T., Seisenberger, S., Reik, W., Pearce, D., et al. (2012). Methylome analysis using MeDIP-seq with low DNA concentrations. *Nature Protocols*, *7*(4), 617–636.

Tang, K., Lang, Z., Zhang, H., & Zhu, J.-K. (2016). The DNA demethylase ROS1 targets genomic regions with distinct chromatin modifications. *Nature Plants*, *2*(11), 16169.

Tang, T., Wen, M., Lin, P., & Wang, Y. (2017). An evolutionary view of the biogenesis and function of rice small RNAs. In *Plant epigenetics* (pp. 69–88). Springer.

Tani, E., Polidoros, A. N., Nianiou-Obeidat, I., & Tsaftaris, A. S. (2005). DNA methylation patterns are differently affected by planting density in maize inbreds and their hybrids. *Maydica*, *50*(1), 19–23.

Teixeira, F. K., Heredia, F., Sarazin, A., Roudier, F. O., Boccara, M., Ciaudo, C., et al. (2009). A role for RNAi in the selective correction of DNA methylation defects. *Science*, *323*(5921), 1600–1604.

Tricker, P. J., Gibbings, J. G., Rodríguez López, C. M., Hadley, P., & Wilkinson, M. J. (2012). Low relative humidity triggers RNA-directed de novo DNA methylation and suppression of genes controlling stomatal development. *Journal of Experimental Botany*, *63*(10), 3799–3813.

Tricker, P. J., López, C. M. R., Gibbings, G., Hadley, P., & Wilkinson, M. J. (2013). Transgenerational, dynamic methylation of stomata genes in response to low relative humidity. *International Journal of Molecular Sciences*, *14*(4), 6674–6689.

Tsaftaris, A. S., Kafka, M., & Polidoros, A. (1997). Epigenetic modifications of total genomic maize DNA: The role of growth conditions. In A. S. Tsaftaris (Ed.), *Genetics, biotechnology and breeding of Maize and Sorghum* (pp. 125–130). Cambridge, UK: Royal Society of Chemistry.

Tsaftaris, A., Kapazoglou, A., & Darzentas, N. (2012). Epigenetics, epigenomics, and implications in plant breeding. In A. Altman & P. Haegawa (Eds.), *Plant biotechnology and agriculture: Prospects for the 21st century* (pp. 207–226). Academic Press.

Urich, M. A., Nery, J. R., Lister, R., Schmitz, R. J., & Ecker, J. R. (2015). MethylC-seq library preparation for base-resolution whole-genome bisulfite sequencing. *Nature Protocols*, *10*(3), 475–483.

van Dijk, K., Ding, Y., Malkaram, S., Riethoven, J.-J. M., Liu, R., Yang, J., et al. (2010). Dynamic changes in genome-wide histone H3 lysine 4 methylation patterns in response to dehydration stress in Arabidopsis thaliana. *BMC Plant Biology*, *10*(1), 238.

Van Oosten, M. J., Bressan, R. A., Zhu, J.-K., Bohnert, H. J., & Chinnusamy, V. (2014). The role of the epigenome in gene expression control and the epimark changes in response to the environment. *Critical Reviews in Plant Sciences*, *33*(1), 64–87.

Varriale, A. (2017). DNA methylation in plants and its implications in development, hybrid vigour, and evolution. In *Plant epigenetics* (pp. 263–280). Springer.

Vidalis, A., Živković, D., Wardenaar, R., Roquis, D., Tellier, A., & Johannes, F. (2016). Methylome evolution in plants. *Genome Biology*, *17*(1), 264.

Wang, Z., Casas-Mollano, J. A., Xu, J., Riethoven, J.-J. M., Zhang, C., & Cerutti, H. (2015). Osmotic stress induces phosphorylation of histone H3 at threonine 3 in pericentromeric regions of Arabidopsis thaliana. *Proceedings of the National Academy of Sciences of the United States of America*, *112*(27), 8487–8492.

Wang, Z., Meng, D., Wang, A., Li, T., Jiang, S., Cong, P., et al. (2013). The methylation of the PcMYB10 promoter is associated with green-skinned sport in Max Red Bartlett pear. *Plant Physiology*, *162*(2), 885–896.

Wen, S., Wen, N., Pang, J., Langen, G., Brew-Appiah, R. A. T., Mejias, J. H., et al. (2012). Structural genes of wheat and barley 5-methylcytosine DNA glycosylases and their potential applications for human health. *Proceedings of the National Academy of Sciences of the United States of America*, *109*, 20543–20548.

Weng, Y.-I., Huang, T. H. M., & Yan, P. S. (2009). Methylated DNA immunoprecipitation and microarray-based analysis: Detection of DNA methylation in breast cancer cell lines. *Molecular Endocrinology: Methods and Protocols*, 165–176.

Wetterstrand, K. A. (2013). *DNA sequencing costs: Data from the NHGRI genome sequencing program (GSP)*. Available at www.genome.gov/sequencingcosts(accessed: January 2018).

Wibowo, A., Becker, C., Marconi, G., Durr, J., Price, J., Hagmann, J., et al. (2016). Hyperosmotic stress memory in Arabidopsis is mediated by distinct epigenetically labile sites in the genome and is restricted in the male germline by DNA glycosylase activity. *Elife*, *5*, e13546.

Williams, B. P., Pignatta, D., Henikoff, S., & Gehring, M. (2015). Methylation-sensitive expression of a DNA demethylase gene serves as an epigenetic rheostat. *PLoS Genetics*, *11*(3), e1005142.

Wu, R., Wang, X., Lin, Y., Ma, Y., Liu, G., Yu, X., et al. (2013). Inter-species grafting caused extensive and heritable alterations of DNA methylation in *Solanaceae* plants. *PLoS One*, *8*(4), e61995.

Xing, M.-Q., Zhang, Y., Zhou, S.-R., Hu, W.-Y., Wu, X.-T., Ye, Y.-J., et al. (2015). Global analysis reveals the crucial roles of DNA methylation during rice seed development. *Plant Physiology*, *168*(4), 1417–1432.

Yamamuro, C., Zhu, J.-K., & Yang, Z. (2016). Epigenetic modifications and plant hormone action. *Molecular Plant*, *9*(1), 57–70.

Yong-Villalobos, L., González-Morales, S. I., Wrobel, K., Gutiérrez-Alanis, D., Cervantes-Peréz, S. A., Hayano-Kanashiro, C., et al. (2015). Methylome analysis reveals an important role for epigenetic changes in the regulation of the Arabidopsis response to phosphate starvation. *Proceedings of the National Academy of Sciences of the United States of America*, *112*(52), E7293–E7302.

Yu, Q., Liu, Y., Li, M., & Yu, B. (2017). Small RNA biogenesis and degradation in plants. In *Plant epigenetics* (pp. 107–127). Springer.

Yu, Y., Yang, X., Wang, H., Shi, F., Liu, Y., Liu, J., et al. (2013). Cytosine methylation alteration in natural populations of Leymus chinensis induced by multiple abiotic stresses. *PloS One*, *8*, e55772.

Zemach, A., Kim, M. Y., Silva, P., Rodrigues, J. A., Dotson, B., Brooks, M. D., et al. (2010). Local DNA hypomethylation activates genes in rice endosperm. *Proceedings of the National Academy of Sciences of the United States of America*, *107*(43), 18729–18734.

Zemach, A., McDaniel, I. E., Silva, P., & Zilberman, D. (2010). Genome-wide evolutionary analysis of eukaryotic DNA methylation. *Science*, *328*(5980), 916–919.

Zhang, B. (2015). MicroRNA: A new target for improving plant tolerance to abiotic stress. *Journal of Experimental Botany*, *66*(7), 1749–1761.

Zhang, X., Bernatavichute, Y. V., Cokus, S., Pellegrini, M., & Jacobsen, S. E. (2009). Genome-wide analysis of mono-, di- and trimethylation of histone H3 lysine 4 in Arabidopsis thaliana. *Genome Biology*, *10*(6), R62.

Zhang, Y.-C., & Chen, Y.-Q. (2017). Epigenetic regulation by noncoding RNAs in plant development. In *Plant epigenetics* (pp. 183–198). Springer.

Zhang, L., Cheng, Z., Qin, R., Qiu, Y., Wang, J.-L., Cui, X., et al. (2012). Identification and characterization of an epi-allele of FIE1 reveals a regulatory linkage between two epigenetic marks in rice. *The Plant Cell*, *24*(11), 4407–4421.

Zhang, C., & Hsieh, T.-F. (2013). Heritable epigenetic variation and its potential applications for crop improvement. *Plant Breeding and Biotechnology*, *1*(4), 307–319.

Zhang, L., Peng, Y., Wei, X., Dai, Y., Yuan, D., Lu, Y., et al. (2014). Small RNAs as important regulators for the hybrid vigour of super-hybrid rice. *Journal of Experimental Botany*, *65*(20), 5989–6002.

Zhang, B., Tieman, D. M., Jiao, C., Xu, Y., Chen, K., Fei, Z., et al. (2016). Chilling-induced tomato flavor loss is associated with altered volatile synthesis and transient changes in DNA methylation. *Proceedings of the National Academy of Sciences of the United States of America*, *113*(44), 12580–12585.

Zhao, M.-T., Whyte, J. J., Hopkins, G. M., Kirk, M. D., & Prather, R. S. (2014). Methylated DNA immunoprecipitation and high-throughput sequencing (MeDIP-seq) using low amounts of genomic DNA. *Cellular Reprogramming (Formerly Cloning and Stem Cells)*, *16*(3), 175–184.

Zhong, S., Fei, Z., Chen, Y.-R., Zheng, Y., Huang, M., Vrebalov, J., et al. (2013). Single-base resolution methylomes of tomato fruit development reveal epigenome modifications associated with ripening. *Nature Biotechnology*, *31*(2), 154–159.

Zhu, A., Greaves, I. K., Dennis, E. S., & Peacock, W. J. (2017). Genome-wide analyses of four major histone modifications in Arabidopsis hybrids at the germinating seed stage. *BMC Genomics*, *18*(1), 137.

Zhu, Q.-H., Shan, W.-X., Ayliffe, M. A., & Wang, M.-B. (2015). Epigenetic mechanisms: An emerging player in plant-microbe interactions. *Molecular Plant-Microbe Interactions*, *29*(3), 187–196.

Zou, B., Yang, D.-L., Shi, Z., Dong, H., & Hua, J. (2014). Monoubiquitination of histone 2B at the disease resistance gene locus regulates its expression and impacts immune responses in Arabidopsis. *Plant Physiology*, *165*(1), 309–318.

FURTHER READING

Eichten, S. R., Swanson-Wagner, R. A., Schnable, J. C., Waters, A. J., Hermanson, P. J., Liu, S., et al. (2011). Heritable epigenetic variation among maize inbreds. *PLoS Genetics*, 7(11), e1002372.

Secco, D., Wang, C., Shou, H., Schultz, M. D., Chiarenza, S., Nussaume, L., et al. (2015). Stress induced gene expression drives transient DNA methylation changes at adjacent repetitive elements. *Elife*, *4*, e09343.

Yuan, W., Luo, X., Li, Z., Yang, W., Wang, Y., Liu, R., et al. (2016). A cis cold memory element and a trans epigenome reader mediate Polycomb silencing of FLC by vernalization in Arabidopsis. *Nature Genetics*, *48*(12), 1527–1534.

CHAPTER TEN

Biotechnologies: The Ideal Victim?

Nayla Farouki[1]
Philosopher and Historian of Science, Grenoble, France
[1]Corresponding author: e-mail address: naylafarouki@hotmail.com

Contents

Abstract

Biotechnologies, in general, and GM plants in particular, suffer from a sort of defiance that has gone beyond ordinary technophobia. Societies, in which all technological advancements are naturally embedded, react in various ways to what industries have to offer. Enthusiasm, slow appropriation, apprehension, rejection seem normal at the consumer's level, and each person ought to have the free choice to consume or not to consume. With plant biotechnologies, this normal evolution of technological offers, and of the free scientific research that goes with it, is now perturbed by overregulation on the one hand and straight and total forbiddance on the other.

These pages are an attempt at an explanation of this phenomenon's specificity. What is it that makes biotechnologies different from other technologies on which there is no consensus, such as nuclear energy or robotics? What are humans scared of? and why?

The answer could reside in the following: biotechnologies, where plants are concerned, that is where agriculture, food and the living environment are concerned, seem to have specific reasons for producing such negative feelings. Among all the "scary" inventions that some want to denounce, could it be that biotechnologies are the ideal victim?

1. INTRODUCTION

When one takes a look, diagonally or in depth, at the chapters of this volume, one cannot refrain from asking why? Why biotechnologies as such and why especially GM plants? What is it that has been happening in the last half century in such a way that all suspicion should converge on the scientific

Advances in Botanical Research, Volume 86
ISSN 0065-2296
https://doi.org/10.1016/bs.abr.2017.11.008

research, the technological research and the industry that allow for the development and commercialization of transgenic plants?

Of course, this is not the only debate in which violent reactions may occur. Nuclear civil energy (fear of radioactivity), emission of carbon dioxide (fear of global warming), chemical industries in all forms and aspects (fear of pollution) are also part of what some people reject with no further analysis.

This generalization of the problem does not aim at downgrading the question of GM plants. After all, this is an area where research itself has been altogether banned in countries like France, where all the "academies" unanimously expressed their dissatisfaction at this situation... to no avail (see for example: http://www.academie-sciences.fr/en/Advice-Notes-and-Reports/french-academies-call-for-freedom-of-research-on-genetically-modified-plants-gmps-to-be-restored.html).

By late, but this has been brewing for some time, the use of animals (for meat or lab research) is under scrutiny, with more or less aggressive encouragement to convert to vegetarianism, veganism, etc.

The need to tackle the GM plants question (Smyth, 2017) is made the more urgent as a potential source of fruitful advances is at worst being dismantled, at best overregulated, a phenomenon that perhaps no other scientific or technological research has seen before.

In what follows, I shall present only a part of the numerous sides of this question. I aim at trying to clarify the picture, rather than to provide solutions. Perhaps, these will eventually show up, but it is important for scientists to understand that GM plant research is facing specific problems, related to the environment, to food ... and to the nature of the sacred.

2. WHENCE THE TECHNOPHOBIA?

First of all, it is important to note that—since antiquity—people (not all, evidently) feel scared of change, of the unknown or unexplained, of the invisible. Therefore, it comes as no surprise to see suspicious reactions to radioactivity, to genetic engineering, as well as to all forms of chemical compounds.

This attitude has further developed with the acceleration of scientific discoveries, with the pervasiveness of industrial products and with the seemingly relentless presentation of new technologies.

Yet, society as a whole has not turned technophobic or antiscientific. Firstly, there is no such a thing as a monolithic society. Social reactions are extremely diverse on any question and frequently go from total rejection to full enthusiasm. Secondly, individuals are in constant contradiction, even with themselves. Example: one could not but notice how antitechnology militants are good at the use of their smart phones and websites.

Nevertheless, when we look at the way biotechnologies are treated, despite their wealth of potentialities, one can see that what is happening does not belong only to the realm of ordinary irrationality. This exists, of course, and it will always be present. The question is therefore: how come some are systematically defiant towards technologies such as biotechnologies, to the point that this defiance ends up affecting policy makers? Did anything happen, different from what existed in antiquity, that made resistance to change so influential?

My attempt at writing something as clear and analytical as possible requires a little bit of history. I hope it will make things clear at the conceptual level for the unfolding of the ulterior issues that presented themselves to industry, to technology, to fundamental science, public or private. We will come to see how anthropology, policy, sociology, philosophy, economics, law and regulation, got so entangled as to create the situation exposed in the preceding chapters.

Technophobia is—by definition—a matter of fear. Fear can be legitimate, genuine, pure fantasy, instrumentally and cynically used. Fear is a powerful emotion (or affect), and once installed, it has a tendency to "colour" all presentations of what is at stake, thus rendering rational arguments powerless. Literature, films, magazines, essays, all use fear as a tool the objectives of which span from amusement to higher audience, to influence. Philosophers, journalists, scientists use it sometimes with a clear interest in its instrumental power. This is, for example, the case of the "heuristics of fear", invented and justified by Hans Jonas in his book *The imperative of Responsibility*, first published in 1979.

Concerning fear, it is important to start somewhere, even if the starting point is arbitrary and one can find earlier references; like for example, the doubts raised about the use of steam power in the 19th century.

I will start with 1962, when Rachel Carson published *Silent Spring* an extremely influential book against pesticides in general and DDT in particular. In a passionate, emotional style, she describes the beauties of nature and expresses her fears that "chemistry" should destroy the harmony of the world. The idea behind the title was that if pesticides continue to be used, the future springs will be silent: no birds. In Carson's book, fear was an outstanding argument as this short paragraph shows:

"As crude a weapon as the cave man's club, the chemical barrage has been hurled against the fabric of life - a fabric on the one hand delicate and destructible, on the other miraculously tough and resilient, and capable of striking back in unexpected ways. These extraordinary capacities of life have been ignored by the practitioners of chemical control who have brought to their task no 'high-minded orientation,' no humility before the vast forces with which they tamper".

Two themes show up that will appear again in the GM plants debates:

- Humans (i.e. arrogant scientists and industrialists, playing God) behave irresponsibly, thus leading to the destruction of Nature,
- Nature (Mother Nature!) will eventually seek revenge.

These arguments are not specific to pesticides or to GM plants. They appear in all debates where the environment is concerned, such as global warming, nuclear pollution, the destruction of species, etc.

Nongovernmental organizations (NGOs) for the protection of the environment started to flourish in the 1960s and became more and more audible and powerful as time went by. WWF was founded in 1961, Friends of the Earth in 1967, followed in 1971 by Greenpeace. Environmental NGOs very intelligently describe themselves as "independent". Independent of what or of whom? No one really knows, especially that it is common knowledge that any organization that works for whatever cause, would also fight... for its own survival.

The policy of fear, since the 1960s, has been used and merchandized deliberately, and in an uncontrollable way: the media (whether social or otherwise) are ubiquitous, rumours and misinformation, added to the expertise of communication science, are easily and effectively available and scientists are mostly unable to fight back in an analogous manner (we shall come back to that later).

The pesticides scare was paralleled by the fear of nuclear warfare during times of cold war and of nuclear risks at industrial civil plants. The accident at Three Miles Island happened 12 days before the release, in 1979, of the film *The China Syndrome* which tells a story of nuclear malevolence. Of course, Hollywood had started the use of technological fears ages ago, but again, in the 1970s, science and technology were highly transmuted into subjects of science-fictional fear, with talent and of course with the emotional impact that goes with it (for a list of films where genetic engineering is part of the (usually scary) story, go to https://en.wikipedia.org/wiki/Genetic_engineering_in_science_fiction).

Environmentalism took several forms and shapes, some justified, some dogmatic, some cynical; and it further developed into "green" parties and organizations making the environment a primarily political issue. The movement started mostly as a reaction against chemical industry pollution, in the United States and in Germany.

In the United States it helped create the legitimate notion of "stakeholders", people whose lives are directly affected by industrial activity, as a complement to "shareholders", people who own the industrial activity (see *Power and Accountability* by Robert Monks and Nell Minow, published in 1992, Harper Collins Publishers). This notion has been distorted later on to mean and to legitimize any collective or individual who feels concerned by a specific question.

In Germany, in its antinuclear form, it helped advance the notion of "risk" related to industry, a notion that was promoted by German sociologist Ulrich Beck, in his book *Risk Society: towards a new modernity* first published in German, in 1986, shortly after the accident at Chernobyl. The book, in its multiple translations, was also highly influential. It presents the consequences of soil pollution (radioactive and chemical) in a sombre, gloomy way and prepares the advances of the precautionary principle.

First stated in the Rio Conference in 1992, article 15 of the Rio declaration notes: "In order to protect the environment, the precautionary approach shall be widely applied by States according to their capabilities. Where there are threats of serious or irreversible damage, lack of full scientific certainty shall not be used as a reason for postponing cost-effective measures to prevent environmental degradation".

Should we consider it amusing or irritating that people in 1992 would require "scientific certainty" in disciplines where the use of the scientific method is essentially based on probabilities and statistics? We know, since the time of Henri Poincaré, that even Newtonian celestial mechanics, though entirely deterministic, is not—cannot be—predicted at large scales with absolute certainty. Evidently, where biology is concerned (whether in plants or in human epidemiology) full scientific certainty does not and cannot exist.

The precautionary principle, as we know, has been used and misused. In a wide context, it has served the advancement of fear. Logically, where precaution is needed, risk must be great. Inversely, where any risk is perceived (imagined? postulated?), precautionary measures had to be taken.

The saga of fear does not stop at environmentalism and precautionary measures. Enter science-fiction not as a literary and cinematographic activity but, taken seriously, as the future of science.

Since the 1980s, some postmodern intellectuals (see later) started stating that science-fiction was the true expression of what the future of science and technology holds. This assertion, coming from Academia, mostly generated by human and social scientists, but also by some physicists or biologists, allowed the layman—reader of newspapers and magazines—to accept without discussion nor critical mindedness the continuity between what is being researched or invented today and whatever nightmarish future science and technology are supposed to bring ineluctably. These days (2017) the principal scare is about Artificial Intelligence, denounced by people as respectable as Stephen Hawking or Elon Musk.

Fear from what science and technology will bring at some moment in the future became a powerful argument in the hands of thinkers whose criteria for extrapolating to what will happen are justified by science-fiction images, and are as fallacious as astrological predictions. When you do not specify the future event, its nature, time and place, you are certain to be right: something "bad" will happen sooner or later. Promises of the sort rival only with ancient scares or modern divination. Most importantly, they do not take into consideration the fact that science and technology do not exist in a no man's land: policies, rules of conduct, modes of control are and will always be there to direct technological innovations, in whatever way people will want it to be.

Technophobia has a dense history and I tried to highlight only a few of its elements. One should add still more major thinkers and publications, the holders of the theory of economic and technological "deceleration" (some today even argue for "slow science") and plenty of websites and webzines to render a complete picture. Sensitivity to fear is such that—as is well studied and well known—some people will only select the information that confirms their apprehension. For them, contradictory rationality and scientific explanation will remain at bay. Worse, if read or heard, all arguments explaining safety measures will be considered further proof that something dangerous is at stake, thus increasing the conviction that fear is legitimate. This, of course, is an old story, and the use of fear, by the few who master the imagination and the emotions of the many, is as old as humanity itself.

3. POSTMODERNISM AND ITS POLITICAL/ACADEMIC CONSTELLATION

In the 1980s, influenced by Nietzsche, Marx, Heidegger and Foucault, a new "theory" came into existence: postmodernism. As the name indicates, postmodernism can be described as what comes after "modernism", a proclaimed nontheoretical antithesis to modernism. This last entity is viewed as an invention of the 18th century enlightenment, and postmodernism aims at criticizing all of the major characteristics of enlightenment values: namely, the idea of progress, scientific rationality and its claim at universality, representative democracy.

The criticism of enlightenment had already started in the 18th century, with Jean-Jacques Rousseau expressing scepticism about the benefits of civilization. Nietzsche lead the war against enlightenment in the 19th century, demolishing practically everything that was brought by rational thought—often called "Cartesian" in reference to Descartes: mainly the idea of the subject ("I think therefore I am"), the use of Reason to discover natural truths. The two major "enemies" of the romantic view of nature were Descartes (who proclaimed that man should be "as Master and Possessor of Nature") and Francis Bacon (who stated that "in order to dominate Nature, we must obey it"). Note that Descartes' quote is often cited out of context. In his own terms, this is what follows: "And this is a result to be desired, not only in order to the invention of an infinity of arts, by which we might be enabled to enjoy without any trouble the fruits of the earth, and all its comforts, but also and especially for the preservation of health,..." (Discourse on Method, chapter VI.)

Philosophers' faith in progress was also heavily put under criticism, as if progress was a pure invention of modernity alone. In the 20th century, Heidegger became the main thinker in the "fear" of technology sphere, through one major text (*The question concerning technology*, published in 1977). This short opus, surprisingly, questions technology via one example: that of the dams on the Rhine, which "control" the river thus symbolizing the "dominance" of promethean mankind over Nature. Nature itself is again presented in a passionate and romantic way as a global harmonious entity that humans disrupt and that may turn against mankind.

Almost simultaneously, in 1975, appeared the famous *Against Method*, by Paul Feyerabend whose attack this time was directly aimed at science, and the scientific method. In brilliant style, and sometimes with arguments worth thinking about and discussing in scientific circles, Feyerabend demolishes the special status of science, and of rationality. His relativism—i.e., science is the production of a specific culture, the Western culture—has been adopted by all those who, for various reasons, need to argue against the results of scientific research.

As he says: "Thus science is much closer to myth than a scientific philosophy is prepared to admit. It is one of the many forms of thought that have been developed by man, and not necessarily the best. It is conspicuous, noisy, and impudent, but it is inherently superior only for those who have already decided in favor of a certain ideology, or who have accepted it without having ever examined its advantages and its limits".

Finally, Michel Foucault ought to be mentioned. His main contribution to postmodernism is to consider that "knowledge is a tool for power", which allows defiance against scientists as being at best themselves tools of political and industrial power, at worst one can imagine them as an influential "community" that seek power for themselves. Suspicion is everywhere: no truth, no universality, no rational method; only conflicts of interest and of power. Doubt was instilled; tools were handed to anyone who wanted to criticize anything scientific. Science became a social endeavour analogous to anything else; scientific expertise was put in doubt.

The postmodernist emphasis was mostly on relativism and on constructivism. It asserts that there is no such a thing as knowledge of the real world, existence as such is a construct in scientists' minds, concepts and lab measurements are only there to justify what scientists had in mind. The attack was led mainly by the proponents of newly formed disciplines such as "science studies", but "gender studies", "colonial studies", etc. were also active at demonstrating that science is a white western male activity, excluding from its realm all of the "dominated" communities: women, coloured people, non-western countries and cultures. Of course, all of these needed to fight back and criticize science for its racist, misogynist, capitalist tendencies.

Years later, synthesizing Feyerabend and Foucault, Dona Harraway, one of the masters of gender studies, expressed things in her own way: "...science – the real game in town – is rhetoric, a series of efforts to persuade relevant social actors that one's manufactured knowledge is a route to a desired form of very objective power".

The relativism that postmodernism defends where science is concerned has lead its proponents logically to the idea that citizens and experts from the social sciences should have a say in science. If the scientific method is irrelevant, if science is a matter of politics like any other matter that concerns people, then why not let these same people have a say in what science does, and what science policy should be? This gave birth to consensus groups, to public debates, to the idea of coconstruction of knowledge and ultimately to participative science seen as a method for integrating scientific research in a democratic construct.

Instead of the classical "public understanding of science", the proponents of "science studies", a group of disciplines focalizing on the notion of "science-in-society" proposed "public engagement with science" and "public debates about science" (see an introduction to science studies: https://en.wikipedia.org/wiki/Science_studies).

The whole idea, at its inception, was based on the so-called arrogance of explanation, which denounces popular science as patronizing... One cannot help but notice that the word "science" here only signifies experimental sciences—physics, chemistry, biology—in so far as they have a relation to technological applications and industry. Yet, it is methodologically and epistemologically absurd to limit "engagement" to natural sciences; those who feel that scientific issues need discussion should also express the same for other disciplines, such as economics, finance, most of the social sciences as well as philosophy, which is rarely advocated by the social scientists. After all, all of the above do have an impact on society, and citizens should both comprehend and—if one follows the logic of "science studies"—have their say in the way these knowledges, theories and hypotheses affect their lives.

The "science studies" community reached its goal. Public engagement was taken seriously by scientific institutions, and when the public is "engaged"—i.e., when the scene is that of a multipartner dialogue, and not that of a conference—what usually happens is that people start by asking for explanations, whether they end up accepting what is being said or not is a totally different matter, and as I said earlier it depends on the turn of mind (apprehensive, defiant, angry, astonished, naïve, sympathetic, etc.) of the auditor. For the huge majority of the public, science is about the real world, and they most eagerly wait for its undertakings, even when not understanding the details of its proceedings.

Public debates on scientific and technological matters, when not preceded by explanations, have mostly failed, especially when the topics discussed are of a passionate nature like GMOs (see Kuntz, 2012). Debates turn fast into posturizing by different "stakeholders", each expressing his own position without listening to the others. The exercise usually ends up in a reaffirming of each position. Science is clearly not welcome in such audiences. Nor is rational argumentation of any sorts. Nor, in general, is the public as such: most people know that they are not knowledgeable enough to participate in so-called democratic or consensus debate about science. Of course, those who want to participate are usually those for whom the stakes exist: this is NOT the general public.

4. WHY ANTHROPOLOGY MATTERS

These systematic attacks on rationality, on the scientific method and on the good faith of scientists themselves, had immediate results that are being prolonged until today. The fact is quite simple: rationality is a thin and fragile veil covering an ocean of affects, of impulses, of social pressures, such as shame and pride, of egotistic shortsighted interests. Unless one remains vigilant and maintains rational argument, objectivity and rational knowledge alive, the above-noted archaisms of the human condition (collective and individual), will come back violently haunting.

In other words, where there is no room for reason, emotion fills the space, with its own categories: fear, superstition, anthropomorphism, symbols and myths, and finally the overriding rule of the sacred (about the "religious" aspects of GM plants rejection, see Tagliabue, 2016). Thus, biotechnologies became the easiest victim of the merchants of fear. In other words, this set of disciplines had all the qualities required to put science, technology and industry in doubt in the minds of citizens in whom, naturally, the sense of the sacred was/is still alive.

Put briefly, the sacred realm is about awe, life and death, fear and respect. Some of the most sacred things in human life are bodies—witness the importance of the rituals of the Medicine Man—and food—witness the rituals of diet and nutrition in all civilizations and cultures. More than ordinary chemistry that pollutes with plastics, more than radioactivity that can be easily spotted with a Geiger detector, GM plants touch on some of the most important elements in everyday life: agriculture, food and now medicine, another major magical element, both adored and abhorred (e.g. debate about vaccines).

Aside from the sacredness of Nature, of Gaïa (https://en.wikipedia.org/wiki/The_Revenge_of_Gaia), the anthropomorphized planet that will eventually turn on us and seek revenge, GM plants seem to be putting at risk what is essential in life, the purity, the beneficial characteristics of what ought to be good to humans and animals alike. When one adds disgust (Curtis, 2011; Sinaceur, Heath, & Cole, 2005) to defiance, the limit of what humans can accept is quickly reached.

I certainly do not mean to say that I agree with this reasoning. But I would like to attract attention to the fact that the easiest form of manipulation by fear, much better and stronger than the danger of dams on the Rhine, is anything that concerns plants and food. And where imagination flourishes, where fear of a sacred nature shows up, also flourish the people in the interest of whom this imagination and fear can be activated. It therefore comes as no surprise that developing countries which usually welcome technological innovation seem hesitant about biotechnologies. Fear is also universal (and so is its deliberate manipulation for political reasons; see Smyth, 2017).

Of course, everyone should be free to accept or reject an innovation of any kind. But it is ruinous for all to constantly debase scientific knowledge, and to deny the existence of scientifically discovered facts (after all, microbes and atoms, a little more than a century ago, were mere hypotheses, highly controversial at that, no one today denies their existence). The ideological elimination of facts, of universal truths, has allowed the spread of the rule of opinion. Democracy-in-science means that all stakeholders are equivalently competent (or not), refusal of expertise means a visceral defiance towards those in the know, accused of hidden interests, distrust in scientific evidence is a patiently constructed ideology, that of the rejection of the notion of "facts".

5. CONCLUSION: REASONS FOR HOPE

Firstly, sympathy for researchers has never waned. In Europe at least, polls suggest that the general public likes scientists and usually trusts them, even when disliking or mistrusting a specific scientific information or a specific technology.

Secondly, scientific culture, in most circles, and at schools, is progressing, especially when compared to what it was one hundred years ago. Democratization of culture has also profited to scientific culture. But culture is a free and open endeavour. People may not feel a necessary need to be knowledgeable in matters scientific, and this indifference may be regretted but ought to be respected. After all, culture, in all fields, is a matter of pleasure and motivation, and that should also be encouraged in scientific education.

Scientists must urgently understand that 50 or 60 years of politicization of the scientific realm, of systematic defiance and mistrust, need to be patiently and systematically unknotted, at schools where the scientific method is not taught, at policy levels where the science needs to be patiently explained and where scientists ought to be more involved, in the news where erroneous or ideologically transformed information should not be left uncorrected. This will be long and difficult. Discussion with ideologues is pointless and it can turn violent: what is essential is not to let fear permeate the minds of ordinary people. And, unfortunately, like all powerful negative emotions, its impact is easily installed and takes a lot of energy to be countered.

The sympathy capital is still there, but work needs to be done. And this does not mean that scientists should carry alone all the time-consuming work. This is not about scientists. This is about Science. It is essential that teachers, journalists, philosophers, social scientists understand that the stakes are too high and that the "value of science" is intrinsic, and should not be tampered with. The price of the war on science is simply too high!

REFERENCES

Curtis, V. (2011). Why disgust matters. *Philosophical Transactions of the Royal Society of London. Series B, Biological Sciences*, *366*(1583), 3478–3490.

Kuntz, M. (2012). The post-modern assault on science. *EMBO Reports*, *13*, 885–889.

Sinaceur, M., Heath, C., & Cole, S. (2005). Emotional and deliberative reactions to a public crisis: Mad cow disease in France. *Psychological Science*, *16*, 247–254. https://doi.org/10.1111/j.0956-7976.2005.00811.x.

Smyth, S. J. (2017). Genetically modified crops, regulatory delays and international trade. *Food and Energy Security*, *6*, 78–86.

Tagliabue, G. (2016). Nature as a totem, GMOs as a contemporary taboo. *North American Journal of Psychology*, *18*(2), 283–294.

SUBJECT INDEX

Note: Page numbers followed by "*f*" indicate figures, and "*t*" indicate tables.

AUTHOR INDEX

Note: Page numbers followed by "*f*" indicate figures.

C

D

E

F

G

H

L

M

P

S

T

X

Y

Z

FSC
www.fsc.org
MIX
Papier | Fördert gute Waldnutzung
FSC® C083411

Kolibri 360 GmbH
Ferdinand-Jühlke-Straße 7
99095 Erfurt, Deutschland
produktsicherheit@kolibri360.de